Praise for *Smarter Than We Think*...

Whilst the need to learn new ideas is exciting, for many it can also be daunting. This book shines a guiding light through the uncertain terrain ahead ... to have one of the country's leaders in mathematics education take what we know from the latest science and weave it into ideas for teachers, parents, and teacher leaders is a wonderful gift.
—*Jo Boaler, Professor of Mathematics Education, Stanford University and founder of youcubed.org*

Smarter Than We Think is a must-read for parents, teachers, policy makers, or anyone who cares about the mathematical education of the next generation. Seeley tackles common misconceptions about students, and she challenges teachers and teacher-educators to engage students by turning their classrooms "upside-down" and to rethink what "everyone knows" about teaching mathematics. She exhorts policy makers to put mathematics first and use this as the guiding principle in making choices about technology in the classroom, restructuring curriculum, and high-stakes testing. Finally, in a set of essays on the mathematical practices in the Common Core State Standards, Seeley reminds us that it is not simply mathematical content that we want students to master. Rather, we should also seek to instill in our students the essential habits of mind that are the hallmark of mathematical thinking, arguably the essential skill for thriving in the twenty-first century.
—*John A. Pelesko, Professor and Chair, Department of Mathematical Sciences, University of Delaware*

Cathy Seeley's sequel to her award-winning *Faster Isn't Smarter* provides another notable source of inspiring and relevant messages; I have been eagerly awaiting the release of it to use with our teachers in our professional development programs. The messages are well-suited for book studies or personal reflections. Our teachers and leaders loved Cathy's first book and benefited from reading and discussing her recommendations and insights. I expect the same effusive, superlative responses to the motivational messages in *Smarter Than We Think*. Cathy is a national treasure.
—*Dr. Anne Papakonstantinou, Director of Rice University School Mathematics Project and Clinical Assistant Professor of Mathematics, Wiess School of Natural Sciences, Rice University*

Parents, teachers, and school leaders all have different concerns and speak different languages about math education. Cathy Seeley speaks fluently to each group with advice and understanding in her brisk and useful book. This is going to be a hit!
—*Dan Meyer, math educator*

The messages in *Smarter Than We Think* focus on the real challenges at work in our schools today: a changing student population and how to support their learning along with teaching, leading, and thinking in a Common Core world. These timely foci are accompanied by five recommendations from Cathy. This book is one I will share and use with our teachers—it will be an invaluable resource and inspiration for their work.
—*Catherine Martin, Director, Mathematics and Science Division of Teaching and Learning, Denver Public Schools*

Smarter Than We Think

More Messages About Math, Teaching, and Learning in the 21st Century

A resource for teachers, leaders, policy makers, and families

CATHY L. SEELEY

Foreword by Jo Boaler

Math Solutions
Sausalito, California, USA

Math Solutions
One Harbor Drive, Suite 101
Sausalito, California, USA 94965
www.mathsolutions.com

ISBN-13: 978-1-935099-36-9
ISBN-10: 1-935099-36-1

Editor: Jamie Ann Cross
Production: Denise A. Botelho
Cover design: Wanda Espana/Wee Design
Interior design: MPS Limited
Cover images: (front) "Super Siblings" © Andrew Rich/Vetta/Getty Images;
   (back) Damon Leo (www.damonleo.com)
Composition: MPS Limited

2 3 4 5 6 7 8 9 10   31   23 22 21 20 19 18 17 16 15 14

To Merv and Kate, who raised me to see, to notice, to think, and to communicate, but mostly who encouraged me to explore and follow my heart in deciding where I wanted to go in my life. I'm still exploring, still following my heart, and still loving the journey, thanks to my parents' good start.

,

Contents

I   Messages About Who We Teach and How They Learn

II Messages About Teachers and Teaching

III Messages About Leadership, Issues, and Policy

IV   Messages About Thinking Mathematically in a Common Core World—Mathematical Practices and More

Foreword

A GUIDING LIGHT

The education world is changing. We are in the midst of a learning revolution and the evidence that is emerging from neuroscience through brain scans and other new technologies is stunning. It is imperative that we get this new evidence out to teachers and parents in practicable, useable forms, and *Smarter Than We Think* is packed full of wonderful ideas that do exactly that. To have one of the country's leaders in mathematics education, Cathy Seeley, take what we know from the latest science and weave it into ideas for teachers, parents, and teacher leaders is a wonderful gift.

One of the most amazing findings that has emerged from the scientific world in the last few years comes from the brain science of mistakes. Message 8 of this book, "Oops! The Value of Mistakes, Failures, and Wrong Turns," takes this and gives it the impact it needs. Brain scans now tell us that when someone makes a mistake in math and they struggle over something, synapses fire (Boaler 2013a). When someone does not struggle and they get an answer correct nothing happens. This new knowledge has huge implications for classrooms as Cathy sets out—it means we want students working on hard, complex problems that encourage them to struggle and even make mistakes. We don't want students working on mathematics questions that they always get correct. Mindset expert Carol Dweck tells parents that when children come home reporting that they got all of their work in math class completely correct, parents should say, "Oh, I'm sorry, you didn't have any opportunities to learn something today." Cathy gives us practical ways to take the latest research on the brain, on mindset, and in mathematics education and turn that research into effective classroom practices.

In the United States, we sit at a critical juncture, with the landscape of teaching, learning, and assessing mathematics changing. The introduction of the Common Core State Standards (NGA Center and CCSSO) and associated assessments means that students will spend more time in classrooms problem solving, reasoning, connecting mathematical ideas, communicating, representing, and a host of other critical activities. These different activities set out in the practice standards of the Common Core are not knowledge to be learned; they are ways of working in which all high-levels users of mathematics engage.

The authors of the Common Core Standards emphasize something important: Mathematical practices should be interwoven into the teaching of all mathematical content. Messages 32 through 39 in *Smarter Than We Think* address these practices and discuss how we can incorporate them into our thinking and into our teaching.

For many teachers, engaging students in more active, mathematical ways of working (Boaler 2013b), rather than learning sets of methods, requires teacher learning of new pedagogical methods. This is an exciting opportunity and the learning of new teaching ideas that engage students actively and at high levels should be the goal of all teachers throughout their career. Whilst the need to learn new ideas is exciting for many it can also be daunting. *Smarter Than We Think* shines a guiding light through the uncertain terrain ahead, taking what we know from research and from Cathy's extensive experience in classrooms to provide paths to follow. You'll learn about ideas like *upside-down teaching* (see Message 12), a term Cathy uses to describe a mathematics classroom focused on engaging students in challenging problems and organized around productive discourse. I am excited for the opportunities this book provides. Every teacher, parent, or researcher of mathematics education who reads it has the opportunity to dive into rich and complex examples of teaching and develop new ideas that will be powerful for students.

—JO BOALER

Professor of Mathematics Education, Stanford University
and founder of youcubed.org

Introduction

It's been five years since the publication of *Faster Isn't Smarter*, the predecessor to *Smarter Than We Think*. During that time, there have continued to be major events in the world of school mathematics, most notably the development of the Common Core State Standards for Mathematics (NGA Center and CCSSO 2010) and the adoption of the standards by most states. In mathematics, we are fortunate that these standards arrived after more than two decades of work from the profession, with the National Council of Teachers of Mathematics having described in some detail the mathematics that should be included in prekindergarten through grade 12 school programs. The Common Core Standards build on the considerable work during this time to continue to refine and update a vision of high-quality mathematics that would prepare every student for a fulfilling life.

Since the publication of *Faster Isn't Smarter*, headlines have continued to report U.S. students not performing as well as we might like in mathematics on international exams. And we have continued to hear arguments over the details of what should be taught in schools and how to teach it, even as we see growing agreement about the broad goals of problem solving, reasoning, and thinking for every student.

Amidst this background, teachers and students have continued to work hard on increasingly rigorous expectations in mathematics. Today's students know more than ever about mathematics, and they continue to grow. That is, they will continue to grow if we can keep politics and other man-made barriers out of their way.

People ask me where I get the ideas for the messages I write. There's no shortage of material out there—since I am far away from the responsibilities of teaching every day or making difficult administrative decisions every day, sometimes it seems so easy and clear how things ought to be done in schools. When things often aren't done the way I envision they should be, I have a seemingly endless supply of topics for potential messages. I don't mean to minimize the seriousness of the work of education. On the contrary, I see improving students' opportunity to learn mathematical reasoning, thinking, and habits of mind as one of the highest-priority goals we can pursue. Education is the most important job on the planet, and mathematics education is a critical focal point for debate and improvement. Against this backdrop, I offer forty new messages for *Smarter Than We Think*.

I appreciate the hard work and commitment of teachers, especially at a time when bureaucratic, administrative, and political challenges sometimes may seem overwhelming. I continue to stand in awe of the teachers and leaders who devote their careers to helping students of all ages expand their minds—helping them become smarter than they think. I hope you enjoy reading these messages as much as I enjoyed writing them.

How to Use This Book

As with *Faster Isn't Smarter*, the forty messages in *Smarter Than We Think* are designed to be read either individually, following an order based on themes that may interest the reader, or sequentially. The book is organized in four sections, each composed of ten messages, beginning with a section of messages about students and their learning, and expanding outward. The second section consists of messages on teachers and teaching, followed by messages on issues, leadership, and policy. A fourth section begins with an overview of mathematical habits of mind, accompanied by eight messages targeting each of the Standards for Mathematical Practice set out in the Common Core State Standards (NGA Center and CCSSO 2010), and concluding with a message on how we can help students develop these mathematical habits of mind. In a special Afterword, I propose five recommendations for the future of mathematics education, accompanied by a table of the messages in both *Smarter Than We Think* and *Faster Isn't Smarter*. The table is organized by the topics reflected in the five recommendations. This table may be useful as another way of individually reflecting on the messages or organizing professional development activities.

Beyond this overall organizing structure, some general themes run through *Smarter Than We Think*, appearing in different messages and from slightly different perspectives. The most important theme is built around intelligence and potential, based on the growing evidence in support of a growth model of intelligence, as well as the influence of how we teach on students' intelligence, as summarized in the title message appearing first in the book. A second theme involves my belief in the importance of a problem-based, student-centered approach to teaching, specifically described in Message 12, "Upside-Down Teaching." A third theme promotes the advantages of professional educators working collaboratively within and across grades, and a fourth reminds us of the importance of paying attention to students' day-to-day learning with informal and formal means of formative assessment. I hope the reader will immediately notice one additional theme—an absolute commitment to equity and the belief that all students need to be provided the opportunity to learn high-level, relevant, engaging, useful mathematics.

I write my messages not simply to offer my opinion, which I love to do. Rather, my primary purpose is to provoke reflection and discussion, and perhaps stretch, challenge, and even push educators, policy makers, and the public to be open to seeing things in different ways or from a different perspective. With that goal in mind, each message is accompanied by questions under the heading "Reflections and Discussion." At the end of each message is a short list of "Related Messages," both from this book and from *Faster Isn't Smarter*. Finally, a list of print and online resources is offered under the heading "More to Consider," so that the reader can choose to pursue the topic further. Resources and related messages are sequenced according to their relevance to the particular message, rather than in numerical or alphabetical order. Whenever possible, I try to include resources that offer different points of view or go into greater depth than my relatively brief messages.

The issues we face in mathematics education are challenging and teaching students the kind of mathematical thinking they need is hard. As we take on the issues and accept the challenge of improving our practice, I hope we will work together collaboratively within and across schools, school systems, and states. The thoughts, questions, and resources offered here are intended to support exactly this kind of collaborative work as we tackle together the biggest problem-solving opportunity we face—helping every student become a mathematical thinker and confident problem solver.

Smarter Than We Think

HELPING STUDENTS GROW THEIR MINDS

Smart is not something you are. It's what you become.

—Salome Thomas-EL ("Principal EL"), Thomas Edison
Charter School, Wilmington, DE (2010)

When I first heard a few years ago about the *Washington Post's* social experiment involving Joshua Bell, world-renowned classical violinist, I immediately thought of the lessons the experiment offers to educators. At the *Post's* request, Bell wore street clothes—jeans, a long-sleeved t-shirt, and a baseball cap—to play his three-million-dollar Stradivarius violin in the L'Enfant Plaza metro station for forty-five minutes during Washington, DC's morning rush hour. The wonderful article in the Sunday edition of the paper described in beautiful detail what nearly all of one-thousand-plus people missed as they walked or ran by on that January day (Weingarten 2007). Reflecting on what those people didn't notice caused me to wonder what talents and abilities we—teachers, families, and other adults—may not be seeing in the young people in our lives.

Why Some Students Don't Succeed

There are many reasons why some students don't reach the high standards we set in mathematics. When I ask teachers to consider possible reasons for a lack of success, many identify factors related to the student (gaps in previous knowledge, attitude, intelligence, motivation, socioeconomic level); factors related to school (superficial coverage of content, shallow or inadequate instructional resources, a repetitive curriculum, lack of relevance of what is taught); factors related to the community or family (low expectations from family, a belief that only some people can do mathematics, societal expectations); and other factors related to equity, opportunity to learn, and so on. While all of these issues and factors may relate to some students' lack of success in mathematics, I would argue that none of them are reasons that make it impossible for those students to succeed. For any reason or cause we might see as an obstacle that keeps a student from learning, I believe that somewhere a student is overcoming that same obstacle and learning mathematics well, possibly to a very high level. And I'm guessing that when we find that student, we can also find at least one teacher who has played a role in that student's learning.

How Students Hide Their Intelligence

Like Joshua Bell in the subway station, our students can disguise their talents and potential in ways that make it difficult for us to see them as smart, hard-working, creative, analytical, thoughtful, or "mathematical." Some students may not dress like our stereotypes of serious mathematics students—perhaps adopting the look of the latest fad, wearing dirty or shabby clothes, or just not being well-groomed overall. Other students may disguise their potential through their behavior or attitude—acting out, behaving rudely, being sullen, or simply remaining silent or behaving in ways so as not to be noticed. Others may not speak English well or may have learning disabilities, diagnosed or not. Others may miss class or arrive late, or show up lacking sleep or food. Some may just be unpleasant or argumentative or just the opposite, placid or accepting everything said without thinking.

The opportunity for teachers is to see through the disguises, to confront our preconceptions and biases about what a good mathematics student looks like or acts like. This is not an easy task, as many of those preconceptions and biases lie below the surface and we may not be aware of them. But if we can move past our unhelpful habits and beliefs, I believe we can open doors for both ourselves and our students.

Understanding Fixed and Growth Mindsets

One of the biggest factors in how well students do in mathematics may lie in their, and our, view of intelligence: Who's smart and how smart are they? Over time, two mindsets have represented the majority of thinking about intelligence. One mindset—a *fixed* or *entity* mindset—is based on the belief that each person is born with certain genes that determine absolutely their intelligence. Based on a fixed mindset of intelligence, nothing in a person's life is likely to change his or her intelligence; the person is as smart now as they are ever going to be, even if they acquire knowledge along the way. The other mindset—a *growth* mindset—is based on seeing intelligence as a malleable quantity. In this view, the genes a person inherits have an influence on his or her intelligence, but genes are not the only influence. A person's intelligence and abilities are also influenced by experience and the kinds of thinking a person does over time.

Consider the implications of these two mindsets. Someone with a fixed mindset, believing that she or he has limited capacity for intelligence, that they were born with whatever mathematics ability they're ever going to have, is far more likely to give up when faced with a hard math problem. *Why bother? This problem is clearly beyond my ability.* On the other hand, someone with a growth mindset realizes that every person can develop his or her own intelligence and abilities. Individuals operating from a growth mindset know that intelligence and ability are related to effort and experience, so they are much more likely to be willing to tackle a hard problem. *This problem looks really hard. I guess I'm going to have to work a while on it.* A growth mindset is also likely to support someone's willingness to take a risk and share his or her thinking in a class discussion, thus contributing to the development of mathematical habits of mind for all involved.

A person's mindset about intelligence is significant and powerful, and it influences even more aspects of life than these examples. In her best-selling book *Mindset,* Carol Dweck (2006) offers an in-depth, accessible discussion of intelligence. The most essential idea related to a growth mindset is that there is increasing evidence and support that a person's experience, especially experience wrestling with challenging academic material, can have a positive effect on that person's intelligence.

This isn't a new idea; support for the contributions of experience to intelligence has been around for more than a hundred years, even before the technology was available for real-time analysis of the brain. In fact, in 1909, Alfred Binet, the creator of IQ tests, stated:

> A few modern philosophers ... assert that an individual's intelligence is a fixed quantity ... which cannot be increased. We must protest and react against this brutal pessimism.... With practice, training, and above all, method, we manage to increase our attention, our memory, our judgment and literally to become more intelligent than we were before. (as quoted in Elliot and Dweck 2007, 105–106)

Today, through brain scans and other recent advances in cognitive psychology, we have support for Binet's point of view. It's fair to say that wrestling with complex tasks in mathematics may make a person smarter, even forming new neural connections in the brain. The flip side seems equally apparent—that not having such opportunities can stagnate, or even diminish, a person's intelligence over time. Since the teacher is the front-line decision maker about the tasks a student sees, the influence of a teacher's day-to-day choices on students' intelligence, not just on what they learn, is significant.

Teaching to Grow Students' Minds

Since the benefits of a growth mindset are so substantial, how do we promote a growth mindset among the students we serve? Teachers— and to a large extent families and other adults—can support students in growing their minds and increasing their intelligence in at least four areas.

BELIEVE IN EVERY STUDENT'S POTENTIAL

Is every student a potential Joshua Bell? Or a Crystal or Richard or Marisa or Jamal or …? The latter are all students I've written about who showed, and thankfully developed, their potential after sometimes many years of appearing to be less capable than they turned out to be. But do all students have equal potential? When I ask this question of teachers after they have been discussing the issues described in this message, they often respond that, while all students may not have the same potential in the same areas, they all likely have far more potential than they have shown, including their potential to do mathematics. I would assert that all students have untapped potential. And the adults around students—teachers, families, community friends, mentors—can have a tremendous influence on helping students unlock that potential and grow their minds.

CHOOSE RICH, DEEP, ENGAGING INSTRUCTIONAL TASKS

If experience influences a person's intelligence, then the mathematical tasks and problems we present to students, and how we present them, take on critical importance. Curriculum developers and authors of instructional materials have a responsibility to ensure that materials include rich, deep, engaging, and relevant tasks that cause students to think and to grapple with challenging mathematics. And, regardless of the curriculum, or for that matter the standards, table of contents, or pacing guide, teachers have a responsibility to make day-to-day instructional choices that ensure that students work with problems that engage their interest and their intellect.

Some teachers, thinking they are helping students by not frustrating them to the point of disengagement, try to protect students from struggling by offering shallow problems or procedural exercises that call for little more than one-step application of recently learned skills. And some parents, reflecting their students' frustration when presented with more challenging problems, advocate only giving students problems they have already learned how to solve. But when we limit students' experience in school to such low-level expectations, we cheat them of the opportunity to wrestle with hard mathematics and to grow their minds, even becoming smarter. In the end, a decision to avoid student frustration may diminish students' confidence and willingness to tackle anything difficult, leaving students ill-equipped to deal with real problems they will face outside of school.

PROVIDE FEEDBACK FOCUSED ON EFFORT

In addition to choosing good instructional tasks, we can monitor the kind of feedback we give students. When we praise students for how smart they are—*You are so smart!*—or unintentionally let them know we don't think they're smart enough to take on a particular problem, we reinforce a fixed mindset and make students feel like the amount of effort they put into something doesn't matter. But when we provide feedback based on effort—how hard students have worked or may need to work to arrive at a solution to a problem—*Your hard work really paid off!*—we encourage a growth mindset. Feedback based on a sincere interest in knowing how a student is thinking can be a powerful stimulus to encourage continued work and learning. In this way, we help them learn to persevere and build their confidence as they expand their learning and increase their intelligence.

MODEL HOW WE HANDLE MISTAKES

By modeling how we handle our own mistakes, we can help students develop a positive disposition toward mathematics and encourage their willingness to tackle hard problems. Many students have come to think that they're supposed to know the answer to a math problem right away. Mathematics can sometimes seem unforgiving, as students view answers to mathematics problems as either right or wrong; they certainly don't want to be wrong! As adults, when we acknowledge that we have made a mistake (whether in mathematics or in life) and as we make adjustments to our thinking or our actions based on what we learn as a result of the mistake, we can help students shift their view and adopt a healthy attitude about what success in mathematics looks like and how mathematical thinking really works.

What Can We Do?

No one's born being good at things; you become good at things through hard work.

—President Barack Obama (2009, Remarks at Back to School Event)

Our job as influential adults in students' lives is to never underestimate the talents and intelligence any young person might be able to develop under the most conducive circumstances, and to express our high expectations and supportive confidence that each one will develop that talent and intelligence. Our job as educators is to make sure every student has the opportunity to wrestle with challenging mathematical tasks and rich, complex problems so that they can grow new neural connections, increase their confidence, build their willingness to persevere, and develop a positive and productive attitude about his or her competence and potential. We can offer feedback based on effort and perseverance and model the importance of mistakes to our learning.

Wouldn't it be wonderful if every student finished their elementary and secondary educational experience knowing they are capable, succeeding in learning high-level mathematics, and having discovered their unique talents and abilities, wherever they may lie? Every one of us can support that goal for at least one student. As we help students develop a growth mindset that lets them know they can become smarter and smarter with hard work, maybe even students will realize they are smarter than they think.

Reflections and Discussion

FOR TEACHERS

- What issues or challenges does this message raise for you? In what ways do you agree with or disagree with the main points of the message?
- In what ways do you teach, or can you teach, that support every student in becoming smarter?

FOR FAMILIES

- What questions or issues does this message raise for you to discuss with your son or daughter, the teacher, or school leaders?

- How can you let teachers know that you support presenting students with challenging, deep problems that cause students to think and wrestle with mathematical ideas, rather than only presenting problems that students have learned specifically how to solve?
- How can you communicate to your daughter or son your confidence that they can become smarter and succeed in mathematics?
- A parent's attitude toward mathematics can have a strong effect on a student's confidence, in either a positive or negative way. How can you let your son or daughter know that you have high expectations for their success in ways that support their confidence and development of a positive attitude about learning, about mathematics, and about himself or herself?

FOR LEADERS AND POLICY MAKERS

- How does this message reinforce or challenge policies and decisions you have made or are considering?
- How do your placement and tracking practices support the growth of every student's intelligence?

RELATED MESSAGES

Smarter Than We Think

- Message 8, "Oops!," looks at the value of mistakes for both students and adults, discusses how we can learn from mistakes, and recognizes the contributions of how we handle mistakes to a growth mindset.
- Message 4, "They Just Aren't Motivated!," examines what motivates students to engage in mathematics.
- Message 3, "He Doesn't Know His Facts," tells the story of a man who overcame an apparent barrier with hard work.
- Message 29, "Finding Great Teachers," offers lessons about supporting teachers in developing their talents.
- Message 13, "Clueless," considers how a student's intelligence may be disguised by what we teach them and how we teach them.

Faster Isn't Smarter

- Message 18, "Faster Isn't Smarter," examines the negative implications of timed tests, potentially disguising how capable some students may be.

- Message 2, "Untapped Potential," advocates allowing all students access to good mathematics and discusses the importance of supporting all students in becoming smarter.
- Message 17, "Constructive Struggling," considers the advantages of offering students challenging problems.
- Message 31, "Do They Really Need It?," tells a story about doing what we say we believe in terms of high expectations.
- Message 30, "Crystal's Calculator," describes a student whose talents emerged after getting past a barrier.

MORE TO CONSIDER

- *Mindset: The New Psychology of Success* (Dweck 2006) is a breakthrough book for both educators and the public about the implications that mindset has on intelligence.
- *Self-Theories: Their Role in Motivation, Personality, and Development* (Dweck 2000) looks at how people work and when they're most functional and successful.
- "Pearls Before Breakfast" (Weingarten 2007) tells the story of the Joshua Bell experiment.
- "Ability and Mathematics: The Mindset Revolution That Is Reshaping Education" (Boaler 2013) presents an overview of the impact of a growth mindset on mathematics teaching.
- *The Immortality of Influence: We Can Build the Best Minds of the Next Generation* (Thomas-EL 2006) offers an uplifting true story of the power of mentoring, high expectations, and support for students apparently unlikely to succeed.
- "Educ115N: How to Learn Math" is a Massive Open Online Course (MOOC) and related website offered by Stanford University and taught by Dr. Jo Boaler for educators, students, or noneducator adults addressing key issues related to learning mathematics, including discussions on intelligence, mistakes, perseverance, problem solving, teaching for student engagement, and other topics.
- *The Mismeasure of Man, revised and expanded edition* (Gould 2012) is an update of the author's concerns about the measurement of intelligence via IQ scores and standardized tests in general, and presents a renewed argument questioning the merits of any view of intelligence as a fixed quantity.
- The Dana Center's website, www.learningandtheadolescentmind .org, includes a nice summary of issues affecting students' success in school, especially in mathematics. It includes background information, teaching suggestions, and resources for further study.

- "Growth Mindset and the Common Core Math Standards" from *Edutopia* (Bryant 2013) looks at a growth mindset as it relates to students developing mathematical habits of mind described in the Common Core Standards for Mathematical Practice (NGA Center and CCSSO 2010).

- "Improving Student Achievement in Mathematics by Promoting Positive Self-Beliefs" (National Council of Supervisors of Mathematics 2010) describes research and includes references related to student beliefs and student learning.

- *David and Goliath: Underdogs, Misfits, and the Art of Battling Giants* (Gladwell 2013) looks at likely and unlikely candidates for success and considers the role of compensation and effort in helping people who might not be expected to succeed overcome their limitations and achieve success.

- *Outliers: The Story of Success* (Gladwell 2008) considers what makes some of the most extraordinary individuals achieve great things.

- *The Smartest Kids in the World: And How They Got That Way* (Ripley 2013) is a reporter's fascinating story of education in the United States and other countries from the perspective of a few students crossing cultures and experiencing education in a different light.

- *Why Don't Students Like School?* (Willingham 2009) is a cognitive scientist's view of fundamental principles about students, the growth of their intelligence, and implications for the classroom.

- *The Path to Purpose: How Young People Find Their Calling in Life* (Damon 2008) discusses how to help adolescents uncover their interests and determine possible future steps.

- *Sparks: How Parents Can Ignite the Hidden Strengths of Teenagers* (Benson 2008) presents ways that parents and other adults can help students uncover their potential.

- *Shattering Expectations Series: Breaking the Glass Ceiling of Achievement for Low-Income Students and Students of Color* (The Education Trust 2013) offers data and recommendations for how we can uncover and develop the potential in all students.

WWW This message is also available in printable format at mathsolutions.com/smarterthanwethink.

2

¿Habla matemáticas?

TAPPING MATHEMATICS POTENTIAL IN ENGLISH
LANGUAGE LEARNERS

If teachers view mathematics as language-free, they likely will
breathe a sigh of relief when mathematics time comes and think,
"Finally! A subject where my second-language learners do not
have to struggle with English!" Consequently, students often end
up working on "word-free" worksheets, practicing only standard
procedures for solving problems, and experiencing mathematics
as something that is done quietly in isolation.

—Rusty Bresser (2003)

I once had an opportunity to visit an elementary school where more
than 90 percent of the students spoke primarily, or only, Spanish
at home. The community is rich in its bicultural heritage, but language issues have a significant impact on students' lives both in and
outside of school. Walking down the school hallway, I couldn't help
but notice the classrooms. One classroom in particular caught my eye.
In the hallway by the classroom door, many examples of student work
were posted alongside a big chart listing all the students' names. Next
to each student's name was a stick-on star for each book the student
had read so far that year. What struck me was not so much the stars
or the work, but the list of the names of the students in the class. Of
the list of twenty-two students, only two of the surnames appeared to
be Hispanic. Looking in the window of the classroom door, I spotted
two or three brown faces amidst a sea of white faces. I was startled; it
was clear just by looking in the window that the demographic makeup
of this classroom was different from the demographic makeup of the
school overall. As I proceeded down the hallway, I continued to see
disparity in the composition of classrooms, not reflecting the overall de-
mographics of the school. When I asked about the classrooms I noticed,
I was told that the classrooms primarily consisting of white, presumably

fluent English-speaking students were in the so-called advanced or upper track. I knew of no research then, nor any now, to indicate that
students who are not yet proficient in English have lower abilities in
mathematics; in fact, evidence abounds to contradict any such belief.
I was discouraged to realize that, in that school at that time, a student's primary language or possibly the location of a student's neighborhood or the color of the student's skin might influence the classroom in
which she or he might be placed. Of even more importance, I continue
to lament the low expectations and limited opportunities afforded to
students whose first language is not English.

Where We Place English Language Learners

How we place students into classes is the first barrier many non-English-
speaking students may face, but certainly not the only one. It can be
difficult to assess how much mathematics a student already knows
when the student first begins school in English, especially if the student's
primary language is not spoken by anyone else at the school. We must
be careful to balance our concern about possibly frustrating the student
with placement into a class he or she may not understand against the student's best interest in experiencing rigorous mathematics. Heterogeneous
(untracked) classrooms may provide the best learning opportunity for
English language learners, as for other students, especially if the classroom is structured for small-group work where every group could include
a strong reader. Pairs or groups are more important than ever when students are engaging in rich problems, so that words don't get in the way
of thinking. Classroom norms can help set the expectation that all members of a group are responsible for ensuring that every group member
understands the problem and the solution. To help develop language
skills, all students in a group, or both students in a pair, can take part in
communicating results, taking into consideration each student's language
proficiency. A non-English-proficient student might be responsible for a
visual representation of the problem or for describing one aspect of the
problem when the group or pair presents their solution. Ideally, the language expectation for a student who is not yet fluent in English will grow
throughout the year as the student develops stronger language skills.

Problem Solving and English Language Learners

Whether a student is in an untracked classroom or a classroom with students thought to be at a similar level of achievement, it can be tempting to

try to protect a non-English-speaking student from words. We may think that the best kind of mathematics for an English language learner should consist of strictly numerical exercises—naked computation. While it's true that the student might find success with the numbers, by depriving a student from the opportunity to engage in complex, interesting, language-rich problems, we deny the student the tools he or she needs to develop in order to solve such problems in the future. The most interesting, challenging and useful problems for learning mathematics are likely to be surrounded by words. Rather than withholding such problems, we owe it to our students to help them develop the academic, mathematical language they need to clarify their thinking and communicate that thinking to others. And we need to do that in the context of engaging mathematics tasks that provide fertile ground for this kind of language development.

Teachers need to monitor how students are doing with their language development, recognizing that not all English language learners will be at the same level of development, either in their use of mathematical language or in their understanding of mathematical ideas. It may be helpful sometimes for a teacher to offer additional supports to students with one-on-one conversation targeted at the particular vocabulary related to a given problem or mathematical concept.

A CLOSER LOOK AT REPRESENTING MATHEMATICAL IDEAS

How a student represents a mathematical idea or procedure can offer challenges and opportunities. All students can benefit from representing mathematical ideas and problem situations with diagrams, tables, or other types of visual representations. Encouraging students to use such representations can help them understand a problem and can also help in giving a basis for building language skills for students to describe the underlying mathematics.

Teachers need to be open to the possibility that a student may know more than he or she shows, but may not use the same representation the teacher has seen before in expressing what he or she knows. I have a friend from Africa who was enrolled in a GED high school equivalency tutoring class for non-English-speaking students. He asked me for help with square roots. I was surprised, since I knew he had studied well beyond square roots in his education in Africa, progressing to a level of about community college. When he showed me what the tutor was doing in class, it was long division. But the use of our typical American division symbol was unfamiliar to my friend, and the closest thing he'd ever seen was a square root symbol. Having used an entirely different schema to represent division with a sort of T-diagram, no wonder he was confused! The way or ways in which we represent ideas and procedures may be different from what non-English-speaking students have seen before, and we need to communicate regularly with them to make sure they see what we write—or what other students write or what is written in a textbook—in the same way it is intended. Otherwise, we may inadvertently plant seeds of misconceptions and undermine students' understanding and confidence.

Representing mathematical ideas and procedures in different ways may be a useful learning tool. Some mathematics resource materials for English language learners suggest the use of multiple representations to help students develop their language and solidify their mathematical understanding. A simple tool like a growing set of index cards or a chart that lists a mathematical idea and shows its representation in words, numbers, symbols, pictures, graphs, and so on can be a helpful tool for students developing mathematical language skills, as well as for all students. For a class in which students may speak several different languages, for example, an educator can create a multicolumn chart with an English word for a mathematical idea and its picture in the first column, with each subsequent column showing the word in a different language (Dong 2009).

Culture and English Language Learners

As educators, it's also important to remember that students who are non-English speakers may also be dealing with varying degrees of new or conflicting cultures or family values. A friend of mine, born in a Spanish-speaking family, spoke only Spanish at home when she was growing up and was forbidden from speaking Spanish at school. For her entire first-grade year, she essentially said nothing. A caring teacher paying attention to students can intervene in ways that help a student deal with difficult times like this.

Moreover, in some cultures, expectations for young people in general, or for girls in particular, may affect a student's participation in class and may inhibit the development and demonstration of language. A student is unlikely to develop academic language about mathematics problems if the student is reluctant to speak in front of the class. For such students, a small-group setting might be more conducive to participation than calling on students in front of the whole class.

Finding Out What English Language Learners Know

Not long ago, I heard a speaker reflect that, for students whose first language is not English, we most likely will never know what they are thinking about advanced, abstract concepts. He noted that most of these students never have the opportunity to advance their academic language in their primary language, since they no longer use that language in a school setting. And they are not likely to reach an advanced level of academic language in their second language to the extent they would need it to communicate complex thinking in mathematics. Thus, even with our best instructional efforts, we may fall short of helping students fully communicate their reasoning or explain their thinking. Nevertheless, we can provide opportunities for students to use and develop their academic language, especially about mathematics, by incorporating extensive opportunities to read, write, and speak as students solve engaging, challenging problems. Teaching centered on such complex problems can help all students—both English speakers and English learners—understand mathematics and learn mathematics more deeply than they would learn in lecture-based or computation-based classrooms.

In terms of assessing what students know, both formatively and summatively, the best way to find out as much as possible is to allow students to explain what they are thinking. Some students may do this best in writing, including the use of pictures or diagrams, and others may do it best by speaking; both opportunities should be provided to students whenever possible so that students continue to develop both their written and spoken language.

Thinking About Labels

Many schools today offer programs for their ELLs—an acronym for *English language learners*—or LEPs or ELPs—acronyms for students with *limited English proficiency*. In education, we tend to overuse acronyms as part of our educational jargon to label students, especially those who may have been identified with special needs, in spite of the fact that acronyms can be off-putting to many people, myself included. Special education students may be called "SPEDs," or students who deal with emotional or behavioral problems may be called "ED" students. But considering all the labels we might use to categorize groups of students, it turns out that the labels probably have little or no relationship to how much mathematics a student might be able to learn. To be fair, acronyms can also be used to label students in ways that indicate recognition seen to assign high status or indicate participation in accelerated programs, such as referring to gifted students as GT (gifted and talented) or GATE (gifted and talented education) or Pre-AP (students taking classes preparing them for advanced placement courses). I recently visited a large urban district where administrators and teachers routinely referred to their ELLs and their ELPs using just one syllable for each acronym. In that city's schools, apparently every student is referred to by a one-syllable label.

Perhaps the first step, then, in helping to tap into the potential of students who are not yet proficient in English is to start thinking and talking about students without using their label, or at least without using the acronym. I often suggest that we advance our goals far better by describing than by labeling. Labeling students can have a negative impact on their learning, making them self-conscious or, worse, making them feel like they are somehow less smart or less able to learn mathematics than other students. Instead of talking about a student or group of students as "ELLs," perhaps we can spend a few additional syllables to refer to them as "students who are learning English" or even to use the words behind the acronym—English language learners. In doing so, we may also become more conscious of the human beings we are helping to learn mathematics.

What Can We Do?

When I taught in Burkina Faso, West Africa, neither I nor my students were operating in our first language in school. Fifty-one local languages are spoken in the country and the national language is French. Students start learning French when they go to school and they are expected to speak French all the time in school. In the large secondary school where I taught, there were at least three, and likely closer to five or six, languages students spoke with their families. Yet they all used French for their academic work. For some students, French was their third language, as their families may have married across ethnic/language groups or may have moved to another region as they were growing up. And most students studied German or English in secondary school as their third or fourth language. As for me, I had studied French in high school more than thirty years before that teaching experience, with additional French studies as part of my Peace Corps training when I arrived. My students were more proficient in French than I was, but we all were dealing with a language other than our primary language as we went through our teaching and learning. At the time, it took almost all of my time and attention to prepare lessons and grade student work in a language I was continuing to learn. Looking back, I realize how much of my energy went toward language issues rather than mathematics issues. I wonder how much of my students' time and energy might have been likewise detracting from their attention to mathematics.

I'll probably never fully know what it's like to try to learn mathematics in a language other than my own. However, I can share ideas I've learned from experience and from experts for tapping into each student's mathematics potential:

- Ensure that any special programs, whether accelerated or remedial, reflect the demographic makeup of the school. This may mean seriously rethinking placement policies and tracking programs, both in terms of classroom placement at the elementary level and in terms of acceleration into secondary school mathematics programs.

- Don't underestimate students who may not be fluent English speakers, as they all deserve an opportunity to explore interesting, engaging problems and they all deserve the opportunity to learn and to excel in mathematics.

- Recognize cultural differences that may affect students' learning or behavior, whether related to their language or not.

- As with all students, use formative assessment regularly, checking to see how well students are learning the intended mathematics.
- Avoid using labels and acronyms when talking about students and the programs we develop to help students learn mathematics while they are learning English.

All of these issues and recommendations can be constructively addressed within the structure of a well-functioning professional learning community. As part of the professional work of the community, educators often address learning issues related to the students they serve. These discussions can be particularly helpful in working with English language learners, since teachers will likely bring different levels of experience and expertise related to the specific issues these students face. Part of the lifelong learning of every educator can include sharing what works and brainstorming what else might work as we continue to learn from professional reading, discussions, conferences, professional development, and, of course, experience.

When we build a relationship with a student and consider the student as an individual, rather than as a member of a group, we can more readily see how to unlock the student's potential. Let's continue to learn as a profession how to help students who are not yet fluent in English, recognizing their considerable expertise in being able to communicate in at least two languages. Let's keep sharing successful strategies that are working in schools to discover more and more stars. Sometimes in Spanish we say, as in English, "Menos es màs": Less is more. But sometimes less is just less. Let us commit to not giving some of our students less—less opportunity to shine or less in terms of the mathematics we offer. Let us commit instead to giving every student an opportunity to flourish and develop his or her intelligence to the maximum.

Reflections and Discussion

FOR TEACHERS

- What issues or challenges does this message raise for you? In what ways do you agree with or disagree with the main points of the message?
- In what ways do adults in your school label students? In what ways might the labels advance or inhibit students' learning?
- How have you successfully helped a non-English-speaking student demonstrate the mathematics he or she knows or learn a mathematical concept or idea?

FOR FAMILIES

- What questions or issues does this message raise for you to discuss with your son or daughter, the teacher, or school leaders?
- How can you support your student for success in a diverse classroom where some students are fluent in English and some are not? What kinds of conversations at home might help your student succeed, perhaps even helping other students?
- Whatever language you speak at home, how can you help your son or daughter learn the mathematical language he or she needs to succeed in school?

FOR LEADERS AND POLICY MAKERS

- How does this message reinforce or challenge policies and decisions you have made or are considering?
- In what ways do programs in your schools label students? To what extent do you refer to students by their labels in working with teachers or other administrators?
- How can you place non-English-speaking students into challenging classrooms where they will be able to learn more advanced mathematics even if they are not yet proficient in English?
- How can you support teachers and parents of students who are learning English and mathematics at the same time?

RELATED MESSAGES

Smarter Than We Think

- Message 1, "Smarter Than We Think," illuminates the notion of a growth mindset and reminds us of the influence of teaching in developing the intelligence and abilities of all students.
- Message 13, "Clueless," describes what happens when we teach students superficially and includes a story about spending time in Costa Rica.
- Message 12, "Upside-Down Teaching," advocates a problem-centered teaching approach that includes students working in groups.
- Message 26, "Leading Change," describes my experience in learning Spanish and offers reflections about facilitating change.
- Message 37, "Communicating Mathematically," describes issues related to precision in language, with implications for English language learners.

Faster Isn't Smarter

- Message 2, "Untapped Potential," reminds us of the potential of every student to accomplish more advanced mathematics.
- Message 31, "Do They Really Need It?," emphasizes the importance of high expectations for all students.

MORE TO CONSIDER

- *The Problem with Math Is English: A Language-Focused Approach to Helping All Students Develop a Deeper Understanding of Mathematics* (Molina 2012) focuses on language development as a key for helping students learn mathematics.
- *Beyond Good Teaching: Advancing Mathematics Education for ELLs* (Ramirez and Celedon-Pattichis 2012) offers insights and recommendations for helping students succeed in mathematics when their first language is not English.
- "Involving Latino and Latina Parents in Their Children's Mathematics Education" (Civil and Menéndez 2010) summarizes research and recommendations related to involving Latino families in the education of their children.
- "Using Two Languages When Learning Mathematics: How Can Research Help Us Understand Mathematics Learners Who Use Two Languages?" (Moschkovich 2009) summarizes research on students learning mathematics in a language other than their native language.
- "Helping English-Language Learners Develop Computational Fluency" (Bresser 2003) discusses the importance of communication and language when students are learning computation.
- *Supporting English Language Learners in Math Class, Grades K–2* (Bresser, Melanese, and Sphar 2008a) provides lessons and other resources for supporting English language learners in mathematics.
- *Supporting English Language Learners in Math Class, Grades 3–5* (Bresser, Melanese, and Sphar 2008b) provides lessons and other resources for supporting English language learners in mathematics.
- *Supporting English Language Learners in Math Class, Grades 6–8* (Melanese, Chung, and Forbes 2011) provides lessons and other resources for supporting English language learners in mathematics.
- *Supporting English Language Learners in Math Class: A Multimedia Professional Learning Resource, Grades K–5* (Bresser, Felux, and Melanese 2009) includes two books, DVD with video, professional development, and coaching resources, and a facilitator's guide for supporting the mathematics learning of students whose first language is not English.

- *English Language Learners in the Mathematics Classroom* (Coggins 2007) offers practical suggestions specific to learning mathematics.
- "Linking to Prior Learning" (Dong 2009) provides concrete teaching suggestions for working with English language learners.
- "Improving Student Achievement in Mathematics by Addressing the Needs of English Language Learners" (National Council of Supervisors of Mathematics 2009a) summarizes research and offers recommendations for helping students learn mathematics when their first language is not English.
- "Supporting ELLs in Mathematics" is a Stanford University site that posts open-source resources for rich tasks that support English language learners in mathematics problem solving. http://ell.stanford.edu/teaching_resources/math.
- "Issues in Assessing English Language Learners' Opportunity to Learn Mathematics" (Herman and Abedi 2004) describes the results of an initial study related to the importance of providing adequate learning opportunities for students whose first language is not English.
- "Seeking Effective Policies and Practices for English Language Learners" (Rennie Center for Education Research and Policy 2007) offers research insights into shifting policies and practices to support English language learners based on results in the state of Massachusetts. Not specific to mathematics.
- "The Critical Role of Vocabulary Development for English Language Learners" (August, Carlo, Dressler, and Snow 2005) presents general research-based strategies for helping English language learners improve their academic language.
- "Who Is an 'English-Language Learner'?" (Lu 2013) discusses issues related to English language learners from a state perspective.

3

He Doesn't Know His Facts

PERSPECTIVES ON MATHEMATICAL FLUENCY
(RICHARD'S STORY)

Oh, the places you'll go!

—Theodor Seuss Geisel (Dr. Seuss)

He was a middle-aged man in a polo shirt seated next to me on the flight to Dallas. We exchanged pleasantries and chatted about odds and ends, and then he asked me what kind of work I do. On hearing the word *math* in my response, he appeared to look for the nearest emergency exit. When it became obvious that neither of us was going anywhere, he quietly confessed, "I was never good at math. I hated those timed tests! I think I have a learning disability or something." Then, after looking both ways, he leaned slightly toward me and whispered, "I still don't know my multiplication facts."

Over the next thirty minutes of our short flight, I discovered a bit more about the man I'll call Richard. Richard came from a strong family who encouraged him to persevere in school in spite of his difficulties with mathematics. With his family's support, and with the help of teachers, tutors, and friends, he made it through elementary school, middle school, and high school. Actually, he made it a bit further. Richard went on to graduate from college with a bachelor's degree in economics. His work experience took him in unexpected directions, so he later chose to pursue a master's degree in physics, which he received. And, eventually, he continued his studies further, earning a Ph.D. in electrical engineering. On the day that I met him, Richard was on his way to a new plant in Asia to help them install manufacturing equipment for a computer chip that he had designed. And Richard still didn't know his multiplication facts.

It's a widespread belief that to be good at math means to be fast at computation. But this belief may in fact do more harm as good. Some of the world's greatest thinkers, scientists, and mathematicians have not been fast at arithmetic, even though they were tremendously successful in working with higher-level mathematics. Dr. Emma King has noted her poor performance on rote arithmetic computation, in spite of her highly recognized work in mathematics and in the scientific field of cosmology (see the "More to Consider" section). Numerous scientists and scholars from around the world could tell similar stories.

Assumptions We Make

With today's emphasis on mastery and accountability, we tend to assume that a person should not continue to move through the mathematics program until having learned all of the necessary prerequisite skills. It seems obvious that if a student doesn't know basic computation, that student certainly should not progress to more complicated computation with longer or uglier numbers, and the student should not be subjected to complex problems involving computation. We decry the practice of moving students on to the next level without a proper foundation in basic skills, and we limit access to calculators only to those students who have demonstrated that they do not need them. There are good reasons behind every one of these assumptions and practices. Without them, too many times we are frustrated with students' lack of prior knowledge, and we see them struggling with facts that should already be firmly implanted in their brains. To accept less than mastery seems too lax, as if we are not expecting enough of our students.

Richard's success in mathematics-dependent fields without strong computational skills may or may not be widespread, but at the very least it should raise questions for educators in terms of decisions we make about which students are allowed to move ahead in mathematics. Is it possible that Richard may not be the only person whose story might challenge our assumptions about mastery, prerequisites, and what it takes to achieve in mathematics or mathematics-related fields? What if it were possible to uncover and develop quantitative talents even when those talents may be disguised by difficulties with memorizing facts or mastering procedures? Perhaps there are ways we can help students develop their strengths as we challenge our own preconceptions about mastery and fluency.

The Common Core State Standards and "Fluency"

One strong emphasis in discussions about mathematics today, reiterated in the *Common Core State Standards for Mathematics* (NGA Center and CCSSO 2010), is the notion of mathematical *fluency*. The standards reinforce the discussion of fluency offered in the National Research Council's important book, *Adding It Up* (2001), where fluency is described as skill in carrying out procedures flexibly, accurately, efficiently and appropriately.

This significant research synthesis emphasizes the importance of fluency and its interrelationship with conceptual understanding. The words *fluency* and *fluently* appear frequently throughout the K–8 *Common Core State Standards for Mathematics*, calling for students to fluently add, subtract, multiply, and divide numbers and solve basic equations. At first glance, it might appear that in order to meet the requirement for fluency, a student must completely master all of the addition and multiplication tables and all procedures. This is an ambitious but important goal. However, I would argue that, even as we maintain our eye on that goal, we should not hold students back, keeping them from getting to the good stuff in mathematics, solely on the basis of isolated gaps or particular stumbling blocks they may experience in computation.

Fluency isn't an all-or-nothing idea, and it may not look the same for every student. Nor does the need for fluency mean that only those who can compute quickly have attained it. Yes, it is important that, when faced with a mathematical problem, students are able to access an effective and efficient method or procedure without having to develop it from scratch, go through an exhaustive guess-and-check process, or search for the nearest calculator. But there may be degrees of fluency that are sufficient for much of the mathematics that lies ahead, and sometimes allowing a student to use a calculator can uncover other mathematical strengths as the student tackles challenging problems that go beyond computation. Some students may be able to make adequate progress (or better), even if they cannot recall a fact or complete a computational exercise as quickly as another student. And sometimes, though not always, the student may fill in gaps as they move into higher-level mathematics. Whether the student does or doesn't fill in the gaps, the importance of working on more complex mathematics may trump the need to continue to pound away on the troublesome area or areas.

Based on Richard's story, and the experiences of others like him, there appears to be much a person can accomplish in mathematics-dependent fields even if there are some gaps in the person's mathematical toolkit. For some students, adequate fluency may involve taking a moment to consider the problem and what is called for. It may involve a quick assessment of how to approach the problem in an alternative way that isn't blocked by some gap in fact knowledge. Among the hundreds of instructional decisions teachers make every day, deciding when a student should go on to something new—how fluent is fluent enough at that point in time—is a critical one.

What Can We Do?

How can we help the Richards in our classrooms? Clearly, some students have tremendous potential that may be disguised by difficulties they have memorizing facts or mastering procedures. Solutions may come from at least two sources. First, we need to structure our mathematics curriculum materials in ways that balance the attention we put on computational proficiency with opportunities to develop understanding of underlying concepts. Second, we need to use computation in ways that may motivate the development of that proficiency, such as organizing instruction around engaging, extended tasks that call for students to develop mathematical thinking and problem-solving skills as they learn mathematical concepts and procedures. Such a balanced approach, where students have many opportunities to develop mental models for the basic operations and the numbers they use, can help students learn their facts and master procedures, while they also develop important mathematical habits of mind.

Our challenge as teachers, then, is to continually look for approaches that tap into students' strengths to help them get past their stumbling blocks. If a student does well with pictures or models, the student should have ample opportunities to use these tools to better understand operations and numbers. If a student is a creative problem solver, the student should have many opportunities to approach problems in different ways. And sometimes it's important just for the student to experience some other aspects of mathematics, particularly engaging in interesting problems or rigorous tasks, even if that means sometimes referring to a multiplication chart or using a calculator for basic facts or procedures we wish they had mastered. For many students, engaging in thoughtful and constructive struggles to solve rich and deep mathematical tasks can itself provide motivation and a foundation on which to finally overcome sometimes longstanding gaps or deficiencies. I often speak of Crystal—an algebra student I taught—who struggled with fraction procedures, but who was a powerful algebraic thinker (Seeley 2009, 159). By building on her success in algebra, she was eventually motivated to go back and learn fraction operations once and for all as a high school student.

Still, gaps in a students' mathematical knowledge can cause difficulties in future studies, and we need to pay attention to a student's gaps, especially if they continue over years. It is particularly important to such students' future success that we engage in communication with next year's teacher to increase the likelihood that perhaps that student will be in a better place to address her or his gaps after a few more months of developing maturity combined with new mathematical experiences that continue to advance her learning. For some students, simply a different approach or some additional time may help them fill in their gaps.

All of us—families, teachers, administrators—need to realize that we may know many Richards. They may not all need the same kind of attention and support, but we can each play a role in helping them succeed. We can all attend to providing the kind of support that helped this particular Richard succeed in life and become an innovator, leaving his positive influence on business and society.

Reflections and Discussion

FOR TEACHERS

- What issues or challenges does this message raise for you? In what ways do you agree with or disagree with the main points of the message?
- How do you deal with students in your classroom who seem to learn differently than most?
- How important is it to you that a student *master* skills from your course or grade level before moving on? How fluent is fluent enough, and is that the same for all students?
- What steps can you take to ensure that any gaps a student carries into next year are acknowledged and/or addressed in the future?

FOR FAMILIES

- What questions or issues does this message raise for you to discuss with your son or daughter, the teacher, or school leaders?
- How can you best support your daughter or son when she or he has not mastered an important skill?
- In what ways can you help your son or daughter learn important skills with which he or she may struggle, while still helping him or her build the confidence to succeed?

FOR LEADERS AND POLICY MAKERS

- How does this message reinforce or challenge policies and decisions you have made or are considering?
- In what ways do your current policies allow or inhibit students like Richard to succeed? How else can you support students like Richard?

RELATED MESSAGES

Smarter Than We Think

- Message 1, "Smarter Than We Think," considers what elements might influence a student's intelligence and capacity to learn mathematics.
- Message 5, "Getting It," advocates helping all students "get" the important mathematical ideas behind the procedures.

- Message 8, "Oops!," looks at the value of making mistakes and learning from them.
- Message 37, "Communicating Mathematically," deals with the role of precision in both mathematical communication and in carrying out computations.

Faster Isn't Smarter
- Message 30, "Crystal's Calculator," tells the story of another student who had difficulty in computation (with fractions), but who went on to fill in her gaps and successfully study more advanced mathematics.
- Message 18, "Faster Isn't Smarter," cautions us about relying too heavily on timed fact tests as a measure of student learning.
- Message 14, "Balance Is Basic," advocates balancing the mathematics program by addressing concepts, skills, and problem solving.

MORE TO CONSIDER

- *Number Talks, Grades K–5: Helping Children Build Mental Math and Computation Strategies, Updated with Common Core Connections* (Parrish 2010, 2014) is a multimedia resource for orchestrating classroom discourse toward building computational fluency and understanding.
- *About Teaching Mathematics: A K–8 Resource, 3rd edition* (Burns 2007) is a practical resource and comprehensive reference for teachers on helping children learn mathematics.
- A fourteen-minute video of Emma King presents a fascinating look at her life and work. She is noted for her work in mathematics and cosmology, after having overcome her dyslexia and difficulties dealing with numbers and computation. (http://vega.org.uk/video/programme/89)
- *Teaching Student-Centered Mathematics: Grades K–3, Volume 1* (Van de Walle and Lovin 2005a) is a comprehensive resource on classroom teaching of mathematics in the primary grades.
- *Teaching Student-Centered Mathematics: Grades 3–5, Volume 2* (Van de Walle and Lovin 2005b) is a comprehensive resource on classroom teaching of mathematics at the upper-elementary level.
- *Teaching Student-Centered Mathematics: Grades 5–8, Volume 3* (Van de Walle and Lovin 2005c) is a comprehensive resource on classroom teaching of mathematics in the middle grades.
- *Helping Children Learn Mathematics, 10th edition* (Reys, Lindquist, Lambdin, and Smith 2010) is a comprehensive resource on teaching mathematics, including the development of computational skills.

- "How Teaching Matters: Bringing the Classroom Back into Discussions of Teacher Quality" (Wenglinsky 2000) discusses the teaching of higher-order thinking skills as a key factor in looking at the quality of classroom teaching.
- *Adding It Up: Helping Children Learn Mathematics* (National Research Council 2001) presents research about what it means to be mathematically proficient.
- Common Core State Standards for Mathematics (NGA Center and CCSSO 2010) presents a balanced program of mathematics within a set of voluntary mathematics standards widely adopted in the United States.

4

They Just Aren't Motivated!

WHAT MAKES STUDENTS *WANT* TO LEARN MATHEMATICS?

It was boring … the teacher just stood in front of the room and just talked and didn't really like involve you.

—High school dropout, *The Silent Epidemic*
(Bridgeland, DiIulio, and Morison 2006)

Why don't all of our students achieve the high standards we set? Why don't some students seem interested in mathematics? What affects students' willingness to participate in the lessons their teachers prepare? When I talk with teachers about these questions, one response often pops up—"They just aren't motivated!" On the surface, it may seem that motivation is a characteristic a person carries as part of his or her genetic makeup or a quality someone may have developed over time based on one's life experiences. Either way, it may seem like motivation is innate in a person and something we just have to live with—someone either is motivated to learn mathematics or isn't.

However, we continue to discover more and more about what motivates students, and the true picture is more complex than a simple either/or condition. Motivation, it turns out, may be more malleable than we may have believed, much like intelligence. And if motivation can change (for the better or for worse), then teachers can do something (positively or negatively) to influence that change.

Understanding Motivation

Is *motivated* a word we can use to describe a person (he's really motivated; she's just not motivated) or is *motivating* something someone does to another (that incredible teacher motivated all of her students to learn)? The answer is probably some of both. What we want as teachers is to contribute to our students becoming motivated to learn mathematics. We want them to enthusiastically, or at least willingly, participate in classroom activities and assignments. As a friend of mine once observed, "Starting is harder than doing." And in a class I took long ago, I learned that action can precede motivation. I didn't fully understand that idea then, but now it seems obvious, based on many experiences since then. Often, if we choose engaging tasks and if we successfully draw a student into doing that task, the student will gain motivation to want to do more. And, if we're lucky, we hope that they will also learn to feel good about mathematics and their ability to do mathematics, perhaps even choosing to pursue opportunities to learn or use mathematics after they leave us. Reaching this goal is not an easy task, but I believe a teacher can play a significant role in creating a positive disposition toward mathematics in the students he or she teaches.

INTRINSIC VS. EXTRINSIC MOTIVATION

Some experts discuss motivation in terms of either *intrinsic motivation*—the internal motivation that comes from how a person feels about himself or herself and about the mathematics or learning being offered—versus *extrinsic motivation*—based on external factors less directly related to the student, such as hoping to please the teacher or wanting to receive a good grade. Extrinsic motivation tends to be less long-lasting and less influential on long-term effects.

In their article "From the Inside Out" in *Teaching Children Mathematics*, Jennifer Fillingim and Angela Barlow (2010) suggest that we can see shifts in students moving from extrinsic motivation to intrinsic motivation when they demonstrate mathematical behaviors by connecting what they're learning to previous material, responding beyond a question asked, or offering a conjecture. Teachers can help students develop these behaviors by posing questions that model what they want students to do: *How does this task relate to the problem we did on …? Can you say more about that? What do you think might happen if …?* Teachers can also acknowledge and reinforce what students say and facilitate conversation among students based on other students' comments, responses, and questions. Communication, both written and spoken, can be a powerful tool in helping students develop mathematical habits of mind that build their motivation.

All Students Are Motivated

Based on their extensive research with teachers and students, Jim Middleton and Amanda Jansen (2011) offer the premise that *all students are motivated*. The issue, they note, is not *whether* students are motivated, but rather, how to create and/or maximize students' motivation specifically to do (and learn) mathematics. They characterize motivation as learned, adaptive, and social, and note that it's subject to shift and evolve based on a student's experiences and current social situation. They suggest that motivation is about what is going on right now, but that it also contributes to long-term attitudes about mathematics. And they propose that success matters in terms of motivating students, but only when students succeed with something significant, even challenging, rather than succeeding with something trivial or routine.

Completing a page of exercises to practice a procedure may be useful in solidifying one's proficiency with that procedure, but success on such a task is unlikely to have a significant positive effect on a student's motivation to want to do more mathematics. On the other hand, working through a hard interesting problem, and finding a solution that works, can build confidence and can help motivate a student to want to do more, or at least to be open to whatever the next task might be. The latter experience is what I would consider *meaningful success*, with the potential to lead to lasting learning whereas the former is more superficial and fleeting.

PRAISE AND REINFORCEMENT

One other factor in spurring motivation is to consider the limited usefulness of rewards and other external ways of reinforcing student performance. While students might perform well for grades or other types of recognition, lasting learning is more often related to various types of intrinsic rewards. Even the use of praise can be tricky; praising students' effort can be more productive than offering general praise for how smart a student is or for praising someone for a right answer. Being frugal with our use of praise may in the end help students. This idea may seem at odds with our natural tendency as teachers to want to encourage our students, but we need to find ways to encourage every student about the student's ability to become smarter with effort on the right type of tasks, rather than reinforcing incorrect ideas about what it means to be smart or capable.

Seizing Opportunities

In education, we are presented with a rare few moments in a student's life when we can seize the student's natural motivation and interest to learn. Too often, we waste those moments in the process of schooling, particularly if we narrow our focus and strip down learning to only those things that are likely to be on *The Test*. Young children start out life curious about the world and eager to learn everything they can; this time in their lives is a precious moment. Children see mathematics all around them and they love learning how numbers and words can offer them tools to talk about the shapes and quantities they see. If we offer them enough opportunities to build on that natural curiosity when they're young, we can help all students develop a foundation for ongoing motivation and for a positive disposition toward doing mathematics.

Likewise, at the beginning of each school year, students may be more open to learning than they will be later in the year. What a difference we could make by beginning each school year in mathematics with a really great task or problem that engages and stimulates students' interest as it offers a glimpse of what is to come, rather than beginning the school year with weeks of review over previously learned skills. I would argue that the former approach is far more likely to generate payoffs throughout the year than the latter, even if students may have forgotten some of what they learned last year. We need not postpone getting to *the good stuff* for the sake of review; if we are clever, we can offer students opportunities to review previous learning in the context of interesting problems as we move forward with new material.

Motivating Every Day

Beyond seizing those too-rare opportunities when students might be more open to tackle mathematics, teachers can help students increase their motivation to learn mathematics and their disposition to do mathematics in the way that they structure the classroom on a day-to-day basis. We can choose engaging tasks that challenge students and allow them to develop their essential and natural tendency to talk with each other and with the teacher about the problem and the mathematics, as they dig deeper and deeper into both the problem and the underlying mathematics. We can nurture and develop our own ability to ask the kinds of questions that push students' thinking (*Why do you think so? Can you convince us? What if ...?*) and help students learn to ask good questions themselves. We can create classrooms where students feel safe to take the risk of speaking up and sharing their solutions in both small groups and with the whole class—classrooms that foster a true sense of community and enthusiasm for learning mathematics among students and with their teacher.

What Can We Do?

Considering recommendations from both researchers and frontline teachers, it seems that we can agree on a few key elements to increase student motivation to learn mathematics. Just as described in 1991 in a beautiful and insightful document from the National Council of Teachers of Mathematics, *Professional Standards for Teaching Mathematics*, the key elements of a rich and motivating classroom continue to lie in the classroom *environment* the teacher creates with students, the selection and use of worthwhile and engaging mathematical *tasks*, and the facilitation of productive *discourse* among students and between students and the teacher. Tasks should cause students to have to think and should tap into their interests whenever possible. It's not necessary for every task to be presented in a nonmathematical context, however, as the inappropriate use of context can occasionally interfere with learning. Sometimes the mathematics in a noncontextual task can itself be engaging. An important addition to considerations of environment, task, and discourse is the development of personal relationships: relationships among students, between a student and teacher, and between a student and mathematics.

As with many other sticky problems in mathematics education, the key to increasing student motivation lies in structuring our classrooms in ways that engage students in tackling challenging tasks where they can grapple with important mathematics and develop mathematical habits of mind. When we do that, we discover new stars—more students who not only are motivated to learn mathematics, but who succeed when they do.

Reflections and Discussion

FOR TEACHERS

- What issues or challenges does this message raise for you? In what ways do you agree with or disagree with the main points of the message?
- How motivated are your students to learn mathematics?
- What is one thing you can do differently to increase the motivation of one student who seems particularly unmotivated?

FOR FAMILIES

- What questions or issues does this message raise for you to discuss with your son or daughter, the teacher, or school leaders?
- If your daughter or son does not seem to be motivated to learn mathematics, how can you help her or him become more motivated?
- How can you encourage your son or daughter to like and participate in mathematics even if you have never liked mathematics or don't see yourself as mathematically inclined?

FOR LEADERS AND POLICY MAKERS

- How does this message reinforce or challenge policies and decisions you have made or are considering?
- How can you support teachers in motivating students to learn mathematics?

RELATED MESSAGES

Smarter Than We Think

- Message 1, "Smarter Than We Think," looks more closely at the relationship among effort, persistence, motivation, praise, and intelligence.
- Message 6, "Loving Algebra as Much as Art," discusses the role of interesting content and challenging, engaging problems in motivating students.

Faster Isn't Smarter

- Message 8, "More Math, More Dropouts?," considers the advantages of expecting more mathematics of all students.
- Message 37, "Boring!," advocates teaching in ways that draw students into the mathematics we want them to learn.
- Message 17, "Constructive Struggling," encourages us to allow students the opportunity to work through hard problems as a way to motivate them and advance their learning.

MORE TO CONSIDER

- *Motivation Matters and Interest Counts* (Middleton and Jansen 2011) relates research on motivation to practical classroom practices that engage students in learning challenging mathematics.
- "From the Inside Out" (Fillingim and Barlow 2010) looks at the source of motivation (internal vs. external) in helping students become "doers of mathematics."
- *5 Practices for Orchestrating Mathematics Discussions* (Smith and Stein 2011) is a simple, elegant description of classroom practice that engages students and helps them develop mathematical thinking.
- *Classroom Discussions in Math: A Teacher's Guide for Using Talk Moves to Support the Common Core and More, Grades K–6, 3rd edition* (Chapin, O'Connor, and Anderson 2014) offers practical strategies for orchestrating productive discourse to engage students in good problems.
- "Reflections of a Dance School Dropout" (Rigsbee 2011) describes the author's experience taking dance lessons to make observations about motivation.
- *Fires in the Bathroom: Advice for Teachers from High School Students* (Cushman 2005), along with *Fires in the Mind* (Cushman 2010) and *Fires in the Middle School Bathroom* (Cushman and Rogers 2008) provides direct quotes from students and offers insights into how we can engage students in learning mathematics.

- *Fires in the Middle School Bathroom: Advice for Teachers from Middle Schoolers* (Cushman and Rogers 2008) is similar to the previous source, but directed at middle school teachers.
- *Fires in the Mind: What Kids Can Tell Us About Motivation and Mastery* (Cushman 2010) offers insights from the students' point of view to help teachers facilitate student learning.
- "Teaching Topics: Interactions with Students: Motivating Students" (Vanderbilt University Center for Teaching) offers suggestions for motivating higher education students; these suggestions can also be applied to elementary and secondary students.

5

Getting It

TEACHING FOR *AHA!* MOMENTS

My heart is singing for joy this morning!
A miracle has happened!
The light of understanding has shone upon my little pupil's mind,
and behold, all things are changed!

—Anne Sullivan

Ever since I first saw the 1962 black-and-white film *The Miracle Worker*, I have remembered one powerful scene. The film depicts a life-changing time in the childhood of Helen Keller, a woman who grew up deaf and blind, with Anne Sullivan, the teacher her family engaged to help Helen learn to communicate. Helen is portrayed as a wild, unmannered child who had never been expected to behave well because of her disabilities. Anne works with the challenging child to help her learn basic manners and also teaches her finger spelling. Over time, Helen eventually succeeds at behaving appropriately and is able to mimic back any word that Anne spells with her fingers in Helen's hands. But Anne's primary goal is to help Helen understand the powerful use of language for communication, an achievement that long continues to elude Helen.

Near the end of the movie, the scene I remember shows Anne and Helen in yet another confrontation. Helen's family has decided that she has advanced so far, they no longer need Anne's help. Helen wildly runs outside and Anne chases after her, eventually catching up to her at the water pump where Helen is frantically using her hands to get a drink. In one last desperate attempt to show Helen what words mean, Anne finger-spells *w-a-t-e-r* into Helen's hands and puts them back under the water. Helen's face just then shows a remarkable and transformative moment of realization—she frantically grabs Anne's hands and finger spells *w-a-t-e-r*, pulling all of their hands into the water. As Anne excitedly pulls Helen's hands to her face, she nods her head repeatedly to let Helen know she's right. Helen then grabs Anne's hand and hungrily runs to everything she can find in the yard, urgently wanting to know the word that goes with tree, grass, doll, and more. She has just experienced a critical moment in her life. Suddenly and forever she *gets* what language is all about, and the world opens up to her.

In mathematics and science, if a person is fortunate, the light bulb will go on with a huge *Aha!* or *Eureka!* as some major discovery suddenly emerges after days, weeks, months, or years of work. In teaching mathematics, one of the greatest rewards a teacher can experience is to see the light bulb come on when a student *gets it*—when he or she understands a central concept or idea with a clarity that opens doors to all kinds of wonderful problems, higher level concepts, and ideas.

The Difference Between Memorizing and *Getting It*

The light of recognition can come to students so gradually that neither they nor their teacher notice. Or it can come suddenly, as an *Aha!* moment, as it did for Helen Keller. In mathematics, sometimes the *Aha!* moment comes as a student internalizes a procedure, such as long division, the addition of fractions, or multiplying binomials, and knows how to do it in many variations. When this happens, it's important for the teacher to know whether the student is *getting it* or has just memorized it. If a student memorizes a rule without *getting it*—understanding the mathematics behind the rule, when to use it, when it works, any limitations in using it, and so on—he or she may eventually forget or misunderstand all or part of the rule and may misapply it. Memorization can be a useful tool, but it's only part of *getting it*—the student needs to internalize what the rule is all about and recognize when it's helpful and when it's not. A student who *gets* multiplication not only knows the multiplication facts but also carries mental pictures of multiplication, like rows of chairs, bowls of marbles, and so on; recognizes when multiplication might be a useful tool in solving a complex problem; and understands the role of place value and regrouping in performing multidigit multiplication.

Teaching to *Get It*

Getting it doesn't happen by telling students facts and showing them procedures. *Getting it* doesn't happen by keeping a checklist of what bits of knowledge or what topics or skills were covered on a given day. *Getting it* happens when students engage their minds in mathematical activity and connect what they are now learning to what they already know or to emerging structures in their minds. We can teach to *get it* in a few key ways.

IDENTIFY THE BIG IDEAS AND FOCUS ON HOW THEY CONNECT

Getting it often comes when a student understands a big idea or connecting thread in mathematics. Ideas like equivalence, proportionality, or the uses of letters to represent unknowns, constants, or variable quantities are all examples of important big ideas in mathematics. Targeting such big ideas as a focus of instruction is essential in helping students understand the importance of the idea and its use as they solve problems or learn other concepts and skills that may build on the idea.

Several years ago a friend gave me a book called *The Important Book* (Brown 1949) that he often referenced when doing workshops with teachers. This children's book has short descriptions of common things, such as grass, an apple, the sky, and so on, listing the author's view of what's really important to know about each thing. For Helen Keller, the important thing about words was that they represented something. We may not all identify the same important thing about a big idea in mathematics, but focusing on something important is far more helpful for students than being distracted by something trivial. For example, we might say that the important thing about numbers is that each one has a particular, fixed relationship to other numbers, rather than being distracted by the shape of the numeral that represents the number. Or the important thing about a cylinder is that its radius is the same at any point of the cylinder, not whether the cylinder is short or tall. (I knew a student in Texas who was distracted by the word *cylinder*, since she heard her teacher's pronunciation to sound like the way she heard the word *slender*. She thought a cylinder could only be tall and thin, not short and wide.)

Helping students attend to what is important can contribute to the student's ability to understand the concept or idea and, perhaps, to truly *get it* in ways that prepare the student for future learning.

The important thing, for example, is not only to be able to solve *this* equality with ratios, but also to understand the bigger idea that in a proportional relationship, as one quantity increases or decreases, a related quantity increases or decreases in a predictable way. When a student *gets* that idea in a deep and meaningful way, the door opens not only to solve a wealth of problems involving ratios and proportionality, but also opens the door to later understand linear relationships and, eventually, a wide range of other functional relationships. Identifying the big ideas and priority topics at each grade level or course is a first step in focusing our teaching on what is important.

BE CAUTIOUS WITH SHORTCUTS AND TRICKS

As teachers, we can be seduced by the temptation to teach students mnemonics to help them memorize a rule or procedure. Tricks and shortcuts can be helpful, as long as they build on deep understanding and, ideally, proficiency. I recently met a man who proudly proclaimed that he had taught his daughter all she needed to know about trigonometry in thirty seconds. He taught her the mnemonic SOHCAHTOA (pronounced in various ways, often sounding like the name of an exotic princess), to help her remember that, in a right triangle, the sine is the ratio of the opposite side to the hypotenuse, the cosine the ratio of the adjacent side to the hypotenuse, and the tangent the ratio of the opposite side to the adjacent side. Aside from potential difficulties in remembering an unusual name (I misspelled it twice in writing this message), obviously this mnemonic does not represent everything a person needs to know in order to *get* trigonometry. Seeing the connection between circular functions and right triangles and realizing the power of periodic motion to model the real world are among the important central themes in trigonometry. A student who has memorized the definitions of trigonometric ratios in right triangles is not likely to *get* these key aspects of trigonometry or to ever experience a trigonometric *Aha!* moment.

Likewise, one student I know believed she knew how to multiply binomials because a friend had taught her the mnemonic FOIL (first, outside, inside, last). Unfortunately, she spent more time memorizing the letters than learning the distributive property. Consequently, she consistently multiplied binomials incorrectly, always thinking that *first* referred to the two terms in the first binomial and *last* referred to the two terms in the second (last) binomial. In both of the above cases, as with learning PEMDAS (order of operations) or the so-called butterfly method (solving fraction or ratio equalities), to name a few others, students might learn the trick but miss the point. If we're going to help students *get* mathematics, and if we have to choose how to spend our precious instructional time, I suggest that it's far better to invest that time in teaching for lasting understanding of the mathematics and working toward those *Aha!* moments than teaching tricks that make students believe they know what's going on when they may not. Let students know a mnemonic if they *get* what's behind it; just don't focus the teaching on the trick at the expense of making sense of the mathematics.

USE WORTHWHILE TASKS TO DEVELOP MATHEMATICAL HABITS OF MIND

Students are far more likely to experience *Aha!* moments when we offer them deep, challenging tasks that elicit mathematical thinking. (See Message 12, "Upside-Down Teaching," which offers a discussion of such tasks, and Appendices A, B, and C, which include resources for selecting and evaluating tasks.) When we structure our teaching around tasks students may not already know how to solve in advance, and when we give students opportunities to discuss their thinking about a challenging problem, students are more likely to engage in thinking about the underlying mathematics and more likely to be open to learn the mathematics we want them to learn.

One of the best tasks I've ever seen for helping ignite students' light bulb appeared in a third-grade class I visited. The teacher had every student make a small book about multiplication. They were to fill the book with things they knew and were learning about multiplication—words or pictures or diagrams that showed what multiplication is, facts they wanted to remember or had already learned, word problems that might be solved with multiplication, puzzles involving multiplication, and so on. I thought this was a great example of connecting mathematics and language arts. And, as students built their books, they were not only learning about multiplication, they were also developing mathematical habits of mind. They represented their growing knowledge in multiple ways and made important connections among words, numbers, tables, diagrams, and ideas. Their little books grew as they learned more and more, and the teacher could easily check each student's developing understanding by periodically looking at what they had added to their books.

What Can We Do?

I'm guessing that every teacher has experienced at least one moment when an inner voice said, "They're not *getting* this...." That voice is trying to tell us that we need to try a different approach. It's not always easy to teach toward *getting it*, and even the best strategies and the richest problems may go astray. It takes time, resilience, and persistence to teach toward big ideas and focus on how those ideas connect, to avoid shortcuts and tricks, to choose worthwhile tasks that engage students in thinking deeply and making connections, and to invest in developing mathematical habits of mind. But all of these elements are central to creating an effective mathematics classroom likely to produce many *Aha!* moments and emerging insights, and as such, they are worth our time and effort.

Helen Keller, of course, went on to become a prolific author and well-known social and political activist, attending Radcliffe College and receiving the first bachelor degree awarded to a deaf and blind person. Her contributions to society are numerous. Perhaps those of us involved in teaching mathematics can also benefit from her story in ways she never anticipated. Imagine if all of our students were able to experience the kind of *Aha!* moment in mathematics that Helen experienced that day with words....

> Once I knew only darkness and stillness ... my life was without past or future ... but a little word from the fingers of another fell into my hand that clutched at emptiness, and my heart leaped to the rapture of living.
>
> —Helen Keller

Reflections and Discussion

FOR TEACHERS

- What issues or challenges does this message raise for you? In what ways do you agree with or disagree with the main points of the message?
- What *Aha!* moments have you ever experienced? How did they shape your experiences that followed?
- Have you ever seen a student experience an *Aha!* moment? What was that like for the student? For you?
- How can you tell when a student has memorized something versus actually *getting it*?
- What structures or pressures from outside your classroom get in your way or keep you from teaching in ways that students can *get it*?
- What is one thing you can do to help students *get it*—understand and deeply internalize an important idea, concept, or skill in mathematics?

FOR FAMILIES

- What questions or issues does this message raise for you to discuss with your daughter or son, the teacher, or school leaders?
- What *Aha!* moments have you ever experienced? How did they shape your experiences that followed?
- How can you help your son or daughter focus on what is most important about the mathematics he or she is learning, rather than focusing simply on getting the answer to the next problem?

FOR LEADERS AND POLICY MAKERS

- How does this message reinforce or challenge policies and decisions you have made or are considering?
- What *Aha!* moments have you ever experienced for yourself? How did they shape your experiences that followed?
- What structures and requirements are placed on teachers that may keep them for teaching towards *getting it*?
- How can you give teachers permission and support to focus on helping students *get it*, rather than focusing on solving the next problem or covering the next topic?

RELATED MESSAGES

Smarter Than We Think

- Message 12, "Upside-Down Teaching," describes a teaching model centered on in-depth problems and discusses the nature of such worthwhile mathematical tasks.
- Message 18, "Finishing Teaching," stresses the importance of teaching for understanding and lasting learning.
- Message 8, "Oops!," looks at the value of learning from mistakes on the road to discovery and learning.
- Message 31, "Developing Mathematical Habits of Mind," presents an overview of mathematical habits of mind.
- Message 40, "Mathematical Habits of Instruction," offers thoughts from across many messages about how we can teach to foster mathematical habits of mind.

Faster Isn't Smarter

- Message 17, "Constructive Struggling," makes a case for challenging students with engaging problems that require them to work hard and wrestle with mathematical ideas.

MORE TO CONSIDER

- *Beyond the Miracle Worker: The Remarkable Life of Anne Sullivan Macy and Her Extraordinary Friendship with Helen Keller* (Nielsen 2010) tells the story of Anne Sullivan and her work with Helen Keller.
- *Little Bets: How Breakthrough Ideas Emerge from Small Discoveries* (Sims 2011) looks at *Aha!* moments in various fields and provides insights into how to stimulate more.
- *The 5 Elements of Effective Thinking* (Burger and Starbird 2012) offers five strategies for becoming a stronger thinker and for gaining insights into challenging problems, including igniting insights through mistakes.
- *Imagine: How Creativity Works* (Lehrer 2012) explores the nature of imagination, creativity, and discovery in business, schools, and the community.
- *A Whole New Mind* (Pink 2006) builds on brain research to look at the kinds of intelligence we need to nurture in students for the twenty-first century.
- Refer to Appendices A, B, and C for several excellent resources on teaching around excellent tasks and fostering classroom discourse, both of which are key ingredients in helping students generate *Aha!* moments.

Loving Algebra as Much as Art

WHAT KEEPS STUDENTS COMING TO SCHOOL?
(JAMAL'S STORY)

School is my way out, into taking care of myself … I really want the college experience.… But out of six of my classes, three are interesting and three are a waste of time.

—Mahogany (student), *Fires in the Bathroom*
(Cushman 2005)

Jamal was a young man sent by his mother from inner-city Detroit to live with his aunt in Texas. His mother worried that Jamal might be headed toward a dangerous future without some kind of dramatic intervention. So Jamal showed up in my algebra class in the semirural suburbs of Austin.

In my class, I was trying out the prototype of a new algebra book developed around the idea that every lesson could begin with an engaging and relevant task or problem. Jamal was a bright young man, always ready to take on the tasks I presented. He participated well in class activities, earned good grades on his work, and increasingly interacted in positive ways with his new classmates.

One day I noticed that Jamal had not been in class for a couple of days. I asked another student where he was, and I was told he was in ISS (in-school suspension). I was shocked, and I asked why. The student told me that Jamal had not been going to class. "Yes, he was!" I contradicted. The student replied, "Well sure, he was coming here. But he was only going to art and algebra."

This response was not what I expected—that any student would choose to participate only in art class and algebra class seemed an odd combination. My ego said it must be because of my great teaching; my common sense told me it was more likely because algebra class, like art class, offered Jamal a stimulating, creative environment. In algebra, Jamal was engaged in the mathematics we were doing—he was hooked into problems that made him think, and he thrived in a learning environment that challenged him.

Why Students Don't Come to Class

When students choose not to come to class, it can be an early warning sign that they may be headed toward dropping out of school. In the 2006 report *The Silent Epidemic*, the authors describe results of surveys and conversations among students who dropped out of school (Bridgeland, DiIulio, and Morison 2006). Almost half of the students reported that their classes weren't interesting and that they were bored and disengaged from school. This was especially true for students with the highest grades and strongest motivation to work hard. And, two-thirds of students who dropped out of school reported that they would have worked harder if more had been expected of them in terms of higher standards and more work. Similarly, an older study of Native American *school leavers* reported of students who spoke of being bored in remedial classes, repeating the same exercises, and studying uninteresting subjects (Deyhle 1989). It seems clear that in many of our schools, we may not be offering the kinds of interesting, relevant, challenging work that both engages and educates students. And, unfortunately, a student's disengagement with school may begin as early as elementary school.

WHEN AM I EVER GOING TO USE THIS?

The work that we were doing in my algebra class that year involved forward-thinking materials and seemed to me to be both engaging and relevant. The program was excellent, but it wasn't entirely unique; an increasing number of programs today focus on rich, challenging tasks and interesting, extended problems as a primary structure for day-to-day teaching. This shifting focus on relevance and engagement is different from the more typical lecture-based teaching model historically seen in mathematics classrooms in past years. But I noticed something remarkable in reflecting back over that year with Jamal's class. In February one student said to me, "When am I ever going to use this?" and that was the only time all year I heard that question. In my experience, mathematics in general—and algebra in particular—can easily give rise to students questioning the relevance of what they are taught. This seems to be especially true when we teach mathematics in more traditional ways that emphasize symbolic manipulation and teacher presentation over students using mathematics to model and solve interesting, relevant problems. When our teaching focuses on modeling and solving such interesting, relevant problems, students simply aren't as likely to ask the dreaded question.

Teaching to Invite Students

What keeps students choosing to come to school? For some students, it can be one or two classes that interest them or one or two teachers who personally connect with them. For others, it can be a particular extracurricular activity. And for some, it can be an opportunity to socialize with friends. Teachers can have a significant impact on a student's decision to come to school in several ways.

TEACH FOR ENGAGEMENT

Beyond personal connections with students, teachers can most directly influence students' interest in coming to school by the way we structure our classes. It may or may not be our job as educators to entertain students, or to cajole, amuse, or trick them into coming to class. But it certainly is our job to engage them in meaningful work that expands their knowledge and, ideally, contributes to their readiness to function as responsible adults after they graduate from school. The good news is that this kind of engagement in challenging mathematics also can motivate students and keep them interested in being in math class.

EXPECT A LOT

Critical to engaging students in mathematical tasks is creating a classroom environment based on expectations for the level of work students will do—asking them to work hard and reach high standards. But simply assigning hard problems or presenting advanced mathematical concepts isn't enough for most students. Helping students understand the nature of intelligence and the contribution of effort to learning can be a critical element in supporting students' perseverance and willingness to engage in doing hard mathematics.

I'd like to see us expect more of students, not only in terms of challenging content, but also by expecting them to think, figure things out, and grapple with hard concepts they may not have seen before. I'd like to see us structure classrooms where students learn to work collaboratively in ways that make each one responsible for learning and where students routinely present their work to classmates or to adults. The design of a fourth-year, twelfth-grade mathematics course developed in Texas and used in many states puts such expectations front and center. In the course Advanced Mathematical Decision Making (Charles A. Dana Center 2010a), also called Advanced Quantitative Reasoning, students are presented applied situations calling for them to research and gather evidence to support solutions to a wide variety of problems in statistics, finance, and numerical reasoning. They work in pairs or small groups, where every student is responsible for contributing, understanding, and presenting results. Overwhelmingly, students report that the course is harder than other mathematics courses they have taken, even if they have taken precalculus, but that it's also more interesting. Both students and teachers seem to be rewarded and motivated by the high expectations that are part of the course. Some other curriculum programs for various grade levels or courses are organized around student engagement with deep problems. And some specialized schools operate this way, such as some of the so-called new tech high schools or specialized interdisciplinary academies.

DEVELOP MATHEMATICAL HABITS OF MIND

Teaching to invite students to be present and participate doesn't need to involve any particular teaching style or strategy. It has far more to do with developing mathematical habits of mind, making sure that every student has the opportunity to engage in worthwhile mathematical tasks that draw students into challenging ideas they can wrestle with. Our goal should be for students to connect with and make sense of the mathematics they are learning as they continually expand their set of mathematical thinking skills.

What Can We Do?

Maybe our job as educators includes doing what we can to give every student a reason to come to school. I think that reason begins with making a personal connection with our students and letting them know they have someone cheering for them. We can also make school a place students want to be by engaging every one of them in interesting, challenging, and often relevant tasks and expecting them to assume increasing responsibility for learning, producing, and communicating about their work. In doing so, we have an opportunity to help them build the knowledge and skills it takes to succeed in life after they finish school.

The day I found out why Jamal wasn't in class, and for several afternoons afterwards, I visited him in ISS, taking a textbook and a graphing calculator with me. Jamal and I had some great interactions around interesting problems during those afternoon sessions. Jamal came back to class for a few days after that, and then left school for good. I never knew where he went from there, but perhaps the relocation experiment hadn't turned out the way his mother wished it would. Or maybe it did, and Jamal grew in a way that his mother saw promising. Maybe he returned to his mother in Detroit and did fine in his school back home. I hope Jamal eventually found himself and I hope he realized what a great mind he has and what a fine adult he could become. I also hope he found more reasons to stay in school besides art and algebra.

Reflections and Discussion

FOR TEACHERS

- What issues or challenges does this message raise for you? In what ways do you agree with or disagree with the main points of the message?
- How can you recognize when a student may begin to become disenfranchised with school or with mathematics class? What can you do if you see a student beginning a downward spiral of disengagement that may lead to lower attendance, poor grades, or worse?
- How can you shift your teaching or the structure of your classroom to more fully engage more students?

FOR FAMILIES

- What questions or issues does this message raise for you to discuss with your son or daughter, the teacher, or school leaders?
- How can you ensure that your daughter or son is attending and participating in the entire school day?
- How can you find out whether your son or daughter may be becoming bored or disenfranchised with school? If you see warning signs, how can you approach your son or daughter and the school to prevent a downward spiral toward lower attendance or poor grades?

FOR LEADERS AND POLICY MAKERS

- How does this message reinforce or challenge policies and decisions you have made or are considering?
- What kinds of supports can you provide for students who may become disenfranchised with school?
- How can you encourage or ensure that all (mathematics) teachers teach in ways that engage students and invite them to be in school?

RELATED MESSAGES

Smarter Than We Think

- Message 1, "Smarter Than We Think," illuminates current thinking about the nature of intelligence and the role of effort in becoming smarter.

- Message 4, "They Just Aren't Motivated," discusses factors related to students' motivation to do mathematics.
- Message 12, "Upside-Down Teaching," describes the benefits of a problem-centered teaching approach.
- Message 35, "Using Math in the *Real* Real World," advocates the development of mathematical modeling through contextual problems, often ill-defined.

Faster Isn't Smarter

- Message 8, "More Math, More Dropouts?," looks at the implications of raising mathematics requirements on students' choices to stay in school until graduation.
- Message 37, "Boring!," considers the importance of engaging students' interest.

MORE TO CONSIDER

- *The Silent Epidemic: Perspectives of High School Dropouts* (Bridgeland, DiIulio, and Morison 2006) describes a large-scale study of students who chose to drop out of school.
- *The Immortality of Influence: We Can Build the Best Minds of the Next Generation* (Thomas-EL 2006) offers an uplifting true story of the power of mentoring, high expectations, and support for students apparently unlikely to succeed.
- *Fires in the Bathroom: Advice for Teachers from High School Students* (Cushman 2005) provides insights from students about their experiences in high school.
- *Fires in the Middle School Bathroom: Advice for Teachers from Middle Schoolers* (Cushman and Rogers 2008) provides insights from students about their experiences in middle school.
- *Fires in the Mind: What Kids Can Tell Us About Motivation and Mastery* (Cushman 2010) is the third book from the What Kids Can Do project, with explicit suggestions from students about how to keep them engaged in school and in their own learning after they leave school.
- "What Is High-Quality Instruction?" (Weiss and Pasley 2004) reports research findings about elements of effective mathematics instruction.
- *Advanced Mathematical Decision Making (Advanced Quantitative Reasoning in Texas)* (Charles A. Dana Center at the University of Texas at Austin 2010b) provides student and teacher materials for a twelfth-grade capstone mathematics course focused on statistics, finance, discrete mathematics, and quantitative reasoning.

7

Families

THE IMPORTANT OTHER TEACHER IN STUDENTS' LIVES

> When I was very young, most of my childhood heroes wore capes, flew through the air, or picked up buildings with one arm. They were spectacular and got a lot of attention. But as I grew, my heroes changed, so that now I can honestly say that anyone who does anything to help a child is a hero to me.
>
> —Fred Rogers (Mister Rogers 2003, 145)

As a seventh-grade student in Mr. Bender's seventh-grade math class, I used a funny little red paperback book that was different from anything I'd ever seen in my previously unexciting experiences with arithmetic. Using the red book, Mr. Bender introduced us to the world of mathematics beyond arithmetic, and my life changed forever. Over the next three years, I was given green paperback math books that addressed algebra, geometry, and lots of other topics. I found out later that all of these books were part of the *new math* movement of the 1960s. All I knew at the time was that I had come to love mathematics. It later turned out that I would choose to spend my career working with it.

As an adult, working with mathematics educators, mathematicians, and policy makers, I've learned about that movement from a different perspective, especially about why some people consider the movement to have not fulfilled its promise of reforming school mathematics. In a recent conversation with an original developer of the SMSG (School Mathematics Study Group) program from that era, he mentioned that one of the mistakes they made at that time was in discouraging parents from helping their students with their math homework. The developers believed that parents would not understand the sometimes radically different approaches students were learning, and that any attempt by parents to help their children with ways of solving problems they themselves had learned might mislead or distract them from the well-ordered path the developers had in mind. In hindsight, it's clear that this was not a useful approach. It's now quite obvious that parents, family members, or other caregivers are crucial partners in every child or young person's education. Rather than ask families to back away, we need to find ways to invite and engage them in supporting their students' mathematics learning.

Families as Partners

Students live in many different types of environments. Most live with parents or grandparents as their primary caregivers, but some may live in different situations. Regardless of what a student's home life might look like, if there is a significant person supporting that student in any way, that person can help the student learn mathematics. While some supporting adults might worry that they aren't strong in mathematics themselves, by working together with the teacher, any adult can find an appropriate way to encourage or support their student. Communicating a positive attitude about mathematics, its importance, and its usefulness is a critical first step, regardless of an individual's own feelings about, or experiences with, mathematics. The most effective teachers and schools know the importance of engaging parents—or whoever supports the student outside of school—in their son's or daughter's education, and they support parents to help them find the best way(s) for parents to support the student.

Reaching Out to Families

Effective schools reach out to parents and families to engage them in the activities of the school and, especially, their students' learning. They use many ways to reach out to parents, making it as easy as possible for parents to stay informed and involved, even if schedules prevent a parent from attending face-to-face events. In today's technology-driven environment, various social media outlets provide ways to connect anyone with a smartphone, tablet, or access to a public computer with the school and, in some cases, the classroom. Email and written or web-based communication like newsletters or updates can keep parents informed not only about important events but also about the direction of their son's or daughter's mathematics program. The most important thing for schools and teachers to do is to invite, invite, invite parents and family members to stay engaged with the mathematics their children are learning.

For parents who are able to spend even a small amount of time in school activities, many schools have had success offering one or more math nights for families and students. These experiences might utilize the resources of the long-standing program Family Math (Lawrence Halls of Science/EQUALS), or might introduce engaging classroom problems to families or showcase student presentations of long-term problems or projects. And when important decisions are being made about the future of the mathematics program, schools should consider involving one or more parents on committees they form.

Families Connecting to Schools

One of the best things a parent can do is to learn more about what's happening in mathematics class, whether the student is in elementary school, middle school, or high school. Examine your beliefs, using the list that follows as a guide, and consider being open to rethinking them based on what we know about how to help students learn to think mathematically. Have conversations about math at home, and look for opportunities to use mathematics in making decisions or in planning family activities. Play games involving reasoning, thinking, and solving interesting problems. Most of all, raise your goals for your son or daughter. Don't let it be enough to learn how to add, subtract, multiply, and divide. Advocate learning how to think mathematically and reason to solve complex problems. This is an appropriate goal for every student, regardless of age or future academic plans.

Learning What to Advocate

The world has changed considerably since most educators and parents learned mathematics, or, in some cases, arithmetic. And like parents, even some educators may need to examine their beliefs about how mathematics can most effectively be taught and learned in a technology-driven world in which thinking, reasoning, and the ability to solve complex problems have become the most critical workplace skills in demand. In particular, consider some of these beliefs and principles underlying today's high-quality mathematics programs:

- **More Than Computation**
  Mathematics involves more than learning how to compute.
  Because learning procedures is only part of the mathematics students need to know, mathematics programs today focus on helping students make sense of mathematics, learn how to reason, explain their thinking, and solve a wide range of problems.
  (See Part IV, "Messages About Thinking Mathematically in a Common Core World" for Mathematical Practice and related mathematical habits of mind.)

- **The Importance of Perseverance**
  Students need to develop a willingness to persevere in solving problems. Solving problems is not about memorizing what

kind of procedure goes with what kind of word problem; the most valuable problems students need to solve do not follow prescribed procedures. Rather, they call for students to interpret the problem, synthesize what they know, and try various approaches to determine which might be most productive. If students have not had experience with this kind of problem solving—tackling a problem without first knowing the rule(s) they might use to solve the problem—they are likely to give up quickly when faced with a new problem, thinking that it's beyond their ability. More and more school mathematics programs offer students opportunities to engage in challenging problems where they need to wrestle with the mathematics and try out possible solution strategies to see what might work. In this way, students learn to think about the mathematical ideas behind the problem rather than trying to remember what rule to use, and they are much more likely to be willing to persevere in solving the next problem they encounter.

Struggling with problems students don't know how to solve in advance is an important mathematics experience. It's not about hoping students will discover the mathematics we want them to learn, but rather about engaging students in thinking about mathematical situations so they connect what they learn to their experience. In this way, students are far more likely to make sense of what they learn and put it into a mental context they can call on when they encounter similar mathematical situations in very different kinds of problems in the future.

- ### The Dangers of Tracking

  Tracking students by so-called ability level does not help them learn mathematics. While it may seem logical to group students together by ability, it turns out that *ability* is not well defined and is not well identified by current measures. Ability is far more influenced by opportunity to learn than we may have thought in the past. Moreover, we now know from brain research that students' actual intelligence can be increased by engaging them in challenging tasks (Dweck 2006). Thus, a better idea than tracking is to focus teaching around challenging problems, as described earlier. I sometimes call this approach—taking on a problem without necessarily first knowing how to perform all the procedures that might help solve it—*upside-down teaching* (see Message 12, "Upside-Down Teaching"). This kind of learning experience, especially when students work with others who may approach problems differently, allows students to develop powerful reasoning and thinking skills. They can advance their learning in ways not possible when students who may have learned particular ways of succeeding in school are simply grouped together.

- It's Not Hereditary

  There's no such thing as a math gene. While it may seem that
  facility with mathematics gets passed on from parent to child,
  today we know that any person without a serious learning
  problem, and even some who may have learning problems, can
  learn mathematics well if taught in ways that engage the person in
  thinking and reasoning. One of the most detrimental things we can
  do with children is to make them think they have inherited either
  the ability to do mathematics (they may give up and feel like they're
  not living up to their potential when they get to a hard problem
  they don't immediately know how to solve) or have *not* inherited
  the ability to do mathematics (they may not even be willing to try
  a problem if they think it's beyond their innate ability). Parents and
  family members can help students considerably by encouraging
  them to work hard and by acknowledging accomplishments rather
  than praising ability or intelligence.

What Can We Do?

Families play a critical role in helping students learn mathematics—a
parent or family member can truly be a young person's *other teacher*,
even if not an expert in mathematics. Working together, families and
schools can reinforce a student's confidence and willingness to persevere
in solving a problem by focusing on hard work, rather than innate ability. Working together, we can help students see mathematics as more
than computation, extending to interesting and relevant problems and
situations they see every day. Working together, we can help students
make sense of the mathematics they see in school and learn to reason
and think mathematically, so that they can solve the kinds of problems
they will face in the future, both in and outside of school.

Reflections and Discussion

FOR TEACHERS

- What issues or challenges does this message raise for you? In what ways
  do you agree with or disagree with the main points of the message?
- What is one thing you can do to invite parents or family members to find
  out more about the mathematics you're teaching?

(continued)

- What kind of technology or social media have you used to reach out to families? What other venues might you consider using or using differently?
- What kinds of experiences might you and your colleagues offer to invite families to directly experience the kind of mathematics their children are learning?

FOR FAMILIES

- What questions or issues does this message raise for you to discuss with your daughter or son, the teacher, or school leaders?
- How open are you to reconsider any beliefs you might have about what is important in a mathematics classroom today?
- Does your school track students—group them for mathematics teaching by identified ability level? Where can you learn more about the negative effects of "tracking" students?
- Have you taken advantage of opportunities the school or teacher may be offering to stay informed about your son's or daughter's mathematics program? If there aren't adequate vehicles for communication, how can you constructively help expand your access to important information?

FOR LEADERS AND POLICY MAKERS

- How does this message reinforce or challenge policies and decisions you have made or are considering?
- How can you help teachers work toward "untracking" their students?
- What outreach vehicles to families and the community, including social media, do you use (or can you use) to communicate important information about school programs?
- How have you involved (or how can you involve) parents or representatives of parents in committees making crucial mathematics program decisions?

RELATED MESSAGES

Smarter Than We Think

- Message 12, "Upside-Down Teaching," describes a teaching model built around problems students may not previously have learned how to solve.
- Message 1, "Smarter Than We Think," presents information about the growth model of intelligence—that any student can not only learn mathematics, but can improve their intelligence.

- Message 32, "Problems Worth Solving," elaborates the nature of mathematical thinking and problem solving called for in today's world.

Faster Isn't Smarter
- Message 17, "Constructive Struggling," promotes the importance of having students grapple with mathematics in meaningful ways while addressing mathematical tasks they don't already know how to solve.
- Message 1, "Math for a Flattening World," reminds us that today's world calls for a different kind of mathematical learning than in previous times.
- Message 27, "A Math Message to Families," offers suggestions for families to stay involved in their sons' or daughters' mathematics programs.

MORE TO CONSIDER

- *Mindset: The New Psychology of Success* (Dweck 2006) is the breakthrough book for both educators and the public about the implications on our lives of our mindset regarding intelligence.
- *A Family's Guide: Fostering Your Child's Success in School Mathematics* (Mirra 2005) offers clear explanations and practical ideas about how families can help students with their mathematics learning.
- "Involving Latino and Latina Parents in Their Children's Mathematics Education" (Civil and Menéndez 2010) summarizes research about the importance of involving parents in the mathematics education of their children.
- *Math: Facing an American Phobia* (Burns 1998) addresses the widespread dislike of mathematics in the United States adult population and offers insights on how to get past any negativity to use math in everyday life.
- "What Should I Look for in a Math Classroom?" (Charles A. Dana Center at the University of Texas at Austin 2012a) offers tips for non-educator adults on characteristics of effective mathematics classrooms.
- *Getting Your Math Message out to Parents: A K–6 Resource* (Litton 1998) offers suggestions for educators in reaching out to families in support of students' mathematics learning, including ideas for newsletters, conferences, and how to involve family volunteers in schools.
- "Doing Mathematics with Your Child" (Hartog and Brosnan 1994) offers helpful ideas for working on mathematics at home with a

child and for connecting with a child's teacher to support his or her learning.

- Common Core State Standards for Mathematics (NGA Center and CCSSO 2010) are widely adopted standards that provide the framework for mathematics instruction in most states; the concisely stated Standards for Mathematical Practice (2010, 4–6) offers insights into the kind of mathematical thinking needed in today's world. (See also Part IV of this book for Messages 32–39 addressing the Standards for Mathematical Practice.)

- *Beyond Tracking: Multiple Pathways to College, Career, and Civic Participation* (Oakes and Saunders 2008) looks at organizing schools to prepare students for life afterwards, including a discussion of tracking.

- "Universal Access to a Quality Education: Research and Recommendations for the Elimination of Curricular Stratification" (Burris, Welner, and Bezoza 2009) summarizes research strongly supporting the elimination of tracking as a strategy to support student learning.

WWW This message is also available in printable format
at mathsolutions.com/smarterthanwethink.

8 Oops!

THE VALUE OF MISTAKES, FAILURES, AND WRONG TURNS

I've missed more than 9,000 shots in my career. I've lost almost 300 games. Twenty-six times, I've been trusted to take the game winning shot and missed. I've failed over and over and over again in my life. And that is why I succeed.

—Michael Jordan

Consider the following views of mistakes, failures, and wrong turns:

- In his fascinating look at the value of working in the trades, Matthew Crawford extolls the virtues of failure (*Shop Craft as Soulcraft*, 2009). He sees failure as a way to help us understand the potential impact of our actions and argues that we will make more thoughtful decisions in the future if we understand that sometimes things can go wrong—sometimes very wrong—and that our actions may contribute to that process.

- Paul Halmos describes his life's work as a well-respected mathematician as being filled with wrong turns on the way to understanding and discovery in *I Want to Be a Mathematician* (1985). His story gives us insights into how mathematical research is done and how much it involves messy work and learning from paths that turn out not to end up where we intended so that, eventually, we can find a different path that yields positive results.

- Peter Sims chronicles the contribution of mistakes to innovation, discovery, and entrepreneurship in *Little Bets* (2011). Sims later wrote about five mistakes Steve Jobs made on his way to becoming one of the world's most successful innovators (Sims 2013).

Successful athletes, entertainers, scientists, artists, business people, workers in the crafts and trades, and those who've chosen many other fields of endeavor almost universally recognize both the inevitability and the value of mistakes and wrong turns, even failures. The inevitability comes with the recognition that each of us is human. More than that, however, it turns out that it may be healthy to accept and not fret over our mistakes, failures, and wrong turns. Rather, if we seize the opportunity to learn from what didn't work, we can advance our thinking. In education, this lesson is important for both teachers and students.

Teaching to Learn from Mistakes

A student might consistently make mistakes in an area of mathematics or on a particular skill. If those mistakes are just handed back to the student marked wrong and the student receives yet another poor grade without the opportunity to learn from and correct the mistakes, the student may simply become discouraged—disengaging from future learning and possibly moving forward carrying a misunderstanding that later will prove to be a significant stumbling block. But a mistake can provide a springboard to a productive learning opportunity if we structure our classrooms to take advantage of mistakes.

It's not enough to simply tell a student to do the problem or exercise over again correctly. By creating a classroom environment where mistakes are part of the learning process, we can help students revisit their errors, discuss their thinking, interact with their peers or the teacher, and approach the problem or exercise in a more productive way. Creating such a classroom environment might involve considerations such as modeling how to handle mistakes, sharing and discussing mistakes as a normal part of learning, using an upside-down teaching model, teaching about the growth model of intelligence, giving students opportunities to fail, and conducting ongoing formative assessment.

MODEL HOW TO DEAL WITH MISTAKES

Don't try to teach to avoid mistakes, either yours or students'. Rather, look for opportunities to teach students how to deal with and learn from their mistakes, wrong turns, and even their failures. A close friend once told me, "If you're not making any mistakes, you're probably not doing much."

When I learned to teach many years ago, I came to think that the best way to offer a lesson was to clearly and logically demonstrate how a particular procedure works. I should prepare my lesson carefully, so as to not make mistakes, and I should deliver it with great enthusiasm so that my students would stay engaged. What I have come to believe over time, however, is that often a student sees a teacher flawlessly show *how* a procedure works (maybe even *why* it works) only to think that the teacher must have a great math mind and that the student is never going to be able to come up with things like that independently.

I now believe that it may help students to occasionally see a teacher go down a wrong path or make a flat-out error in front of them. When students see a teacher make an occasional mistake, notice it, and learn from or correct it, it may well offer them insights into how to handle their own mistakes. More than that, structuring a class where students are allowed to share mistakes and learn from these mistakes turns out to be a powerful instructional approach.

ESTABLISH NORMS FOR SHARING AND DISCUSSING MISTAKES AS PART OF LEARNING

If students are to learn from mistakes, the most important thing is for the teacher to create a classroom environment where sharing, discussing, and possibly rethinking solutions are part of everyday life. Establishing classroom norms where all students' contributions are respected and accepted can, over time, invite students to be part of the conversation, even those students who may initially resist participating. Usually, students become increasingly comfortable and relaxed with sharing and discussing their thinking as they learn to trust that the classroom is a safe place to do so. When we let students work in this way and become accustomed to sharing and discussing both promising and unsuccessful approaches, students feel comfortable taking the risk of sharing, even if they think they might not be correct. And when we work together with colleagues to create that safe place year after year, the long-term benefits can be significant in terms of students' risk taking, perseverance, and learning.

USE AN UPSIDE-DOWN TEACHING MODEL

We can reinforce students' developing understanding of the value of mistakes by structuring our classrooms so that we offer students the opportunity to wrestle, individually and collaboratively, with rich, challenging problems. Using an upside-down teaching model (see Message 12, "Upside-Down Teaching,") we encourage students to share their various approaches, ideas, and possible solution paths. By not focusing primarily on answers right away, but also on how to approach a problem, students become comfortable sharing. Sharing a possible approach that may not pan out in a safe classroom doesn't seem nearly as threatening as raising a hand to share *The Answer* to a problem in a lecture-based classroom.

TEACH ABOUT THE GROWTH MODEL OF INTELLIGENCE

We are learning more and more about the nature of intelligence and the benefits of learning from mistakes. We now know that when a person deals with a mistake and learns from it, that person can grow new dendrites that actually make her or him smarter (Dweck 2006). We can help students not by encouraging them to make mistakes, but by teaching them about a growth model of intelligence, where mistakes are opportunities to learn and grow smarter. Message 1, "Smarter Than We Think," takes a look at a growth mindset of intelligence and how it can help us in teaching and learning.

GIVE STUDENTS OPPORTUNITIES TO FAIL

Teaching students the value of mistakes, even sometimes giving them opportunities to fail on a significant scale, can benefit even our most successful students. I'm convinced that many gifted students and many students who are accelerated through an advanced track in school never have the opportunity to struggle with challenging problems. Often, such students have succeeded by being attentive listeners, good memorizers, and reliable homework-doers. We have perhaps too often covered a lot of material with these students, but may not have given them deep, extended problems that cause them to struggle and persevere through figuring out how to approach a complex, ill-formed problem. In *Shop Class as Soulcraft*, Crawford builds a case for the importance of failing, especially for those who may one day advance to leadership positions. He suggests that a person should lead only if that person understands what can go wrong and what influence he or she has on how things can go wrong: "There may be something to be said, then, for having gifted students learn a trade, if only in the summers, so that their egos will be repeatedly crushed before they go on to run the country." While this point of view may seem extreme and harsh, consider that one of the benefits of learning to learn from mistakes is that we come to realize that mistakes are likely to happen, at least once in a while. We can learn that it's important to try to work carefully and avoid mistakes if possible, especially in high-stakes situations where a mistake might have serious repercussions. But, more than that, we can also learn that not only can life go on after a mistake, but some mistakes may end up benefiting both the person making the mistake and those who participate in learning from it if we take advantage of the opportunity it presents.

CONDUCT ONGOING FORMATIVE ASSESSMENT

There's a big difference between a situation where a student shares a possibly incorrect solution to a problem with the class, with students discussing that solution, versus a situation where a student makes a conceptual error that represents an undiagnosed misconception that will later interfere with other mathematical topics or skills. This critical difference reinforces the importance of a teacher conducting ongoing formative assessment—paying attention on a daily basis to how well every student is learning the mathematics targeted in the instruction.

By regularly keeping up with student learning, mistakes and misunderstandings can be addressed while they are still new and before they have a chance to solidify into misconceptions.

What Can We Do?

A failure is not always a mistake, it may simply be the best one can do under the circumstances. The real mistake is to stop trying.

—B.F. Skinner (*Beyond Freedom and Dignity*, 1972)

When students learn to learn from their mistakes, they become more confident problem solvers. Mistakes can arise from and lead to constructive struggle, and learning how to handle mistakes can help students with persistence. Thus, they become more willing to tackle problems that may take more than a quick look to solve. This willingness to work—to exert effort, to persist—is perhaps one of the most important outcomes we might hope for if we are to adequately prepare individuals for success in their personal life and in their jobs in the future.

I'm told that in some cultures, students applaud discussions that arise after one or more students shares a mistake. The applause reflects the class's appreciation for the mathematical conversation and the learning that can follow a discussion about a mistake, including what may have led to the mistake and what learning can arise from it. I hope that's true. More than that, I hope that we can create classrooms that reflect our commitment to respect and value every student's thinking and to learn from mistakes.

To get there, I suggest focusing on three key elements discussed in this message:

- Choose tasks and structure lessons to invite the sharing of solutions and to include time for learning from solutions, both those that work and those that don't. Establish a nurturing and accepting classroom environment with clear expectations that every student's voice and thoughts have value and where students learn that it's safe to share their thoughts. This may involve establishing ground rules, ideally at the school level, but at least at the classroom level.
- Help students understand intelligence—that making mistakes and struggling with hard problems can make them smarter.
- Examine grading policies to verify that learning from mistakes is valued; that answering correctly on a test is not a once-and-done process with a succeed-or-fail outcome.

Nobody likes to make mistakes. Some mistakes and failures can be painful to experience and painful to witness, and we certainly don't want to go through that pain very often. We can all aspire to making as few mistakes as possible and failing as little as possible in our work and in our day-to-day lives. But accepting and learning from our mistakes in life helps us grow as well-adjusted individuals who know how to deal with bumps in the road. Learning mathematics from mistakes helps us develop mathematical habits of mind and can make us smarter. Creating classrooms where this kind of learning is the norm may help students build into their lives a willingness to struggle, persevere, and learn from their mistakes—important skills for continuing to learn and succeed long after they're out of school.

Reflections and Discussion

FOR TEACHERS

- What issues or challenges does this message raise for you? In what ways do you agree with or disagree with the main points of the message?
- How do you deal with students' computational mistakes? With their conceptual mistakes? With incorrect solutions to in-depth problems?
- What shifts could you make in your teaching in order to maximize the learning payoff from your students' mistakes?
- How safe is your classroom for students to take risks in sharing their thinking and their ideas? What would you need to do in order to make your classroom environment even more inviting for the kind of respectful discussions that incorporate learning from mistakes?

FOR FAMILIES

- What questions or issues does this message raise for you to discuss with your son or daughter, the teacher, or school leaders?
- How can you show your daughter or son that mistakes are a normal part of life and that you try to learn from your own mistakes?

FOR LEADERS AND POLICY MAKERS

- How does this message reinforce or challenge policies and decisions you have made or are considering?
- How well do your grading and placement policies reflect the benefits of learning from mistakes?

- What actions, programs, or policies might help teachers learn how to help students learn from mistakes?
- How can you establish norms for teachers and students at a grade level or within a school or district that support positive, nurturing classroom environments where students feel safe to take the risk of sharing their thinking or their ideas?

RELATED MESSAGES

Smarter Than We Think

- Message 1, "Smarter Than We Think," considers the growth model of intelligence, in which much of recent discussions include the importance of making and learning from mistakes.
- Message 12, "Upside-Down Teaching," describes a problem-centered teaching approach where discussion of mistakes is an integral part of the teaching.
- Message 13, "Clueless," discusses the kind of mistakes we can inadvertently promote by superficial teaching.
- Message 31, "Developing Mathematical Habits of Mind," looks at the mathematical habits of mind mathematicians tend to use that we can help students develop, including the important role of mistakes.
- Message 37, "Communicating Mathematically," offers insights about balancing the importance of precision with the value of learning from mistakes.

Faster Isn't Smarter

- Message 17, "Constructive Struggling," advocates giving students problems with which they may struggle or from which they can grapple with challenging ideas, often making important and useful mistakes along the way.

MORE TO CONSIDER

- *Classroom Discussions in Math: A Teacher's Guide for Using Talk Moves to Support the Common Core and More, Grades K–6* (Chapin, O'Connor, and Anderson 2013) offers practical strategies for organizing mathematics classrooms around worthwhile tasks and productive student discourse about mathematical ideas in a safe learning environment.

- *Little Bets: How Breakthrough Ideas Emerge from Small Discoveries* (Sims 2011) offers insights into how creativity works and the potential benefits of making mistakes and building on them.
- "The Power of a Good Mistake" (Gojak 2013b) is a teacher's perspective on the importance of mistakes and how we handle them.
- *Mindset* (Dweck 2006) offers insights into the role of mistakes in growing minds.
- *The 5 Elements of Effective Thinking* (Burger and Starbird 2012) considers five strategies for becoming a stronger thinker and for gaining insights into challenging problems, including igniting insights through mistakes.
- "CCSS-CTE Classroom Tasks" (Achieve, www.achieve.org/ccss-cte-classroom-tasks) provides a series of classroom tasks relating mathematics to career situations.
- "Educ115N: How to Learn Math" is a Massive Open Online Course (MOOC) and related website offered by Stanford University and taught by Dr. Jo Boaler for educators, students, or noneducator adults addressing key issues related to learning mathematics, including intelligence, mistakes, perseverance, problem solving, teaching for student engagement, and other topics, and including fascinating quotes and interviews from Carol Dweck regarding intelligence mindsets, businessman Ray Peacock regarding the nature of work and needed skills for workers, and others.
- "The Role of Mistakes in the Classroom" (Tugend 2011) discusses the importance of mistakes as part of learning. www.edutopia.org/blog/benefits-mistakes-classroom-alina-tugend.
- *Shop Class as Soulcraft: An Inquiry into the Value of Work* (Crawford 2009) offers a discussion of the merits of work and the role of mistakes in the real world or work.
- *I Want to Be a Mathematician* (Halmos 1985) offers a description of the author's respected work in mathematics throughout his life, including many wrong turns and approaches that didn't turn out.

 Learning to Work

EVERY STUDENT'S NEED FOR REAL CAREER SKILLS
(MORGAN'S STORY)

> The more education you have, the more options you have … as
> long as you get some work skills along the way.
>
> —Willard Daggett ("Preparing Students to Be
> College and Career Ready," 2013)

Morgan's parents, a business education teacher and a police officer, always supported her in following her interests. They wanted her to have a good education, so they chose to live in an area with good public schools and made sure she took the "right courses." In their opinion, the right courses included those she needed to get into college—a given since her birth—as well as fun and interesting electives to help her find and follow her passions, wherever her interests might lie. Accordingly, Morgan took all the academic classes expected of a college-bound student in her large, well-respected suburban high school, including Advanced Placement courses in physics, history, and calculus. She also enrolled in lots of classes involving applications of art, such as jewelry making and photography, and every drafting or engineering class that her school offered. Morgan was always a good student, not necessarily the top student in every class, but always received fairly high grades. She did especially well in mathematics and art. She pursued many hobbies and interests, and she was always involved in activities outside of the regular school day.

As a high school junior, Morgan somewhat surprised herself by essentially knocking the top out of the PSAT. Her high score qualified her as a National Merit Scholarship Semifinalist and opened many doors for potential college scholarships. As she went through her senior year, she started accumulating a wonderful resume, receiving all kinds of awards and recognition. Among them was the Technical Education Student of the Year award for her achievements and participation in various Career Technical Education courses and programs, as well as the prestigious Outstanding Senior leadership award—the highest award given to a senior in each high school by the major newspaper in her large metropolitan area—in recognition of her well-rounded background and diverse achievements. She graduated with a dual diploma in technical education and academic honors. After completing a four-year degree in metallurgical engineering at a highly respected university, Morgan now works as an engineer. Wherever her future path may take her, Morgan's high school education prepared her for a successful and fulfilling life as a responsible adult in the twenty-first century.

College or Career Readiness?

Too often, students aiming at college either ignore or don't have time for practical career-focused coursework. While not all students need to select the courses Morgan did, it may be worthwhile for even college-intending students to consider including in their education one or more courses or experiences where they can learn and/or apply workplace skills.

Certainly, future engineers and those planning on other majors in STEM fields (Science, Technology, Engineering, and Mathematics) should take advantage of whatever calculus-focused courses in mathematics and science are available to them in high school. But all students, whether they are headed toward STEM majors, business majors, other majors, or career paths not calling for four-year degrees, need to be prepared with workplace skills. In today's increasingly globalized and technology-driven workplace, this may include specific training on how to do something—preparation for a profession or business of some kind—or perhaps the skills to enter an apprenticeship in a craft or trade. But true career readiness should also teach students something about the value of work and pride in doing it, as well as preparing them to work collaboratively, tackle and solve complex problems, and communicate a wide range of information in writing and in making presentations.

In an intriguing book called *Shop Class as Soulcraft*, Matthew Crawford (2009) advocates the considerable rewards of skilled hands-on work as a means of developing character and improving one's quality of life. Crawford, who holds a Ph.D. in political philosophy, left his directorship of a Washington, DC, think tank to open a Harley-Davidson repair shop. In reading his compelling narrative, I found myself realizing the limitations of our current focus on the *college* part of *college and career readiness* and constantly wondering how we can do a better job of preparing young people to live productive and rewarding lives. I kept coming up with more questions than answers.... How well are we serving students who aren't interested in, and don't need, a calculus-focused education? Are we overlooking students' talents and interests that may lie in the skilled work of a respectable trade or perhaps may lie in the arts or in a field that blends art and work, such as furniture design, jewelry making, or tile work? How can our educational programs best serve such students, and all students? Can we help students meet adults from different fields as role models who pursue their talents and who find work that stimulates their interests? How can we incorporate into our strongest college-preparation programs the development of workplace skills and a positive work ethic based on the rewards of work?

ACADEMIC-FOCUSED MATHEMATICS LEARNING FOR EVERYONE

In terms of mathematics preparation, we need to ensure that high school course offerings include one or two options other than calculus or precalculus for students' last exposure to mathematics before leaving high school. In particular, rigorous applied capstone courses such as those involving statistics, finance, quantitative reasoning, mathematical decision making, discrete mathematics, and so on, need to be available to all students so that they can learn the skills they need and will use in both STEM and non-STEM career paths. Such courses can have academic integrity and help students learn important, rigorous mathematical skills as they relate to a variety of applied situations.

WORKPLACE-FOCUSED LEARNING FOR EVERYONE

Many workplace skills are currently incorporated into well-designed Career Technical Education programs. But critical twenty-first-century workplace skills—working collaboratively to solve problems, thinking critically about a dilemma, gathering evidence to back up a point of view, writing reports, making presentations—should also be incorporated into courses and programs designed for college-intending students. In today's world, more than ever before, preparing students to be *college and career ready* has to include more than superficial attention to *career readiness*. Too often, school programs use this phrase automatically to describe high school programs focused on preparation for college, sometimes including Career Technical Education courses for those students who aren't "college material," but generally not offering such courses for the college-intending. In mathematics, focusing on college readiness means taking courses aimed at calculus, generally including Algebra I, Geometry, Algebra II, and Precalculus. Workplace-focused learning cannot be just for students who can't make the grade in college-bound programs. On the contrary, students in a high academic track who are busy accumulating AP credits often miss the opportunity to learn how to do a practical skill or to learn what is involved in working in a job. While it was once common for every student to take courses in shop or home economics (although at one time separated for boys and girls), such courses have all but disappeared from school programs today except for students in specialized Career

Technical Education programs. And increased graduation requirements for most students make it difficult to find time to take this kind of elective course outside of basic academic subjects. Yet enrolling in one or more practical or workplace-oriented courses can quite likely benefit all students, regardless of their future direction.

What Can We Do?

Students come in all different kinds of packages, not just *college bound* and *everyone else*. Some will complete a four-year college degree, some may complete a graduate degree, some may complete a two-year degree, some may complete a technical certificate, some may get partway through any of the above, and some may enter the workforce directly. Few students are likely to know as they begin high school where they are likely to be headed a few years later; many who think they know their direction end up someplace else altogether. We have a responsibility to all students to prepare them for a fulfilling future so that every one of them can become a productive, happy, fulfilled adult.

Like all courses in school—especially high school—our mathematics courses could learn a lesson from our Career Technical Education colleagues. Today's marketplace is evolving rapidly, with jobs disappearing or morphing into new, previously nonexistent fields. And new jobs appear every year. Courses should be designed to incorporate emerging workplace technologies and to help students learn adaptable, transferable skills like working collaboratively, solving complex problems, making decisions, and communicating both orally and in writing. And well-designed Career Technical Education courses in a variety of fields should not be reserved only for those students who aren't likely to go to college. Counselors, teachers, and parents can encourage students to expand their thinking and develop practical skills by incorporating one or more such courses into their well-balanced education.

Thanks to the support of her parents and her own motivation and work ethic, Morgan faces a bright future; she knows she's an engineer and takes pride in what she does. It's possible that some day Morgan will shift her interest or her career in another direction. But if and when that happens, I'm confident that with her background of academic knowledge, together with her real-world, practical applied skills, and with her ability to communicate, collaborate, make decisions, and solve problems, she will take pride in whatever she does and she will do it well, or learn how to do it well. She knows the value of work and she understands the value of learning. Maybe some day we can help all of our students understand these concepts and find work that both feeds their souls and also feeds themselves and their future families.

Reflections and Discussion

FOR TEACHERS

- What issues or challenges does this message raise for you? In what ways do you agree with or disagree with the main points of the message?
- Regardless of the grade level or course(s) you teach, to what extent do you incorporate into your classroom the kinds of skills and values called for in today's and tomorrow's workplace?
- How can you find out more about your students' interests and help them discover potential new interests?

FOR FAMILIES

- What questions or issues does this message raise for you to discuss with your daughter or son, the teacher, or school leaders?
- How can you find out more about your son's or daughter's interests and help him or her discover potential new interests?
- How can you best help your daughter or son prepare for the future, both in terms of preparing for education and preparing to have a job?

FOR LEADERS AND POLICY MAKERS

- How does this message reinforce or challenge policies and decisions you have made or are considering?
- How do your programs address the development of twenty-first-century career skills, such as decision making, reasoning, collaboration, and communication? How do they nurture innovation and creativity in both students and their teachers?
- What kinds of apprenticeship programs might you be able to offer to help students experience the kind of work done in professions, such as teaching, business, or health care, as well as in the crafts and trades?

RELATED MESSAGES

Smarter Than We Think

- Message 27, "Fixing High School," looks at how we can modify high school programs to better meet the needs of all students.

- Message 3, "He Doesn't Know His Facts: Perspectives on Mathematical Fluency (Richard's Story)," tells the story of a man who became tremendously successful in education and in the workplace in spite of computational deficiencies.

Faster Isn't Smarter

- Message 1, "Math for a Flattening World," offers thoughts about the kind of knowledge and skills twenty-first-century workers need to thrive in a rapidly changing global environment.
- Message 3, "Making the Case for Creativity," advocates the inclusion of teaching for creativity in mathematics classes and school programs.
- Message 10, "It's Not Just About Math and Reading," encourages us to not eliminate electives and fields of study outside of the *basic* (tested) subject areas of mathematics and reading.

MORE TO CONSIDER

- *Shop Class as Soulcraft: An Inquiry into the Value of Work* (Crawford 2009) offers thoughts about the value of learning how to do work, especially manual work, and the value of learning beyond academics.
- "A High School Under the Hood" (Kilgannon 2007) tells the story of a high school's long-standing curriculum that focuses on automobiles and preparing students to work in the automotive industry, and describes students' challenges adapting to a twenty-first-century workplace and academic demands for college and career readiness.
- *College and Career Ready in the 21st Century: Making High School Matter* (Stone and Lewis 2012) considers how to strengthen connections between Career Technical Education and academic disciplines.
- *Beyond Tracking: Multiple Pathways to College, Career, and Civic Participation* (Oakes and Saunders 2008) looks at organizing schools to prepare students for life after graduation.
- "What Will Future Jobs Look Like?" (Mcafee 2013) considers the kinds of jobs our students may see in their future. www.ted.com/talks/andrew_mcafee_what_will_future_jobs_look_like.
- *What Color Is Your Parachute? For Teens: Discovering Yourself, Defining Your Future, 2nd edition* (Christen and Bolles 2010) helps students think about what kind of work they might like to do and how to find a job.
- *The Path to Purpose: How Young People Find Their Calling in Life* (Damon 2008) helps adolescents find direction and identify possible future directions.

10 Everyone Loves Math

PERSPECTIVES ON HOW WE SEE MATHEMATICS

My license plate proclaiming my personal relationship with mathematics.

Much of my career in education has been spent in roles outside the classroom. I've returned to teaching from time to time, most recently to teach mathematics for two years as a Peace Corps volunteer. I taught mathematics in French in Burkina Faso to one class of middle school students, whose ages ranged from 12 to 14 or so, and three classes of high school students, whose ages ranged from 16 to 19. My seventy-five middle schoolers were energetic, to say the least. Most days, they challenged all of my arguably adequate skills in classroom management backed by years of experience as a parent.

One day, I was talking with my friend and colleague, Leopold, who taught English to this same group of middle-schoolers. We were commiserating about the challenges these particular students presented to us. Leopold said to me, "Yes, they won't participate, they won't pay attention, and they don't do their work!" I replied, "That's not the problem I have with them—they seem to be engaged and they do their work. But they just won't sit still and stop talking." Leopold looked at me with a strange expression and said, "Well, of course they pay attention in your class—it's mathematics. Everyone loves mathematics!"

For a moment I thought I must be in a parallel universe! Everyone loves mathematics? In all my years of education, both as a mathematics teacher and a student, it seemed to me that most people really don't care much for mathematics. I've heard people say: "I never liked math." "Oh, I'm just not a math person!" "I hate math!" One of the worst comments I ever heard came during interviews I conducted for a paper on math anxiety. I spoke to a well-educated woman who had chosen not to apply to law school because she was afraid of the mathematics on the entrance exam. She said, "I see math as a giant pothole in the road of life, and I'm going to do anything I can to avoid it." This kind of negative attitude reflects the strong feelings and misconceptions many people have about mathematics, falling victim to an *I hate mathematics* epidemic or believing they just didn't inherit the right gene.

The *I Hate Mathematics* Epidemic

While it seems to be true that many people everywhere find mathematics challenging, in the United States, people seem to be particularly prone to believing that they can't do mathematics, many developing a negative attitude toward it. Their dislike of mathematics affects personal choices from college courses to career paths. Yet, this pervasive distaste for mathematics is not a universal epidemic. Leopold saw mathematics as an interesting subject and perceived that everyone else felt the same way. In some countries, as in Burkina Faso, while students may see mathematics as hard, the widespread negativity toward the subject that we see in the United States simply is not there. Somehow, in the United States, society has come to think that it's common, even acceptable, not to like mathematics and not to do well in it in school.

A "Math Gene"?

Contrary to some beliefs, there is no evidence of a "math gene." It simply is not the case that some people can do mathematics and some people can't. While it may be true that some people seem to be able to calculate quickly or to grasp some mathematical concepts quickly, people can engage in mathematics in many ways. If we use a variety of teaching approaches and offer engaging problems, and if we help students learn to represent mathematical ideas, skills, and problems in multiple ways, nearly anyone should be able to participate—and succeed—in mathematics. Someone with strong visual skills should be able to use those skills to visualize problems or concepts. Someone else may be a thoughtful problem solver who can ponder the details of a problem and generate a couple of ways to approach it. Someone who is a good communicator has opportunities to talk through a problem or explain his or her thinking as the person tackles a mathematical situation.

Every high school teacher has seen students who were good at algebra but struggled with geometry and students who excelled at geometry even though they may have been only marginally successful with algebra. There's no reason why this should happen. If a student is strong in algebraic thinking or geometric understanding, we should be able to tap into their strengths to help them learn other parts of mathematics. The more we connect across the threads of mathematics, and the more we engage students in rich problems that can be approached in more than one way, the more likely it is that students who may not be traditionally successful mathematics students will be able to learn mathematics well.

The Impact of Negativity Toward Mathematics

The effects of widespread negative attitudes about mathematics can reach into discussions about school improvement or reforming the mathematics program. Sometimes it seems that almost no-one enjoyed mathematics when they studied it, yet almost everyone resists changing anything about the way we teach mathematics in school. Negative attitudes about mathematics and reluctance to talk about anything new in mathematics are widespread in society and show up any time a family or community engages in budget discussions or any time a school, district, or state considers raising mathematics requirements for high school graduation, for example. My theory is that this unlikely coupling of not liking mathematics and not wanting to change it may stem from at least three causes:

1. People don't like change any more than they like mathematics.
2. Some folks believe that if they had to suffer through it, then future generations should as well.
3. Dealing with issues related to mathematics may mean having to deal with mathematics, which may expose a person's gaps or weaknesses. People may feel insecure about discussing mathematics because they aren't sure how well they understand it.

Whatever the reasons, teachers have an opportunity to break this cycle. We know that some adults love mathematics, and early childhood educators tell us about young children's natural interest in numbers, shapes, and quantities in the world around us (Copley 2010). As teachers, then, how can we create classrooms that build on this natural interest and foster positive enthusiastic feelings about mathematics with all students?

Catch a Good Attitude

First of all, teachers, like other adults around students, can help students feel positively about mathematics by demonstrating their own positive feelings toward the subject. In an ideal world, every teacher who teaches mathematics would love mathematics and help their students love it as well. In the real world, a teacher who doesn't yet fall in this category can still contribute to a positive attitude using a technique called "acting as if." Basically, this means that a teacher who realizes she or he may not yet love mathematics can *act as if* she or he does. In other words, every teacher has a responsibility to act in ways that demonstrate a positive attitude about mathematics—in setting up a class activity, in responding to how students engage in the task, in the enthusiasm the teacher shows during and outside of mathematics class. Attitudes are quite contagious, both positive and negative ones. Don't we owe it to our students to help them catch a good attitude?

Embrace an Upside-Down Teaching Model

My personal experience is that students seem to develop positive attitudes when they are engaged in challenging and interesting tasks that require them to grapple with mathematical ideas and think hard. An upside-down teaching model (see Message 12)—where we structure learning opportunities arising from students' work on extended problems, rather than teaching them specific procedures first and then having them solve one- or two-step problems where they apply what they have just learned—gives students a chance to connect with mathematical ideas as they push themselves to figure things out. In this kind of situation, students explain their thinking to others and work together, talking with each other as they move toward possible solutions. When teachers facilitate this kind of interaction, they help students connect personally with the mathematics, contributing to a positive feeling on the part of the student that he or she has something to offer. We cannot underestimate the power of this internal personal connection as a much stronger motivator than, for example, relying on a teacher's praise or other kinds of external rewards to encourage students to do their work.

Encourage Family Support

Supporting positive attitudes toward mathematics is an area where communities and families can make a significant contribution. PBS, as part of their Cyberchase program, offers tips for families in helping develop positive attitudes about mathematics with their children that can offer us some ideas for how to help:

- Learn to learn from mistakes ... talk with students when something doesn't turn out right and help them see that it's OK to make mistakes and that sometimes we can learn new things from a mistake.

- Demonstrate a willingness to try and do math when the opportunity arises ... avoid talking about never having been good at math yourself or saying you can't balance your checkbook. Use your child's learning to motivate you to try these things yourself.

- Talk about mathematics with your child ... look for opportunities to show your child when you are using math to measure or count something.

- Model how to solve problems ... think aloud with your child when you tackle problems of all kinds. Take a break when you get stuck or discouraged. Try breaking the problem into smaller pieces.

- Ask thinking questions ... How did you do that? What were you thinking when you decided to use addition? Can you draw a picture to show me what that means?

- Focus on *how* as well as *what* the answer is ... The process a child uses is as important as the answer he or she gets. It's not always necessary to do a problem in a certain way, and sometimes there may be many ways to arrive at a good answer. It can be very helpful for a child to explain what they did and why they chose to do it that way.

What Can We Do?

I'd like to think that every person can develop a personal relationship with mathematics and can see himself or herself as a *doer* of mathematics. I proclaim my personal relationship with mathematics on my license plate. It reads *DO MATH* and my license-plate holder reads *Do math and you can do anything*. Let's all commit to helping every student form a personal identity as a doer of mathematics. Let's not let another generation go by missing the opportunity to build mathematical success and proficiency on a foundation of enthusiasm and confidence.

Leopold's comment was a wonderful reminder that a person's attitude about loving or not loving mathematics does not arrive at birth, nor does it need to stay fixed over time. We may not live in a culture where loving mathematics is the prevailing belief. But we do have the potential to help every student, and perhaps even adults, come to see mathematics as an interesting and relevant discipline and see themselves as doers of mathematics.

Reflections and Discussion

FOR TEACHERS

- What issues or challenges does this message raise for you? In what ways do you agree with or disagree with the main points of the message?
- How much do you like mathematics? How do you think your attitude toward mathematics plays out in your interactions with students?
- What can you do differently in your classroom in order to help students develop a positive attitude toward mathematics and a willingness to tackle good mathematics problems?

FOR FAMILIES

- What questions or issues does this message raise for you to discuss with your son or daughter, the teacher, or school leaders?
- Have you believed that you or others in your family either had or did not have a "math gene"—that you were born to be able to do math or not? If so, are you open to considering that someone might be able to succeed in mathematics, even if that person has not been successful in the past?
- How much do you like mathematics? How do you think your attitude toward mathematics affects how your daughter or son feels about mathematics?
- How can you help your son or daughter enjoy mathematics or develop a positive attitude about it?

FOR LEADERS AND POLICY MAKERS

- How does this message reinforce or challenge policies and decisions you have made or are considering?
- How much do you like mathematics? How do you think your attitude toward mathematics affects your interactions with teachers and students and the decisions you make regarding the mathematics program?
- How can you foster the development of positive attitudes toward mathematics among the teachers, administrators, and students with whom you work?

RELATED MESSAGES

Smarter Than We Think

- Message 12, "Upside-Down Teaching," considers the use of a problem-centered approach in engaging students and developing positive attitudes about mathematics.

- Message 31, "Developing Mathematical Habits of Mind," describes the mathematical habits of mind we want students to develop.

- Message 8, "Oops!," helps us think about how we handle mistakes and how we can help students come to view mistakes as helpful in the learning process.

- Message 4, "They're Just Not Motivated!," looks more closely at what motivates students to do mathematics.

Faster Isn't Smarter

- Message 18, "Faster Isn't Smarter," looks at the negative impact of timed tests on some students' attitudes toward mathematics.

- Message 27, "A Math Message to Families," provides an overview of shifts in mathematics classrooms that can help students in both achievement and attitude.

- Message 37, "Boring!," considers the impact of teaching in ways that don't engage students' interests.

MORE TO CONSIDER

- *Math: Facing an American Phobia* (Burns 1998) addresses the widespread dislike of mathematics in the United States adult population and offers insights on how to get past the negativity and use math in everyday life.

- *Overcoming Math Anxiety, revised edition* (Tobias 1995) is a landmark book that discusses the impact of extreme negative attitudes toward mathematics and offers insights into how to overcome them.

- "Helping Kids Develop Positive Math Attitudes" offers suggestions for families on helping children develop positive attitudes about mathematics as part of a larger collection of multimedia resources for parents. Cyberchase, www.pbs.org/parents/cyberchase /math-fun-more/using-math-at-home/positive-math-attitudes/.

- *Motivation Matters and Interest Counts* (Middleton and Jansen 2011) looks at the importance of students' interests in motivating their learning of mathematics.

- *Smarter Together! Collaboration and Equity in the Elementary Math Classroom* (Featherstone et al. 2011) looks at the value of having students work collaboratively on in-depth problems as a tool for equity and for developing positive attitudes about mathematics.

- "From the Inside Out" (Fillingim and Barlow 2010) describes the kind of mathematical thinking involved in helping children become "doers of mathematics" in and outside of school.
- "Educ115N: How to Learn Math" is a Massive Open Online Course (MOOC) and related website offered by Stanford University and taught by Dr. Jo Boaler for educators, students, or noneducator adults addressing key issues related to learning mathematics, including intelligence, mistakes, perseverance, problem solving, teaching for student engagement, and other topics.

PART

Messages About Teachers and Teaching

11

Peaks and Valleys

NAVIGATING THE UPS AND DOWNS OF
TEACHING AS A NEW TEACHER

I touch the future. I teach.

—Christa McAuliffe

In 1999, after thirty years in education, I decided to follow a lifelong dream to join the Peace Corps. I was assigned to teach mathematics in French in a small country in West Africa for two years. Reflecting on my life as a new Peace Corps volunteer, I see clear parallels between that experience and the early days, months, and years of a teacher's entry into the teaching profession.

An African Adventure

I can still remember the exhilaration of my trip to Burkina Faso—flying over the vast expanse of Africa on a clear day, touching down once in Bamako, Mali to see Malian families watching near the runway, and proceeding on our way to Ouagadougou. I could not believe I was experiencing this grand adventure! On the four-hour bus ride from Ouagadougou to Bobo-Dioulasso, I could not stop leaning out of the window, waving and smiling at people amused by a busload of foreigners as we passed through their roadside villages. A few of the children returned my waves and called out to us.

We enthusiastically jumped into our training—a combination of language, cultural awareness, health care, safety lessons, and an introduction to the Burkinabé system of education. As one of the older volunteers, I was feeling pretty good that I carried some remnants of my high school French, some knowledge of other cultures gathered over the years, and a lot of experience with teaching mathematics. But I was operating in a very different environment than any I had experienced before. We arrived at the end of the hot season, and the rains were overdue. Bobo is in the part of the country that can be humid, even after months without rain. As Peace Corps volunteers, we had no air conditioning, either at our training site or at the homes of the families who hosted us during our training. Temperatures ran routinely above one hundred degrees Fahrenheit and rarely fell below ninety, even at night. We slept on thin foam mattresses placed on simple wooden beds made by local builders unaccustomed to tall Americans (I'm six foot one). Every night my roommate and I carefully tucked in our mosquito nets in the concrete-block room we shared. I found it hard to sleep in this unfamiliar place, but lack of sleep seemed a small price to pay in exchange for the adrenalin and adventure.

After a few more nights of tossing, turning, and sweating, however, sleep deprivation set in and my enthusiasm decreased. I found it increasingly hard to concentrate during our training sessions. On the fifth day, I had my first mini-meltdown. During a training session on dressing appropriately for school, I felt tears coming and had to leave the room. Finding me sobbing in the hallway, our Peace Corps nurse, Yemi, gently tried to find out what was wrong. I did not have to be coaxed to blurt out my tale of woe: "It's so hot and I can't concentrate and it's so hot and my bed is too short and it's so hot and I haven't slept in a week and it's so hot and I just don't know how I'm going to survive without sleep!"

Yemi reassured me that I was not having a midlife crisis or a nervous breakdown, but that my problems would probably work themselves out, especially if I could get some sleep. She accompanied me on a short walk to the medical building. There she took me to a back room where a fan circulated cool air from a nonstop air conditioner, the first I had seen since my arrival in Burkina Faso. She tucked me into a bed that seemed almost long enough, and stated that I would be much better off with a long nap than with whatever topics the training was addressing that day. I slept like a baby until it was time for dinner.

Yemi's intervention marked a turning point. A few days later, when I arrived home from training at the end of the day, I was surprised to find a longer bed awaiting me (as did two of my tall fellow volunteers). Thanks to improved sleeping conditions, things went along more smoothly, although there were occasional and sometimes predictable ups and downs—my first trip to the market with the obligatory negotiation over the purchase of a pair of flip flops (arguing over what turned out to be about five cents); learning about the French educational system (a grading scale of 20); having clothes made by the local tailor (the fabric had drawings of telephones on it); experiencing the aroma of food from street vendors (fried caterpillars were the most challenging); and taking a field trip to the countryside (mangoes as far as you could see), among other day-to-day adventures. Some days were better than others, and often I reached that high of marveling at my good fortune to be living this dream.

One day well into our three months of training, I was conversing with several volunteers about the ups and downs of our training. To my surprise, nearly every one of them had a similar story of mood swings—ups and downs, peaks and valleys. We all experienced them in slightly different ways, and we noted how thankful we were that the cycles between the valleys were gradually stretching out. One of the volunteers had actually kept a mood graph to record her ups and downs. We all noticed that, over time, we were increasingly moving toward balance. It was reassuring to look back and notice the improvements.

Peaks and Valleys in Teaching

Perhaps entering the world of teaching is not quite as dramatic a change as going to the other side of the world and learning to live with a new lifestyle, language, culture, climate, and educational system. But becoming a teacher is a significant change from being a student, and in some ways, teaching carries with it the need to learn a new language and a new culture. A teacher is the person who creates the educational experience of students, for better or worse, and it can feel like a huge responsibility, which it should. (Maybe new teachers should keep journals or mood graphs to chronicle their adventures and notice their progress!)

Most teachers enter the profession full of idealism and optimism. They want to make a difference in students' lives, and they believe they can. Yet, more than one new teacher has prepared for the first school year by getting their classrooms arranged, organizing students' materials, and putting up inspirational bulletin boards, only to find themselves wondering what to do next, how to start. When teachers actually face students during the first days of school—when they get past the thankfully predictable routines of bookkeeping, attendance, and logistics—teachers can find themselves feeling like they are constantly making it up as they go along, just one step ahead of their students, no matter how much they have planned in advance. What seem like great lesson plans rarely accommodate the realities of the classroom, nor can plans alone provide answers to what a teacher should do given the unpredictable actions and responses of students.

When I started teaching, I gave myself three years to become a wonderful and effective teacher. I thought that was a reasonable amount of time to work through the inevitable challenges of embarking on a new path and to fine-tune my teaching techniques. To say that this was a naïve point of view is an understatement. The ups and downs of those first three years were substantial—emotionally, intellectually, and physically. By the end of my third year of teaching, I was able to make it until spring break before becoming totally overwhelmed by the frustrations I faced and my perceived inability to make a difference. I persevered, learning and improving every year I taught.

I continue to learn and, I hope, improve every year I work in education, whether in the classroom or supporting others. Teaching is a journey, and nowhere is the need greater for a commitment to our own lifelong learning. It is not possible to learn everything you need to know about effective teaching before you start your career. That task will last as long as you continue teaching.

Ultimate Peaks: A Teacher's Legacy

Several years ago, I had the honor of presenting Texas's Presidential Award for Mathematics Teaching at the state's annual mathematics conference luncheon. The recipient of the award that year was Kathleen Walker Murrell, a wonderful high school teacher from Pasadena, Texas. The event was a celebration of excellent teaching in general, with a handful of teachers especially recognized for their work. As part of the event, I read aloud excerpts of letters from students, parents, and fellow teachers recounting the ways in which these outstanding teachers had improved the lives and futures of the students they taught.

Shortly after the conclusion of the awards luncheon, Kathleen approached me to recount an incident that occurred immediately after the ceremony. As she stepped down from the head table, she felt a tap on her shoulder and turned to see her own former mathematics teacher and mentor. Kathleen was ecstatic to be able to tell her teacher that she was the reason Kathleen had chosen to pursue a career as a mathematics teacher. As they were reminiscing, Kathleen felt a tap on her other shoulder. She turned to see a former student, now herself a teacher, who told Kathleen that she, Kathleen, was the reason she decided to become a mathematics teacher. What a wonderful legacy from one teacher to another! A teacher's legacy may be something obvious right away, something the teacher discovers in a year or two, or something that can pass from generation to generation.

Valleys Will Come

My experience in Burkina Faso was powerful and positive, with the peaks far outweighing the valleys. Not everyone in my Peace Corps group made it through the planned two years of service, and a few ended their two years without having had a positive experience. There were many reasons for both outcomes, including factors related to particular assignments or to individuals, or sometimes simply a mismatch of an individual to an assignment. Teaching is much the same, both in the possibility of a potentially positive and powerful experience or in the possibility of something less fulfilling. Some people may need to find a different teaching environment in which they can thrive, and some people may be better suited for a career path other than teaching. But it would be a mistake to decide you have chosen the wrong path simply because you encounter difficulties, frustrations, and even apparently insurmountable challenges.

What Can We Do?

We are confronted with insurmountable opportunities.

—"Pogo" by Walt Kelly

When the inevitable valleys present themselves, it's important to remember the peaks as well. The ways in which you influence students will sometimes be obvious, but can often be invisible. Be assured that teachers do change lives. You'll have to decide whether the peaks are worth the occasional valleys, whether the time between valleys is getting longer, and whether your teaching assignment is a good fit for your talents and philosophy. Maybe you only need a small change of accommodations to find your groove, like I found with my new bed. Only you can know whether you're moving in the right direction. The keys are your commitment to continuing to learn, your actions in implementing what you do learn, and your patience with your progress. Getting plenty of sleep doesn't hurt, either.

So enjoy the journey. If you are a new or early career teacher of any age, welcome to the most important profession on the planet. Over the years, you will experience amazing moments that will stay with you forever. You will have incredible influences on students that you may never hear about. And, if you're lucky, someday someone will tap you on the shoulder and let you know a small part of the influence you've had.

Reflections and Discussion

FOR TEACHERS

- What issues or challenges does this message raise for you? In what ways do you agree with or disagree with the main points of the message?
- If you are a new teacher, what are the top three steps you want to take next?
- If you are an experienced teacher, how can you support a new teacher joining your staff or team?

FOR FAMILIES

- What questions or issues does this message raise for you to discuss with your daughter or son, the teacher, or school leaders?
- How can your experiences with the peaks and valleys of parenting help you constructively interact with your son's or daughter's new teacher so that you work together toward a shared goal of success in mathematics?

FOR LEADERS AND POLICY MAKERS

- How does this message reinforce or challenge policies and decisions you have made or are considering?
- What support structures do you have in place (or might you implement) to help new teachers of mathematics provide the best mathematics education possible to their students, grow as professional teachers, and thrive in their choice of teaching as a profession?

RELATED MESSAGES

Smarter Than We Think

- Message 20, "What's That Goat Doing in My Classroom?," takes a look at dealing with day-to-day challenges in teaching.
- Message 10, "Everyone Loves Math," reflects on an African perspective related to attitudes about mathematics.

Faster Isn't Smarter

- Message 28, "So Now You're a Teacher," looks at issues facing beginning teachers.

- Message 29, "The Evolution of a Mathematics Teacher," considers a teacher's journey.
- Message 38, "Ten Kinds of Wonderful," reflects on ten roles teachers can play to help students learn mathematics.
- Message 41, "Thank You, Mr. Bender," recognizes and appreciates great teachers.

MORE TO CONSIDER

Personal Stories

- *Teachers with Class: True Stories of Great Teachers* (Goldberg and Feldman 2003) is a collection of true short stories about teachers who made a difference in someone's life.
- *The Courage to Teach: Exploring the Inner Landscape of a Teacher's Life* (Parker 2007) offers insights into teaching from a personal perspective.

First-Year Teacher Tips

- *Success from the Start: Your First Years Teaching Secondary Mathematics* (Wieman and Arbaugh 2013) provides practical tips for planning effective lessons and for managing every day in the classroom.
- *Empowering the Beginning Teacher of Mathematics in Elementary School* (Chappell, Schielack, and Zagorski 2004) discusses tips and ideas for new teachers.
- *Empowering the Beginning Teacher of Mathematics: Middle School* (Chappell and Pateracki 2004) offers tips and ideas for new teachers.
- *Empowering the Beginning Teacher of Mathematics: High School* (Chappell, Choppin, and Salls 2004) provides tips and ideas for new teachers.
- Also see Appendix D: Essential Library for a list of titles that might inform your choices for a beginning professional library for teaching mathematics.

12 Upside-Down Teaching

STARTING WITH A GOOD PROBLEM AND SHIFTING TO *YOU-WE-I*

> The hardest part of teaching by challenging is to keep your mouth shut, to hold back. Don't *say;* ask! ... Keep asking "Is that right? Are you sure?" Don't say "no"; ask "why?"
>
> —Paul Halmos (*I Want to Be a Mathematician,* 1985, 272)

In 2013 the United States had almost 4 million job openings, and yet more than 7 percent of those looking for work were unemployed (U.S. Department of Labor 2013). The problem with this continuing disconnect is not geography—where the jobs are compared to where the workers live—but rather that workers simply do not have the skills required for today's available jobs at any level, whether blue-collar or white-collar or requiring a high school education, technical certificate, two- or four-year degree, or graduate study. All workers in jobs today need to be able to think, reason, and solve problems that haven't been solved before, often working in a team or with a small group of individuals contributing different areas of expertise.

Preparing for this kind of future demands a different kind of education, especially in mathematics, than most schools have offered in the past. Throughout most of the twentieth century, it was enough for the educational system to focus on helping students acquire knowledge. But as the century drew to a close, it was becoming obvious that knowledge alone was not enough to secure future employment. Nor was knowledge enough to help communities and the nation address their challenges and thrive. As we made the transition into the twenty-first century, report after report called for ramping up our academic expectations and incorporating significant attention to reasoning, thinking, creativity, and high-level problem solving (see, for example, Friedman 2007; National Center on

Education and the Economy 2008). Those calls continue today, both from within the ranks of mathematics educators and from outside. In particular, the Common Core Standards for Mathematical Practice (NGA Center and CCSSO 2010) put a strong emphasis on reasoning, thinking, problem solving, and communication, and the few states not adopting these standards reflect the same priorities in their own.

So how can we create the classrooms we've been calling for during more than two decades? Maybe we need to turn our traditional teaching model upside down if we're going to prepare students to thrive in their future, rather than our past.

The Traditional Right-Side Up Model

Many mathematics classrooms today reflect the teaching model I experienced years ago as a student, a model that would become my basis for teaching in the early years of my career. That model—what I call the "right-side up" model—involved preparing a lesson thoroughly so that I could clearly explain to my students the specific procedure or concept to be covered next. I was encouraged to fill my explanation with enthusiasm and energy so that my students would stay with me and absorb what I was telling them. Then, after we practiced the procedure together, I would give students exercises to first practice the procedure and then eventually to apply the procedure to solve a few word problems. One way to characterize this teaching approach is *I-We-You*. In other words, *I* (the teacher) will present the mathematical concepts and rules for the lesson; then *We* (students and teacher) will do some guided practice, where we walk through some examples of those concepts and rules, perhaps including word problems involving these same concepts and rules; finally *You* (students) will practice on your own and later do homework on what you have learned.

This method is too often accompanied by several hurdles. First, some students don't learn well from a teacher-delivered explanation; many become bored and, thus, disengage from what's going on. Some students also see an error-free teacher explanation as further proof that they (the students) simply don't have the "math gene." They believe that mathematics is something only some people can do, as demonstrated by their teacher's explanation and by the few students who seem to be able to master the particular concept or procedure being demonstrated. Most of all, when we primarily present students with problems for which they come to expect that they will apply the procedure they have just learned, we withhold perhaps the most important experience students need. We deny them the opportunity to dig into a problem, get a sense of what mathematics might be involved, constructively grapple with the underlying mathematical ideas, try out possible solution

approaches, and learn from mistakes they make in the process of coming to actual solutions. That opportunity represents the heart of *upside-
down teaching*.

The Upside-Down Model

Teaching upside down involves choosing to first present to students a
problem they are expected to mess around with for a while, without
having first taught them the particular rules or procedures they could
use to solve the problem. Engaging students in this way helps them
interact with the mathematics and sets them up to learn the mathematical content the teacher intends.

Rather than the *I-We-You* structure used in many mathematics
classrooms today, this model could be characterized as *You-We-I*: *You*
(students) will mess around with a task for a while, ideally engaging
in some thinking, trying things out, and generally wrestling with or
constructively struggling with mathematics arising from the problem;
then *We* (students and teacher) will discuss the different approaches students tried, with students explaining, questioning, clarifying, and further grappling with the mathematics; finally, *I* (the teacher) will connect
this work and the class's productive discourse around the problem and
related mathematical ideas, facilitating the whole process and ensuring that students come away with the intended mathematics learning.
Sometimes, students' learning may emerge naturally from their engagement with the task. Other times, it may involve the teacher directly telling students a key point or working through an explicit example. Even
when such direct instruction may be called for, students' engagement
with the task and participation in the resulting discourse sets them up
to also take in what the teacher presents.

The way that I learned to teach—clear explanations, shared practice, application of what was just learned—represented a very teacher-
centered approach. The upside-down model I'm advocating here is
more difficult to implement well, calling for considerable time and
teacher skill in orchestrating and managing the classroom—a teacher-
structured approach focused on student engagement, rather than a
teacher-*centered* approach with students playing a more passive role.
Teaching in this way allows students the opportunity to push their
thinking as they constructively struggle with problems that may go
beyond more predictable one- or two-step word problems typically
found at the end of a lesson or chapter in a textbook. And by drawing students into thinking about the problem, students are more likely
to attend to the intended mathematics than they would be if listening
more passively to a teacher explanation.

Choosing Problems to Turn Upside Down

Several years ago, in *Professional Standards for Teaching Mathematics* (1991), the National Council of Teachers of Mathematics suggested organizing mathematics teaching around three key elements: worthwhile tasks, productive discourse, and a safe and supportive learning environment. These basic elements offer timeless recommendations for helping students learn to think and reason on their own and make sense of the mathematics they are learning. The process is centered on *worthwhile tasks*, described in Standard 1 of that document. This standard offers such a clear and beautiful description of the importance of and nature of the tasks we select, I'm inserting the direct text of the standard here.

The teacher of mathematics should pose tasks that are based on:

- sound and significant mathematics;
- knowledge of students' understandings, interests, and experiences;
- knowledge of the range of ways that diverse students learn mathematics;

and that

- engage students' intellect;
- develop students' mathematical understandings and skills;
- stimulate students to make connections and develop a coherent framework for mathematical ideas;
- call for problem formulation, problem solving, and mathematical reasoning;
- promote communication about mathematics;
- represent mathematics as an ongoing human activity;
- display sensitivity to, and draw on, students' diverse background experiences and dispositions; and
- promote the development of all students' dispositions to do mathematics. (NCTM 1991, 25)

Finding such tasks is not always easy. However, the increasing availability of online resources, especially those addressing common standards, makes it more likely than in the past that a teacher will be able to organize a lesson around a rich, deep, challenging, and engaging task. Another place to look for good, worthwhile tasks may be the supplementary materials that come with many textbooks; often, good problems are included as project suggestions or extensions to textbook lessons. And, of course, some curriculum materials themselves are organized around rich tasks. (See Appendices A, B, and C for resources for selecting and evaluating tasks for upside-down teaching.)

Not all tasks offer the same level of opportunity for student engagement and thinking; individually evaluating tasks can be a time-consuming job. But finding and considering such tasks for classroom use can provide an excellent opportunity for collaboration and discussion within a professional learning community, grade-level team, department, or any group of colleagues. And sometimes, a potentially good task can be made even better with the addition of a question or a slight modification, something that might arise in such a collaborative discussion.

Considering Contexts

Problems need not always to be in real-world contexts in order to be effective in upside-down teaching. Some straightforward problems posed in a purely mathematical context can offer nice opportunities for discussing, struggling, thinking, and learning. In *Fostering Geometric Thinking* (Driscoll, DiMatteo, Nikula, and Egan 2007) the authors present the following geometric problem:

> *Two vertices of a triangle are located at (0,6) and (0,12). The triangle has area 12.* (2007, 47)

The authors then describe the kinds of questions that can engage students in deep thinking and discussion:

> *What are all the possible positions for the third vertex?*
> *How do you know you have them all?*
> *How many of the triangles you form are isosceles?* (2007, 47)

It can also be useful to organize a lesson around a task presented in a context outside of mathematics. Choosing contexts should be done carefully so as not to distract students from the mathematics, but rather draw them into it. When we look for problems in contexts outside of mathematics, it simply is not possible to find tasks in which a context will resonate with all students. Students come from different backgrounds with different experiences and interests, and every student will find different tasks engaging or interesting. It's unrealistic and frustrating to eliminate any context that might be unfamiliar to one or more students in a class. Rather, the teacher can help optimize the use of a context by discussing that context with the class in setting up the problem at the beginning of the lesson. For example, a lesson about numerical reasoning based on tire sizes might start out with a discussion of the numbers on tires, perhaps even bringing a tire into class for students to see. Even nondriving students or students who have never looked at the tires on a car can see the numbers printed on the tire and deal with a real-world context from which to explore the mathematics. In the process of doing so, they not only deal with the mathematical content, they also expand their familiarity with that context just a little.

Overall, perhaps the best outcome in terms of teachers' choices of relevant, interesting tasks is that students will engage with enough problems in a wide enough variety of contexts that they come to see mathematics as something actually used in the world outside of school. And, if we're lucky, we can hope that every student's particular interests are piqued by enough problems over time that they come to develop a personal identity with mathematics as relevant to that student's life. Samuel Otten, in a rich discussion of cautions related to real-world contexts, suggests that the most important thing for students to notice about problems posed in contexts is that the mathematical processes they use—the thinking and reasoning skills they develop—are what carry over and apply to a multiplicity of situations (2011, 20–25).

What Can We Do?

Shifting to a problem-centered You-We-I teaching approach, described here as upside-down teaching, involves both instructional time and planning time. Some students, as well as their parents, may complain that "You're not teaching us!," meaning that you aren't telling them every step they should take in solving a problem. Students, parents, teachers, and administrators need to understand the benefits of organizing instruction around good problems that students don't know in advance how to solve. Taking the extra time called for with this teaching approach is an investment in student learning with tremendous potential for positive returns. If we are successful, students not only learn the content they need, they also develop mathematical habits of mind like perseverance, thinking, reasoning, discussing, justifying their point of view, considering variations of a problem, and developing a positive disposition toward mathematics. These habits of mind pay off over and over again—students not only build on their understanding with new content connected to what they have learned, they will also have learned exactly the kinds of skills employers are looking for in filling millions of open positions in the twenty-first-century marketplace.

When I taught mathematics in Burkina Faso I used an upside-down, problem-centered approach. About two-thirds of the way through my first year there, one young man came to me after class. He said to me, "Madame, I know you like these problems of yours. But, you know, we have a program to cover." Although he would never have spoken to a Burkinabé teacher this way, he had seen a few American television shows, and so he believed that Americans were more open to such conversation. He continued, "Perhaps you could do your problems on Fridays and the rest of the time we could cover the program." Looking ahead to the major test students took at the end of high school, he was concerned that the class would not cover all of the material. I thanked him for his suggestion, and continued teaching around problems. Needless to say, he was not happy to find out that I would also be his teacher for the following year. Nevertheless, he came to my house with a group of students on the day that I was leaving to return to the United States at the end of my two years of service. He took me aside, and with a sheepish grin on his face, he said to me, "Madame, I think I learned more mathematics with your problems than I would have learned otherwise." He went on to complete a university degree and became a teacher.

Upside-down teaching seems to have worked in turning this one student's thinking upside down, and I'm sure I was a better teacher by using that approach. Maybe it's time for upside-down teaching to become the new right-side up model for mathematics classrooms.

Reflections and Discussion

FOR TEACHERS

- What issues or challenges does this message raise for you? In what ways do you agree with or disagree with the main points of the message?
- In what ways does your current teaching approach compare to upside-down teaching?
- Do you believe there are certain groups of students for whom upside-down teaching might not be effective or certain topics or courses for which you don't think this kind of teaching would be possible? Why or why not?
- If you don't already teach primarily using a problem-centered approach, what challenges do you see in trying to move closer to upside-down teaching? How might you (and your colleagues) address those challenges?

FOR FAMILIES

- What questions or issues does this message raise for you to discuss with your daughter or son, the teacher, or school leaders?
- How open are you to your son or daughter not being shown all the steps necessary to solve a problem before he or she is asked to deal with the problem? What might be the benefits of such an approach? What might be the drawbacks?
- How can you best support your daughter or son if she or he complains that the teacher isn't "teaching," but rather is expecting students to figure things out?

FOR LEADERS AND POLICY MAKERS

- How does this message reinforce or challenge policies and decisions you have made or are considering?
- How can you best support teachers in developing student thinking using an upside-down model, even if students and parents complain that the teacher isn't "teaching"—meaning that the teacher isn't telling students everything they need to know before giving them a good problem?

RELATED MESSAGES

Smarter Than We Think

Many of the messages in this book advocate an upside-down teaching model; below are a few examples.

- Message 32, "Problems Worth Solving," considers the nature of problems and what is called for from students to solve them.
- Message 31, "Developing Mathematical Habits of Mind," addresses the mathematical habits of mind that characterize real understanding and proficiency.
- Message 16, "Let It Go," offers thoughts on focusing the curriculum through instructional decisions.
- Message 4, "They're Just Not Motivated!," considers motivating students with engaging problems and opportunities for discourse.

Faster Isn't Smarter

- Message 17, "Constructive Struggling," emphasizes the importance of students being challenged to solve mathematically worthwhile problems.
- Message 1, "Math for a Flattening World," makes a case for the kind of thinking and reasoning workers of the future will need.
- Message 33, "Engaged in What?," considers the importance of students engaging in meaningful mathematics while participating in engaging activities.

MORE TO CONSIDER

- *What's Your Math Problem? Getting to the Heart of Teaching Problem Solving* (Gojak 2011) considers the importance of giving students rich, nonroutine problems without having first taught students exactly how to solve them and offers classroom strategies for helping students learn mathematics meaningfully through their work with such problems.

- "Delving Deeper: In-Depth Mathematical Analysis of Ordinary High School Problems" (Stanley and Walukiewicz 2004) suggests how to consider high school problems from a deep mathematical perspective.

- "Takeaways from Math Methods: How Will You Teach Effectively?" (Bay-Williams 2014) offers three big ideas for teaching toward student thinking.

- *Teaching with Your Mouth Shut* (Finkel 2000) advocates a variety of ways to teach without telling (not specific to mathematics).

- "Student-Centered Learning Approaches Are Effective in Closing the Opportunity Gap" is a series of four case studies supporting student-centered learning. https://edpolicy.stanford.edu/news /articles/1137.

- "Connecting Research to Teaching: Shifting Mathematical Authority from Teacher to Community" (Webel 2010) advocates rich, engaging teaching practice based on research about what works with students.

- "The Role of Contexts in the Mathematics Classroom: Do They Make Mathematics More 'Real'?" (Boaler 1993) discusses the use and limitations of real-world contexts in problem solving.

- "Cornered by the Real World: A Defense of Mathematics" (Otten 2011) offers a thought-provoking perspective on issues related to using real-world contexts in problem solving.

- *Motivation Matters and Interest Counts* (Middleton and Jansen 2011) discusses building on students' interests, including a discussion on the use of real-world contexts in selecting tasks.

- *The World Is Flat 3.0: A Brief History of the Twenty-First Century* (Friedman 2007) offers a view of the changing world and the importance of educating twenty-first–century workers for creativity, innovation, and the ability to work together to solve problems.

- *That Used to Be Us: How America Fell Behind in the World It Invented and How We Can Come Back* (Friedman 2011) makes a renewed call for investing in education that prepares workers of the future to think, analyze, and create, among other twenty-first-century skills.

- *Tough Choices or Tough Times: The Report of the New Commission on the Skills of the American Workforce* (National Center on Education and the Economy 2008) lays out the needs for citizens to be educated in more powerful higher-level skills, including creativity, communication, problem-solving, and the ability to conduct research, work in teams, and present findings, and makes recommendations for the education system to accomplish this goal.

- See also Appendices A, B, and C on selecting and evaluating in-depth tasks and Appendix D for several resources on teaching around problem solving listed as part of the Essential Library.

Related Research Briefs from the National Council of Teachers of Mathematics

- "Why Is Teaching with Problem Solving Important to Student Learning?" (Lester and Cai 2010) summarizes research findings about the role of problem solving in the mathematics classroom.
- "What Does Research Say the Benefits of Discussion in Mathematics Class Are?" (Cirillo 2013b) describes how research findings support the importance of offering students opportunities to discuss their work on mathematical tasks.
- "What Are Some Strategies for Facilitating Productive Classroom Discussions?" (Cirillo 2013a) offers research-based techniques in support of student discourse around mathematical tasks.

Resources Related to Specific Problem-Based Curricula

- "A Designer Speaks: Challenges in U.S. Mathematics Education Through a Curriculum Developer Lens" (Lappan and Phillips 2009) offers insights into effective mathematics teaching through the eyes of the developers of the Connected Mathematics Project.
- "The Consequences of a Problem-Based Mathematics Curriculum" (Clarke, Breed, and Fraser 2004) describes results of research on the effectiveness of IMP.
- "Teaching Sensible Mathematics in Sense-Making Ways with the CPMP" (Hirsch, Coxford, Fey, and Schoen 1995) describes results of research on the effectiveness of the Core-Plus Mathematics Project.
- *Advanced Mathematical Decision Making* (Student and Teacher Materials) (Charles A. Dana Center at the University of Texas at Austin 2010b) includes video of the lesson referenced in this message. This resource provides materials and resources for teaching this innovative twelfth-grade capstone mathematics course.

WWW This message is also available in printable format
at mathsolutions.com/smarterthanwethink.

13

Clueless

THE TROUBLE WITH CLUES AND KEY WORDS

Hidden within the language are numerous landmines that can sabotage student learning, arising from the complexities of English itself to poor practices entrenched in traditional instruction.

—Concepcion Molina (*The Problem with Math Is English*, 2012)

In my high school French class, Madame Boissevain regularly told us (in French) that it was better to learn to think in French than to try to translate word for word what we were thinking in English. I don't know how she managed it, but somehow I more or less accomplished that goal. More than thirty years later, when I found myself in the Peace Corps, assigned to teach mathematics in French in Burkina Faso, much of my French from my years with Madame Boissevain came back to me.

More recently, I decided that I wanted to learn to speak Spanish. To jump-start my Spanish learning, I spent two weeks immersed in Spanish in Costa Rica, living with a family and taking Spanish classes four hours a day. I was fortunate to be matched with Teresa, a wonderful teacher who again encouraged me to think in Spanish rather than translate word for word. It's not easy to achieve this goal, but I'm moving in that direction, and I continue to marvel at how much I learn and progress when I turn off the translating switch.

Thinking about these experiences has made me wonder what the payoff might be with mathematics students if we could help them learn to think mathematically when they encounter a problem instead of trying to translate each word into the "correct" mathematical operation.

Why Not Teach Clues and Key Words?

A search on the Internet for *math key words* or *math clue words* leads to list after list of words to look for in word problems to know whether to add, subtract, multiply, or divide. On the surface, this may seem like a good idea and an appropriate strategy. After all, if a student understands addition, for example, then we might think the student should associate addition with words like *total, sum, altogether,* or *increase.* But this strategy comes with a high price.

THE PROBLEM WITH WORDS

The first problem with teaching clues and key words is that this approach focuses on solving *word* problems instead of solving problems. It is true that many of the most interesting and relevant problems worth students' time and energy involve words. However, the key to these problems lies not in the words themselves, but in the mathematics underlying the situation the words present. And the mathematics is not the same for all problems that include the word *decrease,* or for all problems in which the word *twice* appears, or for all problems using the words *shared among.* For any clue or key word on any list, most teachers can think of, or find, a problem that uses the word in a different way from the use commonly presented.

In a compelling video interview shown in Marilyn Burns's Mathematics Reasoning Inventory (MRI), a teacher asks a student, Marisa, to solve the following problem and then to explain how she did it.

> There are 295 students in the school. School buses hold 25 students. How many school buses are needed to fit all of the students?

Marisa looks at the problem, pauses for a moment, then adds 295 and 25 to determine that 320 buses are needed. When the teacher asks how she figured it out, Marisa confidently explains that she listened for information to decide whether to add, subtract, multiply, or divide. She heard the word *all,* so she knew she should add.

WASTING TIME AND WORSE

Clearly, time spent teaching Marisa a list of clue words did not pay off for her in terms of learning how to solve problems. One of the most serious problems with teaching students a list of magic words is the immense waste of instructional time and the waste of students' learning time. Time spent teaching Marisa these words meant that she did not have more time to develop her understanding of numbers and operations. She may not have had adequate opportunities to develop mental pictures for addition, subtraction, multiplication, and division that might have helped her solve the problem by thinking mathematically. She spent her energy memorizing words that turned out to not be useful, instead of working with these operations in meaningful ways in a variety of situations.

MATHEMATICAL MISCONCEPTIONS

Even worse than wasting time, in the process of learning the list of words, students may develop mathematical misconceptions. What is intended as a strategy to help students deal with word problems may actually interfere with students' mathematical understanding. Because Marisa had learned that any problem involving the word *all* called for addition, she could not recognize a situation that might be associated with division or possibly multiplication. This kind of teaching may reinforce the notion that mathematics is a bit like magic—if you know the key word, you can get through it. Such a strategy fosters a very different, often negative, disposition toward mathematics than what might be developed in a classroom that embraces the development of mathematical habits of mind—a classroom where the teacher uses mathematically substantive tasks to engage students in wrestling with mathematical ideas, becoming increasingly strong problem solvers as they develop mathematical ways to think about problem situations.

Connecting Language to Understanding

There's nothing wrong with having discussions with students about what words mean mathematically. This is an important part of learning how to communicate with and about mathematics. But mathematical communication needs to be connected to mathematical understanding. Students are far more likely to be able to take on interesting and challenging problems if they have strong mental pictures of what addition, subtraction, multiplication, and division are, what numbers look like in many different forms, and how numbers and operations can be modeled in different ways. In order to develop these mental pictures, students need many opportunities to model numbers and operations with objects and in situations, combined with a lot of conversation about the mathematics they are using. It is only when words are used in reference to a particular situation or idea that they become a useful tool in mathematics.

What Can We Do?

Let us not waste teachers' and students' precious time or brain cells teaching students lists of words they are supposed to memorize at the expense of teaching them the mathematical ideas, themes, properties, and structures that will prepare them to deal with problems mathematically. Rather, let us spend our instructional time and energy helping students develop a deep understanding of mathematical concepts and the meanings of numbers and the operations they perform on numbers. Teaching for this kind of understanding requires a deep knowledge of mathematics beyond the level being taught. It may not be practical for all teachers to receive adequate professional development and support from coaches to become experts in mathematics. An alternative is to organize mathematics classes so that mathematics is always taught by math specialists.

One of the interesting things I learned in my Spanish class is that there are different words for the different ways in which we use the English word *key*. The word *llave* is used for a physical key to a lock, and the word *clave* is used to describe the idea of a key, like *the key to success*. I would argue that there are different kinds of keys in mathematics as well. We can choose to clunk along, teaching kids how to use the hardware—the keys and clues in the words—or we can teach them the real key: the mathematical tools, ideas, and representations they can use when they encounter a problem.

The family I lived with during my Costa Rican Spanish experience often laughed that they were always misplacing the key (*la llave*) to the outside gate. Let us not spend our time trying to give students keys they will misplace or misuse as they try to remember words on a list. Instead, let us invest in giving them the key of mathematical thinking (*la clave*) that will help them develop mathematical habits of mind to become problem solvers willing to tackle and able to solve challenging problems.

Reflections and Discussion

FOR TEACHERS

- What issues or challenges does this message raise for you? In what ways do you agree with or disagree with the main points of the message?
- Have you had experiences with students like Marisa who focus on the words in problems instead of the mathematics? How did you help the student shift his or her focus to the math?
- How can you help students make sense of the words used in a problem without relying on a memorized list?
- What strategies have you used, or can you use, to help students make sense of and represent the mathematical operations used in a problem?

(continued)

FOR FAMILIES

- What questions or issues does this message raise for you to discuss with your son or daughter, the teacher, or school leaders?
- Have you had experiences with your daughter or son in which a word in a math problem caused a misunderstanding? What happened?
- What kinds of questions might you ask to help your student make sense of word problems brought home for homework?

FOR LEADERS AND POLICY MAKERS

- How does this message reinforce or challenge policies and decisions you have made or are considering?
- How can you help teachers and those who support teachers, especially if they may not be experts in mathematics, understand the trouble with teaching students clue words? How can you help them promote more productive strategies?
- If you work with elementary teachers, how do you use (or can you use) math coaches to support classroom teachers who may not be experts in mathematics or how can you organize classes so that all students are taught mathematics by teachers who are highly knowledgeable and positive about mathematics?

RELATED MESSAGES

Smarter Than We Think

- Message 12, "Upside-Down Teaching," presents a teaching model focused on connecting language to understanding using in-depth problems.
- Message 37, "Communicating Mathematically," looks at the importance of correct language in mathematics.
- Message 14, "Effectiveness and Efficiency," looks at teaching concepts and skills well the first time, so that students can make sense of mathematics problems and avoid developing misconceptions.
- Message 2, "¿Habla matemáticas?," considers language issues for English language learners in learning mathematical concepts and skills and solving mathematics problems.

Faster Isn't Smarter

- Message 17, "Constructive Struggling," makes a case for students to wrestle with problems, rather than being given rules and procedures to memorize.

- Message 14, "Balance Is Basic," emphasizes the importance of teaching a balanced program of concepts, skills, and problem solving so that students can make sense of mathematics.

MORE TO CONSIDER

- *The Problem with Math Is English* (Molina 2012) considers issues and strategies related to the connection between language and mathematics, especially for English language learners.

- "How to Learn Math" (Boaler 2013), an online course convened in 2013 and expected to be offered from time to time in the future, offers adults experiences, information, and insights into the nature of mathematical thinking and offers suggestions for effective mathematics teaching; information and resources can be found at www.youcubed.org.

- "Effective Teaching for the Development of Skill and Conceptual Understanding of Number: What Is Most Effective?" (Hiebert and Grouws 2007) presents a summary of research on developing understanding and proficiency with numbers, with suggestions for effective research-based classroom practices.

- *Focus in High School Mathematics: Fostering Reasoning and Sense Making for All Students* (Strutchens and Quander 2011) provides discussion and recommendations for teaching mathematics around reasoning and sense-making in high school.

- "Math Reasoning Inventory," created to explore formative assessment, offers insights into student learning via straightforward strategies, including videos of student responses such as Marisa's story described in this message. (To see the video of Marisa, go to mathreasoninginventory.com and click on *Learn More*.)

- "The Role of Elementary Mathematics Specialists in the Teaching and Learning of Mathematics" (Association of Mathematics Teacher Educators, Association of State Supervisors of Mathematics, National Council of Supervisors of Mathematics, and National Council of Teachers of Mathematics 2010) identifies the critical need for mathematics specialists so that all children can be taught mathematics by someone who understands mathematics deeply and who likes the subject.

- "Mathematics Specialists and Mathematics Coaches: What Does the Research Say?" (McGatha 2009) offers research on the need for and preparation of mathematics specialists and coaches.

14

Effectiveness and Efficiency

INVESTING IN TEACHING MATHEMATICS WELL THE FIRST TIME

An investment in knowledge pays the best interest.

—Benjamin Franklin

We've heard for decades that the American mathematics curriculum is "a mile wide and an inch deep." The traditional wisdom of spiraling the curriculum by revisiting topics and skills in more depth every year simply has not worked. By trying to revisit many topics and skills every year, the curriculum at every level is so crowded that teachers lack the flexibility or time to teach some topics in depth, even though they know their students would benefit.

Most state standards in the twenty-first century have attempted to tighten the spiral by addressing topics fewer times across the grades. The Common Core State Standards for Mathematics (NGA Center and CCSSO 2010) support this direction, at least through grade 8. But the only way a more focused curriculum works is if we teach topics well the first time.

Teaching Effectively and Efficiently

Effective teaching is teaching that works. If we teach effectively, students learn what we intend them to learn.

Efficient teaching is not about saving time by teaching quickly; it's about saving time by not having to redo what we've already done. If we teach efficiently, we don't need to spend time with extensive review of previously learned material at the beginning of each year or at the start of a new unit. Nor do we need to build into our curriculum intentionally spiraled repetition that essentially repeats previous content, but perhaps with a longer divisor or an uglier denominator. We can teach both *effectively*—so students learn—and *efficiently*—so that they retain what they learn—by focusing on a few key strategies.

IDENTIFYING PRIORITIES

Teaching effectively *and* efficiently begins with identifying the most important priorities at each grade level or course. Much work has gone into recent standards to identify focus areas at each grade or key strands across high school courses. But even in a focused curriculum, teachers need to agree on where to spend the most time. This needs to be done in collaboration with colleagues for the same grade level or course as well as colleagues from other grades and courses. If, for example, fraction operations or quadratic equations are to be a priority at one level or course, that should influence decisions about priorities at other levels or courses. In our crowded mathematics curriculum—even in the more focused K–8 curriculum, but especially at high school—priority topics represent those things that students absolutely need to know. Whether important outcomes in themselves, or prerequisites for what comes later, priority topics and skills particularly call for effective and efficient teaching.

FOCUSING ON DEPTH AND UNDERSTANDING

Spending the time to teach well when a priority topic or skill is first addressed (it may have been briefly introduced earlier) is an investment, not an expenditure. To invest properly, we need to focus our teaching on depth and understanding, giving students opportunities to engage deeply with the content they're learning. They need to mess around with engaging tasks so that they are drawn into the content, wrestling with their emerging mathematical ideas as they work through tasks designed to make them think. The teacher's role is critical in ensuring that students are learning the intended mathematics. Content is uncovered and discussed as students think about relationships or search for strategies to answer provocative questions. Mostly, students need to have time to solidify—to finish—their learning. In this way, they build a critical foundation that is likely to serve them well in the future, perhaps even surviving a summer away from school.

AVOIDING TEACHING "TEMPORARY MATHEMATICS"

Part of investing in the future involves not teaching things to young children that will later need to be untaught, or retaught. For example, adults sometimes tell children that you can't take a larger number from a smaller one. Yet, in middle school, students learn—if they haven't suspected before then from what they see around them—that there are in fact a lot of numbers on the other side of zero and that, among other things, those numbers might show up when subtracting a larger number from a smaller one. Likewise, we might teach children about number sentences and later rename those sentences as *equations* or *inequalities*, or we might use language like *borrowing* and *carrying* to label key elements of mathematical operations. While we may think we are helping children by simplifying mathematical language, we may in fact be helping them develop misconceptions. Children who can learn the names of dinosaurs could probably just as easily learn the word *equation* instead of *number sentence*. When they hear the word *sentence* at a time when they are learning how to read, some children may struggle with making sense of how to capitalize and punctuate that mathematical sentence. And describing the steps in a computation in terms of *place value* and *regrouping*, rather than using labels like *borrowing* and *carrying*, reinforces the connection of the steps to the meanings of the operations.

There are many other examples of how we may inadvertently teach this kind of temporary mathematics that can later interfere with lasting learning—for example, always showing geometric shapes with a particular orientation (usually with a flat bottom); teaching students that an exact answer is always called for in a computation problem then later trying to teach them estimation skills; or even teaching students specific problem-solving strategies and later hoping they can figure things out that may not fit the strategies they've learned.

Keeping mathematical structure in mind can help us notice these issues and perhaps make different decisions when developing curriculum materials or planning the language, examples, and tasks we use with students. Where will this mathematics eventually lead? Will later definitions and properties make sense with what I am teaching today? We don't need to teach young children abstract or complex mathematical definitions like those mathematicians use. But teachers do need to know mathematics well beyond the level they teach and they do need to use definitions that have mathematical integrity appropriate to the grade level (for example, *an equation uses an equal sign to show two things that have the same value*).

FORMATIVELY ASSESSING AND PAYING ATTENTION

One way to know whether students may be developing harmful misconceptions is to constantly monitor how their learning is progressing. Teaching for lasting learning means paying attention to whether students are actually moving in the direction you want them to move. Are they learning what you're teaching, and are their underlying mathematical understandings sound? Formative assessment plays a critical role in teaching effectively and efficiently. It can take many forms, including:

- observations as the teacher moves among students or groups;
- informal conversations with students to verify that they understand key points;

- questions that require students to think about what they're doing;
- short quizzes;
- solutions that students present during a class discussion; and
- a "ticket out the door," or exit ticket—a problem or question students have to answer and hand in as they leave.

Effective teachers use these or numerous other creative ways to keep track of what and how students are learning. The most important formative assessment strategy is to talk with students and listen to them explain their thinking. The Mathematics Reasoning Inventory (MRI) provides many examples of the powerful insights we can gain simply by asking students to describe how they solved a problem.

What If Students Still Forget?

Of course, sometimes students forget something they've learned, even if they learned it well, especially if they haven't used it recently. We can minimize such forgetting by selectively incorporating previously learned content into new problems that also advance or expand their knowledge, making sure that students maintain what they've already learned well. Most of all, when a student appears to not know something, we need to probe a bit deeper with conversation or informal formative assessment strategies to determine whether the student never learned an important concept or skill or whether the student once knew it and could likely retrieve it with a very brief reminder or refresher. In this way, we can avoid the time wasted on extended review, especially at the beginning of the school year. Time is too precious to use inefficiently for review that may not be needed or could better be accomplished while moving on to new content.

What Can We Do?

Teaching effectively and efficiently involves more than learning a set of instructional strategies. It involves making professional decisions that might at times challenge pacing guides and timelines. To teach effectively and efficiently, we need to be clear about what mathematics is the most important investment for the year and then do whatever it takes to ensure that students learn it. We need to focus on sound mathematical structure, including using definitions and properties with both mathematical integrity and developmental appropriateness. We need to avoid teaching "temporary mathematics" that may later need to be undone or, worse, that may cause lasting misconceptions. This kind of effective and efficient teaching calls for teachers who know mathematics deeply across the grades and beyond. In the elementary grades, this either means finding and educating many more mathematical teacher experts than we currently see in schools, or using mathematics specialists, as called for by multiple organizations (AMTE et al. 2010).

Most of all, we need to implement mathematics teaching that deeply engages students in worthwhile mathematical tasks. An increasing number of resources are available to describe and elaborate what this kind of active teaching model looks like (see "More to Consider," page 110). When we invest in this kind of deep, engaging teaching the first time concepts and skills appear in depth in the curriculum, students reap the benefits.

Reflections and Discussion

FOR TEACHERS

- What issues or challenges does this message raise for you? In what ways do you agree with or disagree with the main points of the message?
- What kinds of tasks can you use to help students engage deeply with the mathematical content they're learning?
- How well do you know mathematics beyond the grade level or course(s) you teach? What steps can you take to strengthen your mathematical understanding so that you don't accidentally teach temporary mathematics that may need to later be untaught?
- How can you integrate review into your teaching without spending unnecessary instructional time?

FOR FAMILIES

- What questions or issues does this message raise for you to discuss with your daughter or son, the teacher, or school leaders?
- Do you try to "do it right the first time" in the work you do at home or in a job? What strategies have you found helpful in supporting you to do things right the first time?
- How can you support your son or daughter in remembering mathematics? Are there ways you can incorporate review on important mathematical ideas into your day-to-day family activities? (For example, depending on the age of your daughter or son, you might look for opportunities to discuss the length of nearby objects, estimate how

many cookies are in a box, figure out how much three boxes of cookies will cost, offer an opinion about the most economical package to buy, evaluate an advertisement for a discounted phone plan, etc.)

FOR LEADERS AND POLICY MAKERS

- How does this message reinforce or challenge policies and decisions you have made or are considering?
- How can you support teachers in not spending too much time on review?
- How well do your standards and curriculum planning documents support teaching well the first time, including providing adequate time? How can you support this kind of teaching even more?
- How well do your hiring, staffing, and professional development policies ensure that every student is taught mathematics by a teacher who knows mathematics well, including mathematics beyond the grade level or course being taught? How might you improve in this area?

RELATED MESSAGES

Smarter Than We Think

- Message 18, "Finishing Teaching," addresses the need to teach priority topics until students have finished learning them.
- Message 12, "Upside-Down Teaching," describes a problem-based teaching approach such as that mentioned in this message.
- Message 38, "Building Things," considers the role of mathematical structure in teaching for lasting learning.
- Message 37, "Communicating Mathematically," examines the importance of precise mathematical language at all levels, avoiding the need for later unteaching of "temporary mathematics."
- Message 19, "How to Know What They Know," discusses assessment, including formative assessment.
- Message 13, "Clueless," is based on insights gained with formative assessment, including noticing the kind of teaching that might lead to "temporary mathematics."

Faster Isn't Smarter

- Message 34, "Forgetting Isn't Forever," looks at the need to maximize learning time and minimize instructional time spent reviewing.

MORE TO CONSIDER

- *The Young Child and Mathematics* (Copley 2010) offers research-based strategies and activities for developing mathematical understanding and proficiency in young children.

- *INFORMative Assessment: Formative Assessment to Improve Math Achievement, Grades K–6* (Joyner and Muri 2011) provides an excellent set of resources for monitoring and adjusting teaching based on how well students are learning the intended mathematics.

- "Math Reasoning Inventory" offers insights into student learning via straightforward formative assessment strategies, including videos of student responses. www.mathreasoninginventory.com.

- *Learning and Teaching Early Math: The Learning Trajectories Approach* (Clements and Sarama 2009) looks at the development of mathematical content across the early years and offers recommendations for effective mathematics teaching for young children.

- "Effective Teaching for the Development of Skill and Conceptual Understanding of Number: What Is Most Effective?" (Hiebert and Grouws 2007) presents a summary of research on developing understanding and proficiency with numbers.

- "What Does Research Say the Benefits of Formative Assessment Are?" (Wiliam 2007b) presents research findings on the benefits of formative assessment in mathematics from one of the world's experts.

- "Five 'Key Strategies' for Effective Formative Assessment" (Wiliam 2007a) offers research-based classroom strategies for incorporating formative assessment into effective mathematics teaching.

- "Mathematics, Mathematicians, and Mathematics Education" (Bass 2005) provides insights into the role of mathematical structure in school mathematics, educators, and developers.

- *A Splintered Vision: An Investigation of U.S. Science and Mathematics Education* (Schmidt, McKnight, and Raizen 1997) compares U.S. standards with those in other countries.

- See also the resources under "More to Consider" in Message 38, "Building Things," for strategies on using mathematical structure for lasting learning.

15

You Can't Do *B* If You Don't Know *A*

CHALLENGING THE LINEAR MATHEMATICS CURRICULUM

Mathematics is not a careful march down a well-cleared highway, but a journey into a strange wilderness, where the explorers often get lost.

—W. S. Anglin, mathematician

At a mathematics conference several years ago, I found myself in an elevator with two teachers having a conversation about some of the sessions they had attended. They were discussing the appropriateness of the strategies described at a previous session. One teacher said to the other, "Well, you just can't do fractions if you don't know your multiplication tables." I knew what the teacher meant—in order to write equivalent fractions, a student multiplies (or divides) the numerator and denominator by the same number. As often happens when I hear a statement like that, I first thought about whether I agreed with the statement. I decided I probably did not. Then I started wondering how often in school mathematics we organize our teaching around the notion that "You can't do *B* if you don't know *A*." I decided it's probably fairly often.

Moving from Point A to Point B to Get to the *Good Stuff*

There are many reasons why we may hold students back from rich, challenging problems or more advanced mathematics, including spending too much time on review, redundancy in the curriculum from year to year, inadequate or inappropriate placement policies, or beliefs about a student's level of intelligence or motivation. But the biggest barrier to letting students go on to higher-level mathematical content and problems seems to be the belief that mathematics is a linear discipline, with knowledge and skills—especially skills—arranged in a clear, logical progression from lower-level skills to increasingly complex skills. Too often, we accept the view that a student must therefore follow a linear path—moving on to the next topic or skill only if he has mastered whatever came before, especially in terms of computational skills. Yet there is no evidence whatsoever that basic computational skills are a necessary prerequisite for dealing with geometry; using measurement; gathering, organizing, and interpreting data; or solving complex problems, even those that may involve computation.

PATHS TO SUCCESS

Mathematics is much richer than a limited and rigid view of a well-ordered set of skills to be mastered. For some students, access to complex, challenging mathematics builds on their ability to compute efficiently and accurately. But other students with gaps in their computational skills may also be able to access more challenging mathematics. Not only does mathematics include strands other than computation (for example, geometry, measurement, algebra, and data analysis), but even numerical problems often do not fall neatly into a sequence of "First you must know this before you can do that." In today's technological world, with a growing set of tools and the availability of more and more examples of high-quality, instructionally compelling extended problems for both teaching and testing, the mathematics curriculum may be much less linear than we once thought. For example, there is nothing to keep a student who hasn't mastered multiplication from exploring fractions within a well-designed learning experience. And a student who has difficulty with decimal computation might be able to tackle a problem involving decimals if she uses a calculator or works with a partner. If the teacher structures and monitors these learning experiences appropriately, these students may not only succeed at the *good stuff*—rich, engaging problems or higher-level thinking—but these students might also gain insights or motivation to help them overcome their other numerical stumbling blocks.

What's Really Essential?

Not everything in the mathematics curriculum can be absolutely necessary before moving on—learning time is too precious and the school year too crowded with topics to demand that students finish learning every skill or concept included in the curriculum. For some topics or skills, just an experience or an introduction may be enough, especially if the topic is not a priority at that grade level. And not even the most important skills or concepts may need to be fully mastered in order for a student to go on if requiring mastery holds the student back endlessly.

For many topics, I suggest that we can find other ways to allow students access to what comes later, even if they may be stuck on something that comes sooner.

A PROFESSIONAL DEVELOPMENT EXERCISE

Building on the opening elevator conversation, an interesting exercise for professional reflection among colleagues might include taking a hard look at our assumptions about what is essential for students to know before moving on. Consider the open-ended statement:

Students can't do *B* if they don't know *A*.

Try completing the statement with as many *A/B* pairs as possible, for example, "Students can't do *decimals* if they don't know *fractions*"; *percent/decimals*; *word problems/basic computation*. Then, for each *A/B* pair, consider whether *A* is absolutely necessary in order to do *B*. Could a student do *B* with the aid of a calculator or a fact table? Could a student learn *B* in the context of an interesting problem, even if he or she did not know *A*? Could *B* be an accessible topic or skill if the student were working in a small group or with a partner? Could *B* motivate the eventual learning of *A*?

There may be some skills for which the answer above is, "Yes, *A* is absolutely essential in order to do *B*." If this is truly the case, and if *B* is a priority for the grade level or course, then we need to do whatever it takes to make sure that students have learned *A* before going on to *B*. Or at least, we need to make sure that students do not move on with misconceptions. Understanding fractions well, perhaps using the concepts of unit fractions presented in the Common Core State Standards for Mathematics (NGA Center and CCSSO 2010), may fall in this category of absolutely essential for many skills and problems that follow. Or understanding place value, or knowing the meaning of an operation may fall in this category, at least up to a point.

Thinking About Mastery

Mastery is fine as a goal—we want students to learn the content for their grade level or course so that they will be able to do well at the next level. But sometimes we can embrace mastery too rigidly and, in the process, inadvertently limit students' learning. When a student falls behind or is unable to master what is being taught, we are often tempted to provide more practice on the content he or she has not mastered. Many such students are relegated year after year to skill repetition based on the assumption that they cannot handle higher-level mathematics without lower-level skills. And regardless of how much practice students go through, they never seem to get to the higher-level mathematics they were preparing for, and often they emerge from their extended remediation just as behind on the targeted skill as when they started, and much further behind in terms of grade-level or course expectations.

MASTERY AND TESTS

I would like to offer two assertions about mathematical mastery. First, mastery cannot be shown on a single test. A teacher needs to look at more than one measure and even more than one type of measure, over time, to determine whether a student is proficient in a skill or has a fairly solid knowledge about a topic. And second, passing a test may not indicate absolute mastery and certainly doesn't indicate mastery for life. A student with only superficial learning may do well on a test but not retain much. On the other hand, a student with strong proficiency, lasting knowledge, and deep learning might make a careless mistake on a test or might do well on the test but some day forget something previously learned. In fact, this is likely to happen. However if that student's learning is deep and solid, the student will be able to retrieve the previously learned knowledge with a brief refresher.

So, mastery may not be the absolute quantity some assume. And mastery is complex and difficult to measure with certainty. Rather than always working toward mastery, I'd like to suggest that a more appropriate goal for all students might be to work toward learning content deeply so that students are able to apply what they have learned to solve all kinds of interesting, challenging, engaging problems—what I consider the *good stuff*.

EXPANDING OUR THINKING ABOUT WHAT COMES FIRST

Every year a teacher faces students who come from different experiences. They won't all bring the same background or understanding, even if they come from the same school or the same previous teacher. And in some schools with high mobility rates, most students are likely to have been somewhere else the year before. Thus, we can seriously limit our teaching and our students' opportunities to learn by automatically assuming the necessity to have mastered what came before or by relying too heavily on identified prerequisite skills and knowledge.

What Can We Do?

One of the most important instructional decisions a teacher makes is when a bit more time would help a student and when it's time to allow the student to move on to something different, without losing sight of the gap that may persist. Talk with next year's teacher about not only any gaps that may remain for a student (to be sure they are addressed at some point), but also any strengths the student brings that can help with what comes next. When next year's teacher builds on the student's strengths, it not only gives the student confidence, but also allows the student to develop other skills and knowledge that might eventually lead to overcoming the particular stumbling block.

Meanwhile, let's avoid starting each year with lengthy review and instead be sure that all students have the opportunity to use previously learned knowledge and skills as they engage in new content and compelling problems worthy of their extended attention. Structuring the classroom to allow collaborative problem solving and incorporating appropriate uses of technology can help students access these problems whether or not they have fully mastered lower-level skills. And sometimes, in the process of working on rich, challenging, engaging problems, students may even learn some skills they were lacking before.

We need to find ways to let students progress to interesting problems and challenging mathematics, even as we attend to any gaps or difficulties from what came before. It's encouraging that a growing number of teachers now identify different prerequisite abilities than they might have identified several years ago. Today, an increasing number of teachers cite the following characteristics as critical: number and operation sense, the ability to approach a problem from more than one way, representation skills, critical thinking, and the ability to ask good questions or explain one's reasoning. It may be far more important that a student has experience developing these critical mathematical habits of mind than mastering the procedures or skills from the previous year. Perhaps the best outcome possible is that, when we think of the statement "You can't do *B* unless you know *A*," the *A*s will be mathematical habits of mind even more often than specific skills or procedures.

Reflections and Discussion

FOR TEACHERS

- What issues or challenges does this message raise for you? In what ways do you agree with or disagree with the main points of the message?
- Do the exercise under "What's Really Essential" in this message. What assumptions do you make about topics or skills you think students absolutely must know before they can do something else? What do your responses tell you when you make a list of *A*s and *B*s for the statement *Students can't possibly do 'B' if they don't know 'A'*? How might your

(*continued*)

answers shed light on which essential prerequisites call for our serious attention and which topics or skills might be something students could work around in order to get to the *good stuff?*

- How do the ideas in this message support or conflict with ideas in Message 18, "Finishing Teaching"? If you think there are areas of conflict, how can they be resolved?

FOR FAMILIES

- What questions or issues does this message raise for you to discuss with your daughter or son, the teacher, or school leaders?
- How might you let your son or daughter tackle interesting problems that arise if he or she hasn't yet mastered all of the computation skills needed?

FOR LEADERS AND POLICY MAKERS

- How does this message reinforce or challenge policies and decisions you have made or are considering?
- How flexible, objective, equitable, and successful are your policies for access to high-level mathematics for students who may have potential but who may not fit narrow placement guidelines?

RELATED MESSAGES

Smarter Than We Think

- Message 12, "Upside-Down Teaching," advocates a problem-centered teaching approach that might engage students in challenging mathematics even if they have not mastered all of the expected prior skills.
- Message 1, "Smarter Than We Think," considers how students' intelligence may be positively affected by their engagement with hard problems.
- Message 31, "Developing Mathematical Habits of Mind—Practices, Processes, and Proficiency," describes the kinds of mathematical thinking and reasoning that should be a goal in every mathematics classroom.
- Message 3, "He Doesn't Know His Facts," looks at how one person lacking basic multiplication facts went on to succeed in advanced mathematics.
- Message 16, "Let It Go," reminds us that some topics and skills we have always taught may no longer be necessary or appropriate in order to gain focus under new standards.

Faster Isn't Smarter

- Message 30, "Crystal's Calculator," tells the story of one student who thrived in higher-level mathematics without lower-level mastery. Available for download in printable format from mathsolutions.com/fasterisntsmarter.

- Message 34, "Forgetting Isn't Forever," reminds us that sometimes students forget what they've learned and advocates incorporating brief review as students go on to higher-level mathematics, rather than excessively reviewing at the beginning of the year.

- Message 17, "Constructive Struggling," advocates providing all students with challenging problems that allow them to develop mathematical thinking even if they may not have mastered all prior skills.

MORE TO CONSIDER

- *Mindset: The New Psychology of Success* (Dweck 2006) discusses the implications of a person's mindset about intelligence on their school work and on their lives and offers insights for teachers about the value of providing challenging material to every student.

- *Professional Standards for Teaching Mathematics* (National Council of Teachers of Mathematics 1991) contains the best description of what a mathematics classroom should look like in order to help all students learn to think, reason, and develop mathematical habits of mind.

- "Connecting Research to Teaching: Shifting Mathematical Authority from Teacher to Community" (Webel 2010) makes a case for engaging students in doing the work of mathematics in a problem-centered environment.

16

Let It Go …

MAKING A FOCUSED MATHEMATICS
CURRICULUM WORK

Things which matter most must never be at the mercy of things
which matter least.

—Johann Wolfgang von Goethe

The traditional American spiral curriculum in mathematics may
finally be dead, and no one is mourning the loss. No longer do
we believe that it makes sense to teach students a little bit about
a lot of things every year, each time sprinkling on a little more information about essentially the same things, but sometimes with one more
digit or an uglier denominator. From looking at effective programs in
the United States and considering success in other countries' curricula,
we now know that focusing the curriculum on fewer topics in greater
depth has the potential to reap significant benefits for students' lasting learning. Some reinforcement of previously learned content may be
helpful if done well and in conjunction with moving forward to learn
new content. But the spiral curriculum has not proven to be an effective
vehicle to provide this reinforcement. We now recognize the unfortu-
nate, unintended consequences of trying to teach too many topics without depth—students don't have a chance to finish their learning, and
without adequate learning and understanding upon which students can
build further learning, teachers find it necessary to repeatedly review,
wasting precious instructional time and losing students' interest.

Finding Focus

Recent mathematics standards, including the Common Core State Standards for Mathematics (NGA Center and CCSSO 2010) have become
more focused. High school standards have not yet made as much

progress toward focus as those in kindergarten through grade 8, but greater focus continues to serve as a goal for mathematics standards across the grades. It's now common to find three to five priority areas identified for a grade level, with a total of perhaps thirty to forty standards or objectives at that level. The identification of a few priority topics or clusters of knowledge and skills and the smaller overall number of targets to be assessed on high-stakes assessments increase the likelihood that teachers can help students learn mathematics with understanding in ways that can last over time.

Letting Go

Standards alone—even focused standards supported by aligned tests—do not necessarily translate into focused teaching and improved student learning. For that to happen, teachers must first be willing to let go of some instructional activities on long-standing topics that no longer fit into their grade level or course. By taking the first step of weeding out what may no longer be necessary or appropriate, teachers can open the door for real change, focusing their energy on transforming their day-to-day teaching around a smaller number of targets and taking the time to help all students finish learning those targets. Letting go opens the door for several shifts in teaching, including using "upside-down" teaching, looking for regularity, and ending endless review.

USING UPSIDE-DOWN TEACHING

The key to helping students learn important topics and skills with lasting understanding and proficiency lies in balancing the teaching of computation, conceptual understanding, and problem solving in a classroom centered around student engagement and discussion on rich, worthwhile mathematical tasks. This kind of instruction—I call it "upside-down teaching" (see Message 12)—draws students into thinking about the underlying mathematics in a problem situation as they wrestle with mathematical ideas, even if they have not yet formally learned the related procedures. As they engage with the task, they sometimes uncover important mathematical concepts and understanding. And regardless of what they uncover, realize, or hypothesize about the mathematics of the task, their engagement with it makes them more receptive to learning what the teacher has in mind than they would be by passively listening to the teacher tell them the intended mathematical outcome, procedure, or concept. These kinds of tasks and accompanying classroom discourse call for more classroom time than a teacher-centered presentation of information followed by predictable word problems. However, the payoff is that students are more likely to understand and remember the mathematics they learn and more likely be able to use it in the future.

Finding the necessary additional classroom time for this kind of teaching may mean letting go of a unit, lesson, topic, or skill that has always been part of what we've taught in the past or thought should be taught. Letting go starts with a hard look at the file cabinet—whether digital or paper—and asking hard questions about whether a unit, lesson, or activity truly supports student learning on the specific goals and priorities now in place for the particular grade level or course.

LOOKING FOR REGULARITY

Another way we can let go of old practices and find time for greater focus is to tap into the notion of repeated reasoning, as articulated in the Common Core State Standards for Mathematical Practice 8, "Look for and express regularity in repeated reasoning" (NGA Center and CCSSO 2010, 8). At the elementary grades, this idea is particularly important in reducing redundancy in the curriculum. For example, when students are learning how to add and subtract two- and three-digit numbers, they apply what they know about place value and the meaning of the operations as they learn the procedure. In the past, students might have revisited this topic as a standard in grades 2, 3, 4, and even 5, each time slightly extending the number of digits addressed. Thus, this content consumed considerable time across several years, adding to the topic burden at each grade level and increasing the likelihood that teachers would not have adequate time to fully develop that or other important topics. Now most sets of standards, including the Common Core, address the development and mastery of multidigit addition and subtraction in grade 2 as one of the priority outcomes for that grade level. In later grades, students are expected to generalize and extend what they know, but not as a priority topic. After grade 2, the emphasis shifts to developing the notion of making generalizations based on students' repeated reasoning and extending the patterns and regularity students see, an important mathematical habit of mind. (See Message 39, "Patterns with a Purpose," for a discussion about the notion of regularity in repeated reasoning, including how it unfolds across the grades into high school.)

The challenge for teachers, however, is that a textbook or district curriculum might still contain remnants of previous standards calling for something that has now been eliminated, such as addition and subtraction of six- or nine-digit numbers in upper elementary school. Because we know that addition and subtraction are critical skills, it's easy to want to include such skills that have now been eliminated from a grade if the next lesson in the textbook addresses them. But if we have done the job well of helping students see the regularity and the pattern in the process, we don't need to spend extended time in higher grades teaching as if it were new content. We might incorporate brief review as students move on to other priorities, but we need not spend precious instructional time reteaching this content. The time saved allows teachers to focus on other priorities.

ENDING ENDLESS REVIEW

Perhaps the most important teaching practice to let go is the pervasive time-eating, boredom-generating practice of extended review at the beginning of each school year. First, if we restructure mathematics classrooms to focus on teaching a topic or skill in depth over fewer years than in the past, and if we use the kind of teaching organized around rich in-depth tasks, we may be able to let go of some of our ideas about the necessity to demonstrate mastery of a long list of prerequisite skills. Many such skills can be addressed in the context of the tasks we select to address new content. In this way, we can incorporate meaningful review of previously learned content into tasks and activities that offer students something new—truly building on what students know.

Most of all, we simply don't have instructional time to spend—to waste—testing and drilling students on skills they may have learned but forgotten over the summer or may not have used recently. Teaching and learning time is too precious to sacrifice weeks of time for such review. Rather, if students have adequate time each year to deeply explore and learn priority concepts and skills, they can reinforce those concepts and skills in the context of new, interesting tasks at the next grade level or in the next course.

What Can We Do?

As with almost every other issue in school mathematics, the solution starts with professional collaboration within and across grade levels and courses. In this case, educators can begin by working together to agree on priority topics, concepts, and skills for each grade level or course. Of even more importance, we can agree to focus teaching on those priorities to increase the likelihood that every student will move on to the next grade or course having learned those priorities well. The only way a focused curriculum works is if teachers have the time to focus on priorities and if they trust that other priorities will be addressed well by their colleagues at other grade levels or in other courses. When teachers have confidence that necessary content will be addressed across the grades, then perhaps they can let go of the need to spend extensive time reviewing at the beginning of each year and move on to spend time on topics and skills most needed by the students they teach.

Teachers can buy the time to teach those topics and skills by taking a hard look at their textbook, their curriculum frameworks, their files of previous lessons and instructional resources, and, most of all, their own teaching practice in years past to see what they might be able to omit. Such a thorough audit may expose time-eaters that might be expendable, given changes in standards and a tightened focus at the grade level. Teachers might ask themselves: *What am I hanging onto that I might let go? What might I be able to eliminate to buy time for priority concepts and skills?* When teachers have identified the most important ways to invest their time, their focus can shift to spending that time engaging students in wrestling with worthwhile problems and helping students see the regularity in the mathematical concepts and procedures they learn.

Teachers almost universally agree that their most precious resource is time. One of the best ways to provide that precious resource may be to let go of something that takes it away.

Reflections and Discussion

FOR TEACHERS

- What issues or challenges does this message raise for you? In what ways do you agree with or disagree with the main points of the message?
- How are the topics and skills in the standards you are now expected to teach different from those you might have been expected to teach a few years ago?
- On what topics or skills at your grade level do you find yourself spending significant amounts of instructional time, even if they are not identified as priority topics in your grade level standards? What would be the benefits or disadvantages of letting any of these topics and skills go?
- How can you work with your colleagues to provide more focus at your grade level or in your course?
- If you are a high school teacher, how can you find greater focus in your overcrowded curriculum?

FOR FAMILIES

- What questions or issues does this message raise for you to discuss with your son or daughter, the teacher, or school leaders?
- What are some things you can do to help your daughter or son not forget mathematics over the summer?
- How can you reinforce the mathematics your son or daughter is learning at school throughout the year?

FOR LEADERS AND POLICY MAKERS

- How does this message reinforce or challenge policies and decisions you have made or are considering?
- How can you provide time and support for collaboration within and across grade levels to reduce redundancy and better focus on priority topics and skills?
- How can you support teachers in having the confidence to not have to teach everything every year?

RELATED MESSAGES

Smarter Than We Think

- Message 3, "He Doesn't Know His Facts," reminds us that sometimes students can learn new mathematical content even if they may have gaps in their computational skills.
- Message 12, "Upside-Down Teaching," describes the teaching model advocated here.
- Message 14, "Effectiveness and Efficiency," considers what it takes to help students learn for lasting understanding and proficiency.
- Message 18, "Finishing Teaching," looks at the need for students to have adequate time to "finish" their learning of priority topics and skills.
- Message 39, "Patterns with a Purpose," addresses the mathematical practice involving looking for regularity in reasoning, a key element in finding adequate instructional time to focus the curriculum on priorities.

Faster Isn't Smarter

- Message 4, "Good Old Days," looks at the illusion that the mathematics curriculum, mathematics teaching, and student achievement were better in days gone by compared to today's classrooms.
- Message 16, "Hard Arithmetic Isn't Deep Mathematics," reminds us that ramping up our expectations of students involves providing opportunities to tackle rich problems, not just harder computation.
- Message 30, "Crystal's Calculator," recounts the true story of a student who failed to learn fractions year after year, but for whom moving on to Algebra 1 provided the platform for her to learn fraction operations once and for all.

- Message 34, "Forgetting Isn't Forever," advocates minimizing beginning-of-the-year review and building in reinforcement of previously learned skills and content as students move forward to new material.

MORE TO CONSIDER

- *Learning and Teaching Early Math: The Learning Trajectories Approach* (Clements and Sarama 2009) offers an important way to focus on mathematical content as it develops across the early years and supports recommendations for effective mathematics teaching for young children.

- *Engaging Young Children in Mathematics: Standards for Early Childhood Mathematics Education* (Clements, Sarama, and DiBiase 2007) presents a variety of resources for early childhood mathematics, including standards for preschool and kindergarten mathematics education and classroom suggestions for developing priority mathematical knowledge and skills.

- "Effective Teaching for the Development of Skill and Conceptual Understanding of Number: What Is Most Effective?" (Hiebert and Grouws 2007) presents a summary of research on developing understanding and proficiency with numbers across the grades, including suggestions for effective research-based practices for the classroom.

- Common Core State Standards for Mathematics (NGA Center and CCSSO 2010) are the standards that form the basis of teaching and learning in many states.

- *The Intended Mathematics Curriculum as Represented in State-Level Curriculum Standards: Consensus or Confusion?* (Reys 2006) provides an overview of state standards showing the redundancy in what was called for in many states during the early twenty-first century.

17 The Journey to Algebra

DEVELOPING ALGEBRAIC THINKING ACROSS
THE GRADES

It is one thing to see threads of algebraic thinking in a student response or to notice gaps in thinking; it is quite another thing to know how to teach the student to develop his or her thinking in productive ways.

—Mark Driscoll (*Fostering Algebraic Thinking*, 1999)

The United States is one of the only countries in the world to teach courses with names like Algebra 1 or Algebra 2. Most schools in the rest of the world teach mathematics, not as separate courses, but as a continuous program from elementary through secondary school. In the United States, some schools offer an alternative to these courses, such as an integrated program that incorporates algebra as a strand blended with geometry and other advanced topics. Others continue to offer a course sequence that includes Algebra 1, Geometry, and Algebra 2. In many states, the study of algebra in some form is required of all students for high school graduation. Regardless of whether a school's secondary curriculum includes separate courses in Algebra 1 and Algebra 2 or offers a more integrated system that incorporates algebra across courses, we can take concrete steps to ensure that students will flourish and succeed as they move into the formal study of algebra. A key to this success is the development of algebraic thinking as a cohesive thread in the mathematics curriculum from prekindergarten through high school.

Charting a Course to Algebraic Thinking

Algebraic thinking includes recognizing and analyzing patterns, studying and representing relationships, making generalizations, and analyzing how things change. Of course, facility in using algebraic symbols is an integral part of becoming proficient in applying algebra to solve problems. But trying to understand abstract symbolism without a foundation in thinking algebraically is likely to lead to frustration and failure. Algebraic thinking can start when students begin their study of mathematics.

THE EARLY GRADES

At the earliest grades, young children work with patterns. From an early age, children have a natural love of mathematics, and their curiosity is a strong motivator as they try to describe and extend patterns of shapes, colors, sounds, and eventually letters and numbers. And young children can begin to make generalizations about patterns that seem to be the same or different. This kind of categorizing and generalizing is an important developmental step on the journey toward algebraic thinking.

Throughout the elementary grades, patterns may not be a formal object of study, but working with patterns can be especially useful as a tool for thinking about a variety of mathematical concepts, especially numbers and numerical relationships. As students develop their understanding of numbers, they can use patterns in arrays of dots or objects to help them recognize what 6 is or whether 2 is greater than 3. As they explore and understand addition, subtraction, multiplication, and division, they can look for repeating patterns that help them learn facts and extend procedures to increasingly large numbers. Patterns in rows and columns of objects help students get a sense of multiplication and see that the facts they're learning make sense. Patterns within the multiplication table itself are interesting to children and can help them both learn their facts and understand relationships among facts. The process of noticing and exploring patterns sets the stage for looking at more complex relationships, including proportionality, in later grades.

THE MIDDLE GRADES

As students move into the middle grades, their mathematics experience connects their work with numbers and operations to more symbolic work with equations and expressions. At this level, the emphasis of the mathematics program should be on proportionality, perhaps the most important connecting idea in the entire prekindergarten through grade 12 mathematics curriculum. This concept should take students well beyond the study of ratios, proportions, and percent to address increasingly complex problems calling for students to notice and use proportionality. A deep and solid understanding of proportionality allows students to connect their experience with numbers and operations to ideas they've studied in geometry, measurement, and data analysis. They begin to get a sense of how two quantities can be related proportionally, as seen on maps, scale drawings, and similar figures, or in calculating sales tax or commissions. As students do so, they build a foundation for understanding linear relationships.

HIGH SCHOOL

A solid understanding of proportionality sets the stage for students to succeed in the more formal study of algebra. From this base, notions of linearity and linear functions emerge naturally. As students explore the use of linear functions to solve problems, the bigger world of functions that may not be linear also begins to open for them. Looking at what is the same and what is different among functions lies at the heart of understanding and using algebraic skills and processes.

The journey doesn't end with a student's first formal study of algebra in high school. Continuous development of increasingly sophisticated algebraic reasoning can provide an avenue into the study of geometry, statistics, and advanced mathematics, whether in courses called Algebra 1, Algebra 2, Geometry, Statistics, or in more integrated courses. In the real world outside of school, these topics are not separated. When more advanced courses regularly incorporate opportunities to build on students' algebraic understanding, they are far more likely to succeed than if the courses present just one mathematical perspective. Thus, algebraic thinking and an understanding of algebra serve as not only goals but also powerful tools to access the broad world of secondary and postsecondary mathematics.

What Can We Do?

The development of algebraic thinking is an ongoing journey, not something that happens in a single course. Making the transition to algebraic thinking is most natural for students when, throughout their mathematics experience, they increasingly move toward thinking abstractly, making generalizations, connecting representations of mathematical ideas (tables, graphs, equations, numbers), and solving problems. This kind of deep, lasting learning cannot be accomplished by telling students things they need to learn. It has to happen in the context of rich tasks, considerable student reflection and discussion, and a learning environment that supports students in making sense of mathematics and learning to think mathematically, even when a problem may not look like others they have seen before. Thus, the most important thing we can do on a day-to-day basis to support students' development of algebraic thinking is to organize mathematics classrooms so that the classes focus on students' engagement with tasks that cause them to grapple with algebraic ideas and engage discourse to help them dig deeper into those ideas.

Longer term, we can continually examine our programs and instructional materials to make sure that the threads of algebraic thinking—multiple representations, generalizations, communication, and justification, among others—are woven into the mathematics program in appropriate ways at every level. This notion of integrating algebra across the grades is just as important at high school, where we need to take a hard look at how long we can justify continuing to offer courses organized strictly around algebra or strictly around geometry. A mathematics problem that students will face outside of school will not be labeled so conveniently—students would be far better served if algebra was a unifying theme across all grades, connecting with geometry, statistics, and other mathematical topics around worthwhile and meaningful tasks.

Reflections and Discussion

FOR TEACHERS

- What issues or challenges does this message raise for you? In what ways do you agree with or disagree with the main points of the message?
- How can secondary school mathematics teaching capitalize on the inclusion of algebraic thinking throughout the elementary and middle grades?
- How can we incorporate algebraic thinking into the prekindergarten through grade 12 curriculum at all levels? At your level?

FOR FAMILIES

- What questions or issues does this message raise for you to discuss with your daughter or son, the teacher, or school leaders?
- How can you help your son or daughter practice representing mathematical ideas in many ways—with numbers, pictures, words, tables, or graphs? What situations in your everyday life might provide opportunities to draw a picture about a mathematical situation or make a table or graph? In what situations can you imagine engaging your son or daughter in describing in conversation a mathematical relationship, for instance, speed or distance while driving, buying something, planning an event, and so on?

FOR LEADERS AND POLICY MAKERS

- How does this message reinforce or challenge policies and decisions you have made or are considering?
- How can you ensure that instructional materials and curriculum frameworks across the grades reflect the importance of developing algebraic thinking in an appropriate way?
- In what ways do your elementary and secondary mathematics programs incorporate notions of algebraic thinking?
- How can secondary school mathematics capitalize on the development of the many aspects of algebraic thinking throughout the elementary and middle grades?

RELATED MESSAGES

Smarter Than We Think

- Message 38, "Building Things," considers the incorporation of mathematical structure, including the use of appropriate definitions and properties related to algebra, into school mathematics across the grades.
- Message 39, "Patterns with a Purpose," emphasizes the importance of noticing regularity in the repeated patterns students see across mathematical topics, an important element in algebraic thinking.

Faster Isn't Smarter

- Message 25, "Pushing Algebra Down," discusses accelerating students into a high school algebra course.
- Message 30, "Crystal's Calculator," relates a true story about a student's transition to high school algebra, even when she couldn't do fractions.
- Message 7, "Not Your Grandpa's Algebra," reminds us that the study of algebra today needs to reflect changing priorities and the goal of increasing the number of students studying algebra in high school.

MORE TO CONSIDER

- *Thinking Mathematically: Integrating Arithmetic and Algebra in Elementary School* (Carpenter, Franke, and Levi 2003) offers research-informed considerations for teaching mathematics to elementary students so they can develop algebraic thinking as they learn arithmetic concepts and properties.

- *Fostering Algebraic Thinking: A Guide for Teachers, Grades 6–10* (Driscoll 1999), a well-known resource from a respected mathematics researcher, helps teachers in middle school and high school support students in developing algebraic thinking.
- *Developing Essential Understanding of Algebraic Thinking for Teaching Mathematics in Grades 3–5* (Blanton, Levi, Crites, Dougherty, and Zbiek 2011) focuses on mathematical properties and relationships in support of the transition from numbers to symbols.
- *Connecting Arithmetic to Algebra* (Russell, Schifter, and Bastable 2011) uses the transition to algebra as a platform for addressing mathematical structure through properties, generalizations, and reasoning.
- *It's All Connected: The Power of Representation to Build Algebraic Reasoning, Grades 6–9* (Van Dyke 2012) discusses the use of a variety of mathematical representations, including symbolic representation, and offers classroom activities to help students make the transition to algebra.
- *It's All Connected: The Power of Proportional Reasoning to Understand Mathematics Concepts, Grades 6–8* (Whitman 2011) discusses proportionality, a critical connecting idea from arithmetic through algebra, including classroom activities.
- *Lessons for Algebraic Thinking: Grades K–2* (von Rotz and Burns 2002) includes sample lessons for integrating algebraic thinking into elementary classrooms. *Lessons for Algebraic Thinking: Grades 3–5* (Wickett, Kharas, and Burns 2002) includes lessons for integrating algebraic thinking into elementary classrooms. *Lessons for Algebraic Thinking, Grades 6–8* (Lawrence and Hennessy 2002) includes sample lessons for integrating algebraic thinking into middle school classrooms.
- *Active Algebra: Strategies and Lessons for Successfully Teaching Linear Relationships, Middle and High School* (Brutlag 2009) offers active learning strategies for helping students learn about linear relationships based on brain research with adolescents.
- "What Do We Know About the Teaching and Learning of Algebra in the Elementary Grades?" (Kieran 2007b) summarizes research on developing algebraic thinking in elementary school and offers research-based insights into classroom practice and curriculum development.
- "What Do Students Struggle with When First Introduced to Algebra Symbols?" (Kieran 2007a) presents research findings about making the transition from numbers to letters as students begin to work with algebraic ideas.

18 Finishing Teaching

COVERING CONTENT DOESN'T CUT IT

You quit teaching too soon ...

—Japanese education official (Daro 2012)

Not long ago I attended a presentation by Common Core Standards codirector Phil Daro in which he referenced a conversation with a Japanese education official who had visited schools in the United States. The official stated, "I know why Japan outperforms the United States on international mathematics exams. You quit teaching too soon and go on to the next thing. We finish." Finishing happens, the official noted, when students have learned the intended mathematics (Daro 2012). In the United States, we clearly focus on covering content, and our target typically is to cover the content that will be on the accountability test near the end of the year. But effective teachers know that covering content and focusing instruction primarily on how to answer questions like those that will be on the high-stakes test are not strategies that lead to lasting learning. In short, covering doesn't cut it. What can we do instead of just covering content?

Postponing Answer-Getting

In the United States, teachers often focus on how to get students to find the answer to a problem. In other countries, such as Japan, teachers are more likely to focus on how to use a problem to help students learn the intended mathematics. When we focus primarily on getting answers, students come to believe that mathematics is a bit of a guessing game, where they are supposed to look at a problem and remember which rule or procedure—probably one recently learned—they should use.

On the other hand, if we organize teaching around deep rich tasks, and if we can learn to postpone or slow down answer-getting (not eliminate it), we can use the task or problem as a springboard for in-depth discussion in which students work through their thinking, consider approaches used by others, try alternative strategies, and have many opportunities to justify and explain what they're doing. As a result, students can not only learn the intended mathematics, but can also develop mathematical habits of mind that can evolve and grow over time as they apply their mathematical thinking in solving other problems.

Answers are not bad; a primary goal of school mathematics is to equip students to be able to solve problems and arrive at answers. But racing to answers and rewarding whoever gets there first can give students an inaccurate and often negative view of mathematics. Worse, focusing solely on answers misses rich opportunities for students to engage with mathematical ideas, wrestle with mathematical concepts, and participate in discourse with other students and the teacher. We also rob them of the opportunity to see the value of making and rethinking their mistakes, ideally coming to see that their mistakes can contribute to their learning. When we use problems as the basis for lessons, especially if we use them to kick off discussion of a new concept or process, we offer opportunities to draw students into thinking deeply about as yet unnamed ideas as they formulate ideas and hypotheses and reflect on their own and others' thinking.

One reason so many teachers focus on answer-getting is because that's what students are expected to do on the end-of-the-year high-stakes test. But students are far more likely to do well on such tests if they develop mathematical thinking and deep understanding. Unfortunately, that's a hard idea to accept if pressures from every direction push teachers to cover content and prepare students to answer test questions correctly. Nevertheless, new types of assessments on the horizon will focus far more on thinking and reasoning about extended problems than most current tests. We can hope that such assessments will liberate teachers and allow them to teach in ways they know help students develop mathematical habits of mind.

Teaching for Depth and Understanding

During the two years I taught in Burkina Faso, from 1999 through 2001, I tried to teach around good problems—the best deep problems I could find (in French) that would push students' thinking. One young man in my high school class, Boukari, approached me about two-thirds of the way through the first of my two years there. He explained to me that he noticed I really liked problems, but that he thought we were spending too much time on problems and not enough time *covering the program*—a complaint often heard from some parents and administrators in the United States. Boukari suggested that perhaps we could do problems on Fridays, but the rest of the time we could cover the program. I thanked him for his input, and I continued to teach using problems. He was understandably disappointed when he discovered that I would also be his teacher the following year. At the end of my two years there, however, as I was preparing to leave for the United States, Boukari came to see me with a group of students to say their good-byes. He took me aside, and, with a sheepish grin, said to me, "Madame, I think I learned more mathematics with your problems than I would have learned otherwise." What he was acknowledging was that, even though we may not have covered the whole program, he finished his learning on the mathematics we did.

In the United States, both students and teachers are also accustomed to covering material. We may think teaching for depth and lasting understanding just takes too much time, even if we know it's valuable. But finishing teaching—teaching until students understand—is an investment, not an expense. Students who have the opportunity to finish learning mathematics deeply are far more likely to be willing to take on problems they've never learned specifically to solve and far more likely to persevere in seeing a problem through to its solution. The mathematical habits of mind students develop can help them think and reason in ways that simply do not emerge in classrooms where we *cover the program*. Unfortunately, most teachers in today's test-focused world will say that they know their students need more time on some topics, but that they don't have time to teach in the depth their students need because they have to *cover the program*. Covering the program, however, may not allow students to learn much of anything well, and, even if they are able to perform satisfactorily on a few multiple-choice questions on the high-stakes test, they are not likely to retain their superficial knowledge for long. And when we cover material, rather than letting students finish their learning, they can often carry misconceptions with them that will cause them difficulties later. If we focus instead on using good problems to teach mathematical concepts and skills and let students finish their learning on those concepts and skills, students may well learn more than they would by simply *covering the program*. Ironically, students may learn more, not less, by covering less, not more.

Making Choices and Identifying Priorities

There probably isn't time to teach everything in the curriculum in the depth we might like. So we have to decide where to put our greatest emphasis. Teaching in depth so that students finish learning means identifying the most important topics, concepts, and skills for a particular grade level or course. This can only be done by working with other teachers across grades and courses. The Common Core State Standards, for example, provide somewhat more focus than standards several years ago in terms of the number of standards per grade level, and priority areas are generally identified. This can help in deciding which topics call for the most in-depth teaching. High school standards, however, tend to still pack quite a bit of material into each course. But even in high schools, by working with colleagues, we can identify the highest priority areas in each course where students would benefit the most by in-depth teaching that could help them finish their learning. When students have the opportunity to finish learning, we can also work together to revisit the time-wasting practice of starting each year with days or weeks of review, instead offering students a chance to remember their "finished" learning in the context of new content and new problems.

What Can We Do?

We don't need to copy what Japan does in mathematics classrooms. Or what Finland, Singapore, or any other country does. Their cultures, contexts, and educational systems vary dramatically from ours. But we can learn lessons by looking at effective approaches or philosophies of others, whether down the hall or across an ocean. Teaching until students have learned—finishing teaching—can have tremendous payoffs. When students have learned well the priority topics, concepts, and skills for their grade, they gain confidence in their mathematical ability and tend to perform well on tests, including high-stakes tests. Most of all, if we prioritize developing mathematical habits of mind, and if we provide enough time for students to develop mathematical understanding as they work fully through rich tasks, what students carry into next year can be a way of thinking, not just a short-term, easily forgotten, set of skills on a checklist of what they covered the previous year. The former will last far longer than the latter.

My mother always told me to listen to that little voice in my head, which is sometimes more difficult than it sounds. When a teacher hears the little voice saying that students aren't there yet—that they need more time or a different approach—I hope the teacher will consider staying longer and teaching until students finish. Imagine the difference if, instead of moving on to the next unit when some students haven't finished learning the meaning of fractions or the concept of a proportional relationship, a teacher chose to invest a bit more time to correct students' misconceptions. Such a choice would pay off year after year after year as students are able to learn more and make sense of what they're learning. By letting them finish, we ensure that they will be able to go even further, building on success, connecting what they're learning to their growing body of knowledge, and developing ever higher abilities to think, reason, and use what they know to successfully take on complex, challenging problems.

Reflections and Discussion

FOR TEACHERS

- What issues or challenges does this message raise for you? In what ways do you agree with or disagree with the main points of the message?
- What are the priority topics, concepts, or skills for your grade level or course? How can you ensure that you finish teaching these topics, concepts, and skills, even if you don't thoroughly cover everything else for your grade level or course?
- How can you help students focus on learning mathematics rather than racing to answers?
- How can you help a student who needs more time to learn a priority topic, concept, or skill?
- How do the ideas in this message support or conflict with ideas in Message 15, "You Can't Do *B* If You Don't Know *A*"? If you think there are areas of conflict, how can these be resolved?

FOR FAMILIES

- What questions or issues does this message raise for you to discuss with your son or daughter, the teacher, or school leaders?
- How can you help your daughter or son focus on learning mathematics rather than racing to answers?
- How can you find out what the priority areas are for deep learning for your son or daughter in mathematics class?
- How can you reinforce important mathematical skills and concepts your daughter or son learns at school?

FOR LEADERS AND POLICY MAKERS

- How does this message reinforce or challenge policies and decisions you have made or are considering?
- How can you give teachers permission not to try to cover everything, but to teach for in-depth learning and more complex thinking?
- How can you give teachers more time to finish teaching?

RELATED MESSAGES

Smarter Than We Think

- Message 12, "Upside-Down Teaching," describes a problem-centered approach where students develop depth and understanding on the way to finishing their learning.
- Message 13, "Clueless," considers what happens when we don't finish teaching all the way to student learning.
- Message 14, "Effectiveness and Efficiency," looks at how to maximize the outcomes from our instructional decisions.

Faster Isn't Smarter

- Message 22, "We Don't Care About the Answer," offers a caution about ignoring correct answers and discusses the importance of balance in teaching.
- Message 31, "Do They Really Need It?," recounts a story from Africa regarding assumptions I made about a class of students and what I learned from the experience.
- Message 17, "Constructive Struggling," advocates challenging students to wrestle with good problems as they develop depth and understanding and finish learning.

MORE TO CONSIDER

- *What's Math Got to Do with It?* (Boaler 2008) looks at important shifts in teaching and learning mathematics for a broad audience of educators and noneducators.
- "The Mathematics Education of Students in Japan: A Comparison with United States Mathematics Programs" (Mastrull 2002) is a personal report that considers a variety of factors in Japanese and American mathematics classrooms, including the notion of postponing answer-getting to focus on mathematical learning.
- "Reassessing U.S. International Mathematics Performance: New Findings from the 2003 TIMSS and PISA" (Ginsburg, Cooke, Leinwand, Noell, and Pollock 2005) looks at findings from these two international tests and provides insights into the apparently low relationship between topic coverage in various countries and those countries' performance on the tests.
- "Informing Grades 1–6 Mathematics Standards Development: What Can Be Learned from High-Performing Hong Kong, Korea, and Singapore?" (Ginsburg, Leinwand, and Decker 2009) considers a variety of topics about standards and curriculum at the elementary grades and includes a discussion of topic coverage.

19 How to Know What They Know

ASSESSING AND SUPPORTING STUDENT LEARNING

By permission of Marshall Ramsey and Creators Syndicate, Inc.

I believe strongly that the purpose of assessment must be to improve student learning. Some might debate the truth of this statement, however, especially in light of the many ways policy makers, school administrators, and the public use test scores in today's culture of accountability. I stand by my statement. Whatever the stated purpose of a test or other type of assessment, if the end result does not in some way contribute to improving student learning, the purpose and nature of the assessment should be seriously reconsidered.

Summative Assessment, Formative Assessment, and Things in Between

Traditionally, assessments fall into one of two categories—*summative* and *formative*. Summative assessments are generally given at the conclusion of some instructional time period or at the end of the semester or school year, like a year-end state test or a semester exam. Such assessments can often be *high-stakes*, meaning that a decision will be made based on the results, such as for student promotion, award of credit, program evaluation or, increasingly today, for teacher evaluation. Formative assessment, on the other hand, is used by teachers to monitor student learning as it is occurring. It is integrated into instruction, with results used on the spot or in the immediate future to adjust learning experiences for students as needed.

For making day-to-day decisions, the teacher is the best person to assess mathematics learning, and the classroom is the best context in which to do so. Teachers today are using an increasing array of practical strategies to find out how well students are learning on a day-to-day basis. If we can catch students' misunderstandings and confusion before they become habits, we can help students improve their learning both now and in the future. This type of assessment is generally considered formative.

Sometimes other tests, like an assessment given at the end of a major unit, might be somewhat summative, in that a grade may be taken that will contribute toward a student's course grade. However, a teacher might use a student's results on a unit assessment to determine what next steps are appropriate to further support the student's learning. So the teacher may use the assessment formatively. Whether an assessment is labeled summative or formative is less important than it is for teachers to recognize the availability of a wide range of resources, tools, and protocols to both adjust instruction and to measure and grade learning.

Assessing and Learning

Ideally, assessment should be seamlessly woven into the fabric of teaching and learning, minimizing interruptions in instructional time and maximizing the immediate impact on students' learning. When assessment is part of the learning process, it need not sidetrack an effective mathematics program. A formative strategy might be as simple as asking a student a question as the student is working on a problem, or engaging a group of students in a probing conversation as they collaborate on a task. The most useful assessment results for influencing students' learning are those that are immediate and help students see for themselves whether they're headed in the right direction. So learning how to ask questions that push students' thinking is key, as well as knowing what to ask next based on a student's response. When we help students see what they are doing well and help them clarify and redirect any developing misunderstandings, they are far more likely to learn mathematics well and have that learning last.

Asking Good Questions

Teachers can use effective questioning strategies in many types of settings, whether informally talking with one or more students or conducting in-depth one-on-one student interviews. The ability to ask good questions also improves the quality of quizzes and tests (*How did you decide to ...? What does the 7 mean in your solution? Can you convince Maria?*). The nature of student responses to productive questions is much richer than if students are asked procedural questions or fill-in-the-blank questions (*If we add three to each side, we get ...? How do I find the greatest common factor of these two numbers? What shape is the graph of a quadratic function?*). When students are asked productive, probing questions that push their thinking, and when we ask them to explain or justify their responses, the level of student reasoning and mathematical thinking goes up. Learning how to ask good questions and develop effective assessments can offer a powerful return on the investment of time and resources in terms of student learning, especially if educators work together within an active professional learning community.

Teaching to *The Test*

When a high-stakes large-scale assessment, such as a state accountability test, is well aligned with the system of mathematics teaching and learning, preparing students to perform well on the test should involve not much more than teaching the mathematics program well. Unfortunately, this is a hard premise for teachers to accept, especially when they feel pressure from both in and outside of school to spend extended periods of time on explicit test preparation. Unfortunately, however, many of the most common types of test preparation strategies may interrupt and actually interfere with learning as they decrease the opportunity for students to learn mathematics with meaning, understanding, and proficiency. For example, benchmark testing—the use of interim tests in preparation for the year-end test like the students in the opening cartoon apparently have experienced—can be a costly interruption to teaching and learning. Too often such tests yield only minimal information, if any, beyond what a teacher could find out using formative classroom strategies. And the costs can be high in terms of lost time and increased student boredom or anxiety about testing.

Real preparation for a high stakes accountability test should involve focusing on teaching mathematics for understanding and proficiency. Students can quickly learn what they need to learn about how to take a large-scale test. So little preparation about the nature of the test itself is quite adequate most of the time. In fact, major advances in large-scale test design, and a commitment to assess meaningful understanding by the major test development consortia in the nation—PARCC and Smarter Balanced Assessment Consortia—hold the promise that large-scale assessments may one day test what we say we value with respect to thinking, reasoning, and solving complex problems. These two consortia developing assessments for the Common Core State Standards for Mathematics (NGA Center and CCSSO 2010) have committed to focusing on assessing mathematical practices, especially modeling, reasoning, and communication. If the assessments develop in ways that actually accomplish this goal, even if it takes a few years and multiple iterations to get there, we could see such tests acting as a major driver toward teaching in the ways advocated by the profession for more than two decades. Students may be expected to engage in tasks that require thinking about multiple, connected mathematical ideas, explaining and justifying complex solutions to problems taking more than just a minute or two to solve. Unfortunately, such tests are expensive to develop and to administer, and scoring can be both expensive and time-consuming. But the payoff in allowing teachers to help students learn to think, reason, and apply the mathematics they're learning can be huge. Whether or not these advances in large-scale high-stakes testing develop as hoped, such rich, deep and meaningful assessments are an appropriate goal for the future.

Even if we see such advances in high-stakes accountability tests, no matter how wonderful a large-scale test may be, it's unlikely that it will ever fully tell us how a student is thinking or provide a comprehensive picture of what the student knows. That job will continue to rest with teachers in their everyday work with students.

What Can We Do?

First, it's important for teachers, supervisors, administrators, and families to learn to be smart consumers of test data, learning what test data do and do not tell us. Large-scale tests give, at best, a snapshot of some of the content for the year. We need to recognize that a single picture taken on a single day may or may not accurately identify students' strengths and weaknesses. Such tests tend to test only some of what is important, often in superficial ways that can be quickly and inexpensively scored. Effective teachers will use the information from such tests to point toward potential areas calling for more evaluation for a student or a group of students. Knowledgeable parents and caregivers can follow up beyond looking at test scores by talking with the teacher to determine how well their sons and daughters are learning the mathematics being addressed.

In classrooms, teachers can continue to learn how to assess student learning as part of instruction and develop skills to design a variety of formative assessment measures that show what students know, whether in daily teaching or for evaluating learning on tests or long-term projects. The commitment to assess well begins with a commitment to pay attention to what students are doing every day and ask questions that push their thinking in appropriate directions.

Reflections and Discussion

FOR TEACHERS

- What issues or challenges does this message raise for you? In what ways do you agree with or disagree with the main points of the message?
- How comfortable are you with the notion of formatively assessing your students?
- What ways have you found to determine how well your students are learning the mathematics expected of them on a day-to-day basis?

FOR FAMILIES

- What questions or issues does this message raise for you to discuss with your son or daughter, the teacher, or school leaders?
- How can you find out what your daughter's or son's state test scores really tell you about the mathematics she or he knows?
- What questions can you ask your son's or daughter's mathematics teacher to find out how well your son or daughter is learning mathematics throughout the year? Is there information sent home about his or her progress, or are there other vehicles for communicating with the teacher?

FOR LEADERS AND POLICY MAKERS

- How does this message reinforce or challenge policies and decisions you have made or are considering?

(continued)

- How can you support teachers' use of effective formative assessment strategies?
- What professional learning opportunities do you, or can you, offer teachers to learn how to be more effective at designing formal and informal ways to monitor student progress?
- In what ways do you, or can you, help teachers and families make sense of large-scale test scores in terms of how well students are learning the intended mathematics?

RELATED MESSAGES

Smarter Than We Think

- Message 28, "Bringing Testing into the 21st Century," discusses shifts in assessment, especially in large-scale testing and offers suggestions for making sense of test results.
- Message 22, "Building Our Work on Evidence," considers the importance of making instructional and programmatic decisions based on evidence and research whenever possible.
- Message 40, "Mathematical Habits of Instruction," includes a discussion of the role of formative assessment in helping students develop mathematical habits of mind.

Faster Isn't Smarter

- Message 35, "Putting Testing in Perspective," discusses the role of testing in learning.
- Message 19, "Embracing Accountability," considers issues related to large-scale assessment and its impact on classroom practice.
- Message 11, "Weighing Hens," looks at the pros and cons of benchmark testing.
- Message 18, "Faster Isn't Smarter," considers the negative side of timed tests.

MORE TO CONSIDER

- *INFORMative Assessment: Formative Assessment to Improve Math Achievement, Grades K–6* (Joyner and Muri 2011) provides resources for monitoring, learning, and adjusting teaching based on how well students are learning the intended mathematics. *INFORMative Assessment: Formative Assessment to Improve Mathematics Achievement, Middle and High School* (Joyner and Bright forthcoming) provides resources for monitoring learning and adjusting teaching based on how well students are learning the intended mathematics.

- *Embedded Formative Assessment* (Wiliam 2011a) offers practical suggestions on how to check daily for student learning. *Sustaining Formative Assessment with Teacher Learning Communities* (Wiliam 2012) is an e-book written to support implementation of the suggestions in *Embedded Formative Assessment*.

- *Good Questions for Math Teaching: Why Ask Them and What to Ask, K–6* (Lilburn and Sullivan 2002) offers practical suggestions on asking the kinds of questions that inform teaching and help students learn to think.

- *Good Questions for Math Teaching: Why Ask Them and What to Ask, Grades 5–8* (Anderson and Schuster 2005) offers practical suggestions on asking the kinds of questions that inform teaching and help students learn to think.

- *Mathematics Assessment: Myths, Models Good Questions, and Practical Suggestions* (Stenmark 1991) is a wonderful, timeless, and easy-to-use professional resource for designing problem-scoring rubrics and for administering and evaluating a wide range of other types of classroom assessments. *How to Assess While You Teach Math: Formative Assessment Practices and Lessons, Grades K-2: A Multimedia Professional Learning Resource* (Islas 2011) offers a variety of resources for formative assessment in the early grades. *How to Assess While You Teach Math: Formative Assessment Practices and Lessons, Grades 3–5: A Multimedia Professional Learning Resource* (Islas and Terry forthcoming) offers a variety of resources for formative assessment in the upper elementary grades. *Investigations, Tasks, and Rubrics to Teach and Assess Math, Grades 1–6* (Lilburn and Ciurak 2010) provides a variety of resources for formative assessment.

- "Are We Obsessed with Assessment?" (Gojak 2013a) considers the downside of overly assessing students.

- "What Does Research Say the Benefits of Formative Assessment Are?" (Wiliam 2007b) presents research findings on the benefits of formative assessment in mathematics from one of the world's experts.

- "Five 'Key Strategies' for Effective Formative Assessment" (Wiliam 2007a) suggests research-based classroom strategies for incorporating formative assessment into effective mathematics teaching.

- "Assessing to Learn and Learning to Assess" (National Council of Teachers of Mathematics 2005) provides a variety of resources on the connections between teaching, learning, and assessment, including publications, online resources, and

professional development support. www.nctm.org/profdev
/content.aspx?id=4420.

- "Assess Teaching and Learning" (Eberly Center at Carnegie Mellon University, www.cmu.edu/teaching/assessment/index.html) offers a concise summary of important points on assessment in higher education, but also useful at the elementary and secondary level.

- "Mathematics Assessment Project: Assessing 21st Century Math" (Mathematics Assessment Resource Service) includes a variety of assessment resources, with an emphasis on extended mathematics tasks. http://map.mathshell.org.

- *This Is Only a Test: Teaching for Mathematical Understanding in an Age of Standardized Testing* (Litton and Wickett 2008) offers strategies for helping students succeed on large-scale tests while teaching for lasting understanding.

- *New Frontiers in Formative Assessment* (Noyce and Hickey 2011) offers essays on formative assessment, including a section on mathematics.

- *Assessment Standards for School Mathematics* (National Council of Teachers of Mathematics 1995) presents standards for assessing student mathematics learning, including examples and vignettes. Also available online at www.nctm.org/standards/content .aspx?id=24282.

- "Smarter Balanced Assessment Consortia" (Washington State Office of Superintendent of Public Instruction) is one of two national assessment consortia developing large-scale accountability tests to accompany the Common Core State Standards. www.smarterbalanced.org/.

- "Partnership for Assessment of Readiness for College and Careers" (PARCC) is one of two national assessment consortia developing large-scale accountability tests to accompany the Common Core State Standards. http://parcconline.org/.

What's That Goat Doing in My Classroom?

LESSONS FROM TEACHING IN AFRICA

Everyday distractions at the Lyceé Yamwaya in Ouahigouya, Burkina Faso

After thirty-five years in education, starting as a teacher and eventually working for many years in roles supporting teachers, I decided in 1999 that I needed to go back to the classroom. I wanted to remind myself what I was asking teachers to do, and see if I could practice what I advocated. I knew that it was a challenging and stressful time to be a teacher in the United States during such a high-stakes era of accountability, politics, bureaucracy, and pressure to teach to *The Test*. So I decided to try teaching someplace "easier" than the U.S.—I joined the Peace Corps. I was assigned to teach mathematics (in French) in Burkina Faso, a small poor country in West Africa. For the next two years, I had a wonderful life-transforming experience, and I learned lessons that will stay with me forever.

Goats and Other Farm Animals

The Lyceé Yamwaya, in the town of Ouahigouya, was a large school, with more than two thousand students in grades roughly equivalent to junior high school through high school. It consisted of nine open buildings, spread out around an open space where students parked their bicycles or the occasional car or cart. In the open space it was not unusual to see various types of animals wandering through from nearby houses. The first time it happened, I was startled to see a goat standing and bleating in the doorway of my classroom. After regaining my composure, I found myself laughing out loud at the unique distraction most teachers in the United States don't have to deal with. Over the course of my two years, I would hear similar sounds from animals just outside the window or door and regularly see and hear chickens moving about nearby.

Other School Challenges

More than the animal distractions, my classrooms were crowded, with seventy-five students in each of my four classes, sitting at desks built for two to three students each; my former colleagues at Yamwaya tell me that class sizes rose to more than one hundred in the first years after I left, and they are relieved that classes are back down to only one hundred now. Windows in the school consisted of fixed open slats, with hot, dry winds routinely blowing the dust from the red earth into the classroom. If the rainy season started early, we occasionally also dealt with rain blowing in. The chalkboard consisted of a piece of wood repainted with chalkboard paint a couple of times a year. Teachers were issued one box of colored chalk at the beginning of the year and one box of white chalk, with one replacement allowed when we had used all of it. Every classroom had a coffee can filled with water and a rag; students were expected to make sure there was always water in the can to clean the board. Teachers typically used the colored chalk to draw diagrams and to highlight key words on the chalkboard. This was important, since students did not have textbooks and they were accustomed to carefully copying verbatim, into their small paper *cahiers*, everything the teacher wrote on the board. I was fortunate to have overhead lights in my classroom and even one electrical outlet, although I never found anything to plug in.

Beyond the physical challenges and the large class sizes, we missed a considerable amount of school—more than one-fourth of the days each year—because of strikes and similar disruptions (Seeley 2009, 166). As a result, I continually had to adjust the year's program and hope to accomplish the same outcomes with students as if we had gone through a complete school year. This was particularly challenging, since some of my classes had only one more year to prepare for their post-graduation high-stakes test, the *Bac*.

One of the hardest issues to overcome was a long-standing school culture of rote teaching, with carefully written-out notes and procedures on the board, followed by the administration of very difficult unit tests. Students were not expected to do well. On the 20-point French grading scale, a 10 was considered a good score, and many students did not achieve that, so they failed. No-one was expected to score a 20 on a test. The expectation was that many students would fail and would need either to repeat the grade or drop out of school, possibly paying to attend a private school if they had the resources. The principal explained to me that there wouldn't be enough room in the school if more students passed. I tried to make a case that there would be plenty of room if more students advanced and actually finished, instead of staying in school to repeat grades. But my argument was lost to long-standing cultural beliefs about students and learning, the principal noting as well that the reason I didn't understand was because I was a woman.

Finally, there were broader societal issues to deal with in terms of poverty and language. Burkina Faso is one of the poorest countries on earth. And there are fifty-one local languages spoken nationwide, with French as the designated national language. Thus, none of my students, nor I, were operating in our native language in school. We all spoke French, but probably thought about mathematics in another language.

Focusing on Priorities

I offer these descriptions of my teaching situation not because I think I had a worse situation than other teachers; in fact, I would much prefer to deal with these types of challenges than with the challenges I see many American teachers facing. Rather, I want to share a slightly different perspective about the challenges of teaching than we usually consider. In this drastically different physical and cultural environment, I had to face fundamental questions regarding my beliefs about teaching and learning. I had to identify what was really important for my teaching and, mostly, for my students' learning.

TEACHING THROUGH PROBLEMS

My first priority, and my first recommendation to mathematics teachers anywhere, is to organize teaching around engaging problems. Students need the opportunity to make sense of the mathematics they're learning, as they wrestle with good problems and constructively discuss those problems. Students can learn to reason and think mathematically as they explain and justify what they do.

In my situation, teaching around problems wasn't always easy, especially with such large classes, and especially in French. But I was determined to do it, and I was able to find enough good problems in a few French math books that I could structure much of the year around problems. I worked hard to have students wrestle with the problems and collaborate with each other as they tried to solve them. I asked them to share their strategies with the class as we engaged in mathematical discourse about the various approaches students had taken and drew from their work as we learned the mathematics of the unit at hand.

ASSESSING FOR THINKING

My second priority and recommendation is to focus on assessments that cause students to have to think and explain their reasoning. I had one particular student in Burkina Faso who was my litmus test as to whether I had accomplished the construction of a good test that really pushed their thinking. He was an incredibly bright student who always seemed to figure things out quickly. I would watch his face as he engaged in each problem on a test to see if he seemed to have to think hard. I never gave a test he couldn't do, but I knew I had pushed his thinking at least a little bit if the light bulb didn't go on until he had wrestled with a problem for a while.

On my tests, I asked students to explain solutions to challenging problems, sometimes giving students problems that might have more than one way to approach them or even more than one possible answer. (I sometimes questioned my choice to do this when grading seventy-five or one hundred fifty open-ended, problem-solving tests involving student explanations in French....) My tests were different from those of my colleagues, who were accustomed to administering hard tests that were essentially procedural in nature. But by the end of the first year, one colleague was routinely asking me for copies of the quizzes and tests I used in my middle school–level class so that he could use them in his classes at that level.

I would also argue that meaningful formative assessment is possible, at least to some degree, no matter what the teaching situation or how large the class. Of course, with a large class, there is less time to formatively assess, or even speak with, every student. But I'm convinced that it's important enough that we need to find time to do it. It's worth asking questions of at least some students some of the time to see if they're getting into the mathematics in appropriate ways. I tried to do this by walking around the classroom while students worked on a problem alone or with a partner, asking questions I hoped would push their thinking (*How did you decide to ...? Have you convinced your partner that works?*)

ADVOCATING FOR RESOURCES

A third priority for me, and a recommendation to educators in the United States, is to advocate—aggressively if necessary—for the resources needed to teach well.

I was not surprised that my very large classes in a very poor community did not have textbooks. In fact, the school library in Ouahigouya had only about twenty or thirty books, accumulated over many years. I was quite surprised, however, when the principal told me that teachers had access to a book room, where teachers could use donated textbooks from France for their own reference. A bigger surprise was when I discovered that, among the books sitting in cobwebs in the book room, there were approximately eighty copies of a textbook appropriate for the course I was teaching to one of my classes and one hundred fifty copies of a textbook appropriate for the course I was teaching to two other classes. Unfortunately, there were indeed no books available for my fourth class of middle school–aged students. But I asked my principal if we could distribute the books we had to students in my three high school classes. The first response was a resounding *no*; if we let students use the books, they might damage them or lose them, he said. After persevering for several days, I was able to persuade him to let me have enough books for half the students in those three classes. While this was hugely impractical with many students living in remote locations, it was better than not having any books. And I managed to squeeze out a few extras so that students who were very isolated might have access to their own books.

Today, most schools in the United States have at least some kind of print or virtual textbook or resource for most students, although many such resources are outdated or out of sync with new standards or with innovative teaching approaches. So we may still need to advocate for appropriate, problem-centered teaching materials, whether in textbook form or in other forms. And we need to advocate for access to technological resources particular to the mathematics classroom, such as calculators (including graphing calculators), dynamic geometric software, or computer algebra systems. Access to these critical tools can allow students to tackle far more challenging and relevant problems than they could take on without the tools, thus raising their level of mathematical thinking in ways that will serve them after they leave school.

What Can We Do?

It turns out that the animal in the doorway was a sheep, not a goat. But in my defense, it didn't look anything like sheep I'd ever seen before, and both sheep and goats make similar sounds. It may be that not many teachers in the United States will see either one of these animals in their classroom doorway. Although I do recall that in my first year of teaching in rural Colorado, there were not enough regular classrooms while a new school was being completed. That year, I taught high school geometry in the ag shop, with animals on the other side of a small wall, eighth-graders in the typing room, and a small class of Algebra 2 students in a large book closet.

The reality is that teachers in the United States today deal with far more difficult challenges than inappropriate rooms and animals in the doorway. Thankfully, class size may not be seventy-five or one hundred in most schools, but it's often greater than thirty, and it's worth working to keep classes smaller than that. Research has shown that significantly reducing class size can make a positive difference in student learning, regardless of the subject area (Chingos and Whitehurst 2011). In mathematics, having smaller classes gives teachers the opportunity to have students work in pairs or groups on extended problems and gives adequate time for the important classroom discourse that helps students learn from this kind of collaborative work. Even when learning computational procedures, teachers can do more ongoing (formative) assessment with smaller classes in order to catch misconceptions before they become entrenched practices.

We can also work to ensure that mathematics classrooms have adequate instructional resources, and that students all have desks and a safe physical space in which to work. Making sure students have these basic necessities is the job of schools and communities, and we all need to advocate investing in our students to this level and beyond.

Most of all, we can organize classrooms around effective teaching and good tasks, not around preparing students for *The Test*. If our emphasis is on effective teaching to develop mathematical thinking, and if we pay attention to how well students are learning from day to day, test scores on the state accountability test will take care of themselves. If teachers are not knowledgeable about assessment, especially formative assessment, this is an important area for professional development. All teachers need to know sound principles for regularly monitoring students' progress and for larger-scale unit and semester tests that show what students have learned.

We can't let the sheep or goats or chickens distract us from helping students learn mathematics. We can't let outside pressures push us to teach to a test if we know it's not the best thing for students. And we can't go back to an outdated teaching model by trying to tell students what they need to know rather than engaging them in meaningful tasks. By focusing on priorities—helping students make sense of mathematics, emphasizing reasoning and communication, and paying attention to students' learning on a regular basis—we can help every student prepare for a productive future in a global society.

Reflections and Discussion

FOR TEACHERS

- What issues or challenges does this message raise for you? In what ways do you agree with or disagree with the main points of the message?
- Have you taught in different types of environments—different grades, schools, communities, states, countries? If so, what lessons did you learn from your experience(s)?
- Are there critical instructional resources (print, technology, or other) you are lacking? If so, what avenues are available to you to advocate obtaining these resources?
- How well do you know how to formatively assess students and how to create good quizzes and tests? Where can you go to learn more and improve your knowledge of assessment?

FOR FAMILIES

- What questions or issues does this message raise for you to discuss with your daughter or son, the teacher, or school leaders?
- Do you know whether your son's or daughter's mathematics classroom has adequate resources? If it doesn't, how can you advocate for providing such resources?
- How open are you to a kind of teaching that focuses on problems your daughter or son might not have been told how to solve? Can you see the benefit of teaching around such problems as a way to help her or him learn how to think and reason?

FOR LEADERS AND POLICY MAKERS

- How does this message reinforce or challenge policies and decisions you have made or are considering?
- How can you work with your mathematics teachers to determine whether mathematics-specific technologies like calculators, graphing calculators, dynamic geometry software, CAS software, and so on, are being used to raise the level of mathematics being taught? If your teachers don't know how to make use of this technology to support improved student learning, how can you help them get up to speed?
- How well do your teachers understand assessment, both formative and summative? How can you support them in becoming more knowledgeable and effective with everyday and longer-term assessment?

RELATED MESSAGES

Smarter Than We Think

- Message 12, "Upside-Down Teaching," describes a problem-centered teaching approach that focuses on students learning to reason and justify their thinking as they make sense of mathematics.
- Message 19, "How to Know What They Know," addresses issues related to assessing student learning.
- Message 14, "Effectiveness and Efficiency," emphasizes the importance of teaching for lasting learning.

Faster Isn't Smarter

- Message 31, "Do They Really Need It?," describes the impact of the strikes in Burkina Faso on my teaching and my expectations.

MORE TO CONSIDER

- *Inside Teaching: How Classroom Life Undermines Reform* (Kennedy 2006) offers insights into daily teaching challenges and their impact on improving student learning.
- "Class Size: What Research Says and What It Means for State Policy" (Chingos and Whitehurst 2011) summarizes research on the effects of lowering class size.
- *Fires in the Bathroom: Advice for Teachers from High School Students* (Cushman 2005) offers thoughts from students in challenging high school environments.
- *Fires in the Middle School Bathroom: Advice for Teachers from Middle Schoolers* (Cushman and Rogers 2008) offers thoughts from students in challenging middle school environments.

Moving from Where We Are to Where We Want to Be

WHAT WE KNOW ABOUT FACILITATING CHANGE (GLORIA'S STORY)

How does change happen? Slowly, carefully, and all at once.

—Theodore Sizer

Most educators will agree that we can improve the educational system in some way. But when we translate *improvement* into its inevitable counterpart *change*, it somehow seems much less desirable or doable. Change is a personal process, and an individual asked to do something different is likely to feel uncomfortable with at least part of that process at least some of the time.

Over the years, I've learned lessons about the change process via many routes, some through formal study and others by living through change in various roles. I've been a participant in change, an initiator of change, and a facilitator of change. I think my most effective role has been when I saw myself as a facilitator of change, working to make the process a little easier for those trying to do something new. An effective leader—an effective change facilitator—understands the change process, can support individuals going through implementing a change and, in the process, help a school, district, state, or organization achieve its goals by learning about change and by paying attention to the people going through it.

Understanding the Change Process

When I started my first job as a mathematics coordinator in the 1970s for a suburban district outside of Houston, my colleague and new friend, Rosemarie, convinced me to take a graduate course with her. The course was called Institutional Change, and eventually it led me to complete a doctoral program at the University of Houston, even though

155

that was the furthest thing from my mind when I enrolled in the course. What was most remarkable about that first course was that Rosemarie and I were learning theories about change every Monday night that we were then using in real life on Tuesday morning ... and Wednesday and Thursday and so on. I had been hired as the district's first mathematics coordinator to lead the implementation of a new mathematics program, and Rosemarie, the language arts coordinator, was my confidant, advisor, and general "partner in crime." As we learned about change together, we talked about how to help the district's more than one hundred teachers of kindergarten through grade 12 mathematics succeed with the new mathematics program the district had adopted.

THE CONCERNS-BASED ADOPTION MODEL

One of the most helpful theories of change I learned then and still rely on today is the Concerns-Based Adoption Model (CBAM), originally developed at the University of Texas (SEDL 2014). I found out many years later that, in my early years as a teacher in Colorado, I was probably one of the subjects of the research that led to the development of the CBAM model as our school district developed a new mathematics program.

CBAM consists primarily of three components, and I think all three can be quite useful: stages of concern, levels of use, and innovation configurations. I encourage every leader to learn more about these components (see "More to Consider" at the end of this message for some suggested resources). In particular, *stages of concern* relates to an individual undergoing a change; in dealing with a transition to something new, a person's attitudes, issues, and concerns progress through fairly predictable stages. The following list summarizes the CBAM stages of concern and includes typical comments or thoughts a person might express as he or she moves through each stage:

- **Stage 0: Awareness:** *I'm not really interested in what you want us to do, I just want to be left alone.*

- **Stage 1: Informational:** Peeking at the classroom next door ... *That looks kind of interesting ... so tell me a little about what this is you're doing ....*

- **Stage 2: Personal:** *They want me to do WHAT? I'm not sure what I'll be expected to do and whether I'll be able to do it. What if I mess it up? What if it doesn't work? I know what I've been doing isn't perfect, but at least I know how to do it ....*

- **Stage 3: Management:** *OK, so show me step-by-step what I need to do, but it feels like I'm going to be spending all my time moving paper and getting ready to teach.*

- **Stage 4: Consequence:** *I can see that my students may be benefitting from this. I'm thinking that they might do even better if I focus more attention on ....*

- **Stage 5: Collaboration:** *How can I relate what I'm doing to what others are doing? Can we work together to be even more effective?*
- **Stage 6: Refocusing:** *I've got an idea how we can make this even better next year .... What if we ....*

By recognizing where a person is in this process, a leader or peer can help the person with an appropriate kind of support or intervention. When I was responsible for facilitating the implementation of the district's new mathematics program, there came a time fairly early in the process when it was clear that most teachers in the district were at Stage 2 (Personal) or Stage 3 (Management) in terms of how they were experiencing the implementation of the new program (I now know that CBAM includes surveys and useful tools for assessing where people are, but at that time I was still tempering theory with intuition as I learned about what I was doing). Rosemarie and I brainstormed how we might best support the teachers and decided that we would hold school-by-school meetings right after the end of the school day, bringing food and chart paper. We wanted to acknowledge teachers' concerns and deal with everyone on a personal level. We also wanted to provide motivation and encouragement to teachers to continue moving through the process, taking into account those who might already be at Stage 3 (Management). At each meeting, we decided to let every teacher share one thing he or she was concerned about or having an issue with, but the teacher also had to offer one thing he or she liked about the new program. We recorded everyone's responses on two lists—Challenges and Benefits. Going through this process visibly changed the tone in the room as people had an opportunity to be heard. But, just as important, we followed up with specific actions targeted at the things people were struggling with. Some required something simple, like making a personal visit to a teacher's classroom or sharing a management idea from another teacher in another school. Others required more work, like customizing the district's record-keeping form to more closely fit district practices. The actions we took were important and directly targeted what people needed. More importantly, our personal acknowledgement of what people were going through helped everyone relax and feel valued as professionals.

There are other models that can help us understand change and several experts who write about how to facilitate implementation of new programs, some more theoretical than others (see "More to Consider" at the end of this message for suggested readings).

Change Takes Time

One of the lessons I learned in that first course, and have relearned several times since, is that *change is a process*. Closely tied to that idea, we know that *change is incremental* and *change takes time*. However, some years ago when I was working at the state level, I was struggling with these ideas amidst the massive, rapid-fire changes that seemed to be taking place in the education arena, especially in my state. I called my graduate advisor, the professor of my first course on change, and I asked her, "How relevant is this vision of change in today's world?" Her wise response was, "While change is a process that takes time, today the world and our circumstances change so quickly that we simply don't have the luxury of taking all the time we might wish or moving as incrementally as might be ideal." It might seem logical, for example, to try out a new set of kindergarten through grade 12 mathematics standards with a group of kindergarteners and see how they do throughout their thirteen years in school before large-scale implementation of the standards. But that's just not practical, and even if it were, the world would have changed so much by the time those students graduated that the original standards would already be far out of date. Sometimes in these days of ever-accelerating change, we have to do the best we can at planning well, collaborating across constituencies, and taking the boldest steps we can in order to move toward goals that are likely to change even as we try to implement them.

Thus, we need to look for any ways we can to buy as much time as possible to increase the likelihood of a successful implementation. Starting the planning process as far ahead of planned implementation as possible is clearly the best strategy. Although planning well in advance is not always an option, sometimes we have to take a stand and say "Stop!" when things are being implemented so fast that they simply are not sustainable. Otherwise, we risk going down a path doomed to failure and frustration.

Considering Scale

Plans should be focused on sustainability, starting with collaboration and continuing with ongoing support for all key players throughout the implementation process. In planning for sustainable change, consider a principle I've learned over the years: *The size and scale of a change should be inversely proportional to the size and scale of the audience being expected to implement the change and the timeline with which it is to be implemented.*

For example, the use of a new form of report card that is mostly based on format, rather than a dramatic change in philosophy, might be effectively implemented on a wide scale at the school, district, or even state level. The change does not involve a lot from each person, and though some teachers might resist or have difficulty with the new format, the change could be facilitated in a reasonable fashion with adequate planning, involvement, and support.

However, if we want to implement new standards or, especially, a new model for teaching, we are asking all teachers to dramatically change what they do and maybe even challenge what they believe. Especially if the timeline is aggressive, we might consider scaling back our expectations of how much each teacher will be expected to change in the first months or over the first year. Additionally, the more ambitious the timeline, the more significant the change, and the broader the scale of implementation, the more support is called for. Many teachers can recount horror stories of one after another initiative where they were expected to change too much too fast without adequate support, with the initiative inevitably failing in the end.

Planning for Sustainability

Some of what I've learned about facilitating educational change comes from an unexpected field. During my service as a Peace Corps volunteer in my fifties, and ever since returning from that service, I have become aware of the literature on *development* (as in providing assistance to developing nations). In particular, I've been intrigued by the attention paid to sustainability—considerations that help a program or initiative last, whether in community development, AIDS prevention, or sustainable energy. Among Peace Corps documents I found during my service was a list of useful considerations regarding the sustainability of any new initiative:

- Information is a starting point, but not enough.
- Fear messages have limited use.
- People are more likely to try behavior they feel capable of performing.
- People are more likely to adopt a new behavior if they have choices.
- Change is more likely to succeed if influential people adopt the change.
- Relapse is expected. Build in ways to continue the forward motion.

These suggestions were credited to Thomas J. Coates, a medical researcher working on AIDS prevention in developing nations, but they have much to offer us in terms of how to work with people as they try something new. The notion of sustainability is one that should be foremost in our minds as we work to implement changes we hope will last. Our history in education is littered with fads that were superficially, often impulsively, implemented with very short lifespans. It should come as no surprise to leaders when teachers resist the *next new thing*, since most of the previous *next new things* didn't last long.

The most important lesson we can take from the literature on sustainability is to plan for sustainable change from the very beginning—before the new program or initiative is ever rolled out. This includes planning a reasonable scale and timeline for the initiative, informing and involving the various stakeholders, and building in supports for teachers and others responsible for implementing the initiative at all stages of implementation.

Informing and Involving Stakeholders

Keeping all stakeholders informed of the plans and progress of implementation of a new program is a critical step for its success. This means starting within the school system to make sure all levels of faculty, administrators, and staff send a consistent message to the community. In addition to this internal involvement, it's important to reach out to teachers, families, community members, and even students through a variety of strategies. We can learn lessons from successful practices used to implement programs in schools in many communities, including traditional communication vehicles like newsletters or information posted on a website or communicating to a broad audience through any of a growing array of social media. This type of technology offers a variety of outreach possibilities that may be accessed by different groups, and they provide an economical way to reach many more constituencies than was possible just a few years ago.

Beyond sending out information, there is much value in convening face-to-face experiences for those who can attend. Whether a regularly scheduled event like back-to-school night or a special event like Family Math Night, sharing information in an active and engaging way can reassure families and community members of the positive aspects and potential benefits of the program.

Communicating with and informing these stakeholders of what's coming is a good step, but it can be even more valuable to substantively involve them in providing input to the plan. Thus, when forming committees to inform or even develop implementation plans, consider inviting representatives of key stakeholder groups, especially parents and community members, as active or advisory members. This kind of personal outreach and involvement builds goodwill among the various constituencies who will be involved in the implementation or affected by its outcomes. But more than that, as they bring their perspective(s) to the thinking and conversation about what will be involved in implementing the new program, someone may raise an issue nobody else had thought about. Anticipating where there may be bumps in the road can significantly improve implementation plans, and in some cases you may be able to avoid the bumps altogether once the issues are raised and addressed within a collaborative environment.

Building in Teacher Support

We know that teachers are the frontline implementers of most new programs. And we know that teachers can benefit from all kinds of support as they go through the process of implementing something new (Hull 2000). That support can come in many forms, but the most important kinds of support teachers seem to ask for and benefit from are time and learning. In terms of time, teachers overwhelmingly request more time than they currently have to plan, teach, learn, implement, and improve. Finding ways to provide working and learning time on an ongoing scheduled basis can be one of the best gifts and strongest supports a leader can offer teachers implementing a new initiative. Giving as much time as possible to see results can relieve the stress and pressure teachers are sure to feel. Most of all, staying the course is a critical way to provide time for teachers to work out the details, try things out, and work together to fine-tune their practice. The worst thing we can do to teachers is to implement a new program only to replace it with something different just a year or two or even three years later. Teachers have seen this kind of short-term thinking over and over again, and it's one of the reasons many teachers resist the *next great idea.*

In addition to providing time, there simply is no substitute for professional development, ideally with the involvement of administrators and any paraprofessional staff who may be involved with implementing the new program. However, professional development will make a difference only if it is done well and part of ongoing learning over time. In a comprehensive study of systemic change initiatives, high-quality professional development structured around high-quality instructional materials yielded significant payoffs in terms of teaching change and student learning to a far greater extent than either professional development or materials alone (Weiss and Pasley 2006).

GLORIA'S STORY

One teacher's story provides insights into how we might help teachers implement a new program. Gloria, a teacher in what some might consider a difficult inner-city middle school, participated in a district-wide implementation of the Connected Mathematics Program (CMP) in the 1990s. I visited her classroom nine years after her first CMP professional development experience. Everyone told me I had to see her classroom; they called her the "Queen of CMP."

Gloria told me her story of transition to the role of CMP Queen. She recounted the first week of professional development as wonderful and challenging. She worked hard and was treated respectfully in a comfortable environment that afforded plentiful opportunities to interact with other teachers, both intellectually and personally. What they learned was invaluable and would clearly benefit their students. Gloria explained to me that when she returned to her classroom in the fall, she wanted to implement what she had learned, but she worried that she wouldn't do it right. And she discovered that it was hard to put into practice the kind of teaching she had learned. She kept thinking that, even though she knew her previous methods of teaching weren't as effective as what she had learned over the summer, maybe it would be okay to go back to the old way because at least she knew how to do it.

But Gloria knew that the instructional leaders in her district were going to come by to visit classrooms once a month to support the teachers implementing the program. Gloria confided to me that probably the only reason she did not go back to the old teaching model was because she was afraid she'd get caught when they came to help her. She laughed as she recalled this. Gloria then told me that she just kept pushing ahead, attending the follow-up sessions, relying on the coaches and leaders who came to support her, and doing the best she could. She said that she wasn't sure when it happened, but at some point a few years after starting this journey, she realized that she had become accustomed to teaching in this very different way. Gloria had transformed her teaching and had seen the results with her students. She knew she could never go back.

When I visited Gloria's classroom, her students were actively engaged in learning mathematics. Every year her students performed great on the state assessment. Her leaders had set the expectation that teachers would implement the program and had made that implementation possible by their frequent and ongoing involvement in classrooms. Gloria had received the kind of support that savvy leaders provide—ongoing professional development, personal one-on-one help, time to learn, and, perhaps most of all, a long-term commitment to continue with the same program long enough to see results.

What Can We Do?

As with much of life, it's all about relationships. If the change process is personal, then facilitating it is all about personal connections. Supporting people as they go through the process of implementing something new is the primary role of a change facilitator, and it can provide tremendous rewards to both the implementers and the facilitator. The best way to serve and support those people is by listening to them, paying attention to what they say and what they do, learning from what you see, hear, and notice, and offering appropriate kinds of help, support, or intervention at the right time. The three key steps in supporting change that I've mentioned above—planning, involvement, and support—all deal with considering and working with people responsible for the on-the-ground implementation of the change and those who will most directly see the results.

Many factors can influence and support positive change. But real improvement in student learning only happens one student at a time with one teacher at a time. What a teacher does and how a teacher interacts with a student contributes directly to what that student learns. The teacher's actions and attitudes may be the most important factor of all in terms of student outcomes. Those of us supporting teachers— whether peers, supervisors, or policy makers—need to do everything we can to give teachers what they need in order to help every student learn what we all want every student to learn. Being a smart facilitator who uses common sense, pays attention, and cares about the people involved can go a long way to providing the support that teachers need in order to improve mathematics teaching and student learning.

Reflections and Discussion

FOR TEACHERS

- What issues or challenges does this message raise for you? In what ways do you agree with or disagree with the main points of the message?
- Can you think of some kind of change in your work or your life that you are dealing with? Which of the CBAM stages of concern do you feel currently represents where you are with respect to that change? What would it take for you to move to the next stage?
- Describe the best experience you've ever had in changing what or how you teach. What contributed to making it effective?
- Describe the least successful or most unpleasant experience you've ever had in being expected to change what or how you teach. What are the differences between your best and worst experiences? What can you learn from these experiences about yourself and about how to get maximum payoff from a new initiative?

FOR FAMILIES

- What questions or issues does this message raise for you to discuss with your son or daughter, the teacher, or school leaders?
- Can you think of some kind of change in your work or your life you are dealing with? Which of the CBAM stages of concern do you feel currently represents where you are with respect to that change? What would it take for you to move to the next stage?

(continued)

- How can you stay informed about what new programs your daughter's or son's school may be implementing?
- If you were invited to be part of a planning process for a new initiative, would you choose to actively engage in the process or would you prefer to just be informed as the process goes on? How might your input help the school in its implementation efforts? If you're interested in participating in such efforts, how can you communicate your interest to the school?

FOR LEADERS AND POLICY MAKERS

- How does this message reinforce or challenge policies and decisions you have made or are considering?
- Can you think of some kind of change in your work or your life you are dealing with? Which of the CBAM stages of concern do you feel currently represents where you are with respect to that change? What would it take for you to move to the next stage?
- For the next (or current) initiative you may be implementing or planning, how can you involve the various stakeholders in planning for sustainable change?
- What are some ways you may not have tried before that you can use to support teachers in implementing a new initiative for lasting success?

RELATED MESSAGES

Smarter Than We Think

- Message 26, "Leading Change," offers insights from learning a new language that can inform how we facilitate change.
- Message 23, "Common Sense and the Common Core," discusses particular implementation issues related to the Common Core State Standards for Mathematics (NGA Center and CCSSO 2010) and also broader issues related to implementing the *next new thing*.

Faster Isn't Smarter

- Message 12, "Beyond Band-Aids and Bandwagons," looks at educational fads and considers how we can make lasting change.
- Message 13, "Seek First to Understand," focuses on the role of collaboration and the importance of establishing personal relationships as we try to improve mathematics teaching and learning.

- Message 27, "A Math Message to Families," offers guidance on keeping parents and caregivers in the loop about how mathematics classrooms might be different today and how those differences might help their students.

MORE TO CONSIDER

- "Concerns-Based Adoption Model" (SEDL 2014) provides an overview of this very useful model of change. www.sedl.org/cbam/.
- *Cultures Built to Last: Systemic PLCs at Work* (DuFour and Fullan 2013) offers insights and practical suggestions for creating and improving the effectiveness of professional learning communities (PLCs) from an internationally known expert on facilitating change and educational improvement (Fullan) and the recognized leader in PLCs (DuFour).
- *Change Forces: Probing the Depths of Educational Reform* (Fullan 2012a) looks at educational change within systems at the broadest level and how what we know about that level of change can inform more localized efforts.
- *Stratosphere: Integrating Technology, Pedagogy, and Change Knowledge* (Fullan 2012b) considers the integration of multiple factors related to education, including technology, and how these factors interact on the change process in schools.
- *Leading in a Culture of Change* (Fullan 2007) provides insights into facilitating school change for leaders.
- "Mathematics Professional Development" (Doerr, Goldsmith, and Lewis 2010) considers the nature of professional development most likely to support improved teaching and learning.
- *The Common Core Mathematics Standards: Transforming Practice Through Team Leadership* (Hull, Miles, and Balka 2012) discusses issues related to improving mathematics teaching, including how to deal with resistance to change, with a focus on how teachers can help students develop mathematical habits of mind.
- *Sensible Mathematics: A Guide for School Leaders in the Era of Common Core State Standards, 2nd edition* (Leinwand 2012) offers practical suggestions for school leaders dealing with implementing new standards and assessments.
- "Scaling Up Instructional Improvement Through Teacher Professional Development: Insights from the Local Systemic Change Initiative" (Weiss and Pasley 2006) provides research insights into using professional development and high-quality instructional materials for effective improvement of mathematics teaching and learning.

- *Taking Charge of Change* (Hord, Rutherford, Huling-Austin, and Hall 1987) remains the frontline educators' manual on leading change, with a good overview of the Concerns-Based Adoption Model (CBAM).

- *Implementing Change: Patterns, Principles and Potholes, 2nd edition* (Hall and Hord 2005) considers the change process, including the CBAM model.

- "Teachers' Mathematical Understanding of Proportionality: Links to Curriculum, Professional Development, and Support" (Hull 2000) describes research on implementation of the Connected Mathematics Project.

22 Building Our Work on Evidence

MAKING SENSE OF RESEARCH AND DATA

The plural of *anecdote* is not *evidence*.

—Unknown

At a recent meeting of researchers and practitioners, a researcher remarked that the plural of *anecdote* is not *evidence*. Yet much of what passes for research, data, and evidence in schools seems to be based on urban myths, superficial information, and wishful thinking.

Policymakers and the public have appropriately raised a call to base teaching and educational decisions on a solid evidence base—to use research and data to guide our practice. As educators, we embrace this call. After all, what we care about most is how well our students learn, and we want to implement practices and programs with the highest likelihood of providing strong student learning. Yet, how can we do this in a world spinning so fast that by the time we ask the question calling for research, we already need decisions?

The Need for Research-Informed Practices and Programs

I am not a researcher, but throughout my career in the classroom and at the district, state, and national levels, I have needed to know what research had to tell me so that I could become a more effective teacher or help others become more effective teachers. I have also yearned to connect to the research community and let them know what questions I wanted answered that I heard every day from teachers and other educators: *What teaching approach is most effective? Should we track our students? Are there ways to use calculators to support my students' learning? How should the curriculum be structured to maximize students' learning?*

It's easy to become confused or frustrated trying to implement research-informed practices, when we see differences of opinion even within the research community about what *counts* as *real* research. Some experts would have us define research in increasingly narrow ways focused on pure experimental studies with a control group and an experimental group. Yet for many questions in education, as in some fields of science, this specific type of experimental research is simply not practical or may not be possible at all. A great deal of high-quality and informative research in mathematics education may not fit the experimental mold, yet might yield key insights or could support significant changes in instructional practice. Further complicating the landscape, sometimes studies seem to give us conflicting results. And sometimes studies focus on studying a particular phenomenon or practice but aren't intended to directly influence classroom practice—even though such studies may inform future research or development efforts. And sometimes research reports are just difficult to understand.

The fact is that we need to know more about many things. We have much to learn about not only classroom practices, but also what it takes to implement a program well and how we can measure results with respect to the fidelity of implementation according to the original design of the program. But there are also many things we do know. Fortunately, we now have available a growing collection of practical, readable, and usable research briefs and summaries from professional mathematics groups like the National Council of Teachers of Mathematics and the National Council of Supervisors of Mathematics. And there are many recent books and resources that describe research findings related to mathematics teaching and learning written for a nonresearch audience (see "More to Consider" at the end of this message). Yet we long for more research results on questions we most care about in a format and style that is easy to grasp. For the research we do find, nonresearchers may need help making sense of what we see reported, both in scholarly venues and in the media.

Being a Smart Consumer of Research

Research comes in many forms: formal experimental studies; small case studies; targeted studies of particular practices; meta-analyses synthesizing results of multiple studies related to an issue or question; longitudinal studies reporting data collected over a period of time, often years; surveys of current practices and preferences; and many other forms of studies reporting data and evidence to answer particular questions. In interpreting the results we read or hear about, it's important not to be seduced or sidetracked by dramatic headlines or oversimplified conclusions. There are many ways to be fooled or misled in the name of *research*. A smart consumer of research knows to look at reported research from several perspectives.

DISTINGUISHING AMONG FACTS, OPINIONS, AND CONCLUSIONS

Interpreting research findings and reports begins with getting clear about what the facts are and what may be the opinion of the person(s) reporting or relating the results. The interpretation of results may be influenced by the particular point of view of the person or organization doing the interpretation. Even if not intentional, it can sometimes be difficult to separate the facts from someone's opinion about those facts, especially if original findings are reported by someone other than the researchers, such as a blogger or social media site. It may be useful in interpreting conclusions presented in the media or in research reports to consider what organization or institution conducted the study and also to consider what organization, institution, or individual is reporting the results. If a study was conducted by a commercial entity about their own program, look carefully to determine whether they involved unbiased researchers or perhaps worked with a respected university to conduct the research. Or if a report comes out of a political organization, for example, consider any likely point of view the researchers or authors may have brought to their interpretation of the results.

Furthermore, conclusions may not always be justified by the study being reported. Over the past several years, we have seen reports that offered conclusions such as, "Because this group performed differently from this group, therefore this factor is the cause." A conclusion as to cause and effect is extremely difficult to back up with research about performance. Within the many facets of educational systems from the classroom level to nationally, a multitude of factors influence educational performance, and very few studies are designed to authoritatively rule out some factors in order to attribute performance solely to one or two. Thus, we may hear about the performance of students in the United States on an international study, but determining what causes their performance is a complex challenge dependent on far more types of data than are readily collected as part of such a study.

Small differences in performance or responses may not be considered significant based on statistical principles, yet we may read conclusions into the reporting of such performance. Many types of studies, especially those involving survey data, report results with a stated margin of error. One television reporter recently presented results of a study that stated that "only 29 percent" of the sample performed in one way, but "almost a third" performed in another way, with a note that there was a 4 percent margin of error. Given that margin of error, both groups represented a statistically similar percent of the population involved, and no inferences were justifiable based on these results alone. Yet, much of the population seems to be intimidated by statistical data, and it can be hard to avoid believing that, between two different numbers, one is better or worse than the other, regardless of disclaimers about margin of error.

BEING SKEPTICAL OF GENERALIZATIONS

Many studies are conducted with small or very specific samples or in specialized situations, yet results may be generalized beyond the original sample or setting either by those reporting results of the study or by those interpreting results. A study of a particular type of mathematics instruction conducted with a small number of special education students in a certain community should not be considered evidence of how the same type of instruction might perform in a larger setting, different type of classroom, or even a different community. Many popular education reforms that catch on as fads can have their roots traced to small-scale or specialized studies. This doesn't mean the programs aren't good; it only means that we can't assume that the research supports their use more broadly. A well-known historical example is the use of IQ tests developed by Alfred Binet in France during the early years of the twentieth century. Binet's charge was to develop an instrument to determine what type or types of alternative education might benefit students we today refer to as "learning disabled." Binet himself cautioned against generalizing from the results of his test and described at length its limitations. Yet the IQ test has been used for years for purposes far different from those it was developed to address, and some argue that the test is the basis for most or all of today's standardized tests (Gould 2012).

In general, we need to be careful not to generalize from studies done outside of mathematics, studies conducted without adequate attention to sound principles of research and statistics, studies done in particular settings different from our own, and results interpreted beyond their intended purpose. Being a smart consumer of research means paying attention not only to the headlines, but to the substance of the study and to factors surrounding the study.

NOTICING LIMITED OR CONFLICTING EVIDENCE VERSUS AN ACCUMULATION OF EVIDENCE

As healthily skeptical consumers of research, we want to look for accumulated evidence, rather than a single study, to support any significant change we consider making in a classroom, school, or school system. Ask for and look for follow-up studies, replication studies, or studies conducted in different types of settings that might support the claim being made. At the very least, look carefully at the question the stated research is attempting to answer or illuminate and determine whether the setting and experimental variables match the setting in which you hope to apply the results. It can be helpful to look for meta-analyses that summarize results from multiple related studies whenever possible.

Sometimes studies seem to offer conflicting findings. There's a real danger in misunderstanding findings from a single study or in mistakenly believing that such results represent absolute truth. One study might find, for example, that class size doesn't matter, and another might find that it does. Looking more closely, we might notice that the intended purposes of the studies or the desired outcomes in terms of student learning could influence the findings—if we want to know how well students remember information presented in a lecture setting, it may not matter whether there are twenty or forty students in the class. If, on the other hand, we want to know how well students succeed at tackling challenging problems calling for exploration, possibly wrestling with ideas or concepts, the teacher's role as facilitator and questioner becomes central. In such a classroom, class size might make a difference in terms of student success. In this case, an appropriate study to determine whether class size matters might look different from such a study for the lecture-based outcomes. In either case, it might make a difference what discipline was being studied—does class size make a difference in teaching reading, art, physical education, public speaking, and so on.

Often, reports summarizing multiple studies can be helpful to practitioners, provided that authors adequately note limitations and factors outside of the studies that might be important to an audience of consumers of research. (See "More to Consider" at the end of this message for a summary of some of the key research—with appropriately noted limitations—on class size [Chingos and Whitehurst 2011].)

Gathering Your Own Data

Closer to home, consider gathering data from the beginning of the implementation of a new program or practice in your own classroom, school, or school system. Even if a new program or practice is supported by existing research, it's important to determine its effectiveness in a new setting by gathering data and making any necessary adjustments supported by the data. In the classroom, keep track of how students do and how well they learn what you hope they will learn when you try out a new practice or new materials. Talk with colleagues about differences you see compared to results you've seen in previous years and possible differences from what your colleagues see.

For large-scale change, consider starting with a small-scale trial, such as a couple of classrooms. If possible, connect with knowledgeable researchers at a university or research organization to plan how to determine the success of what you have in mind. Gather appropriate kinds of data to make future decisions about how to make adjustments if results aren't immediately as positive as you hope—which they rarely are in the early stages of something new. Even after a program is underway, we may be able to modify what we're doing if we collect the right kind of data or evidence or if we conduct or locate one or more relevant research studies. Dylan Wiliam, an expert in assessment, recommends "Decision-Driven Data Collection" as a counterpoint to the widely used notion of "Data-Driven Decision Making" (Wiliam 2013). Wiliam's point is that we don't need more tests just to gather more data. Rather, we can identify the kinds of decisions we want to make and gather selective data that can inform those decisions. There's a delicate balance between using data to inform decisions and drowning students and teachers in the data-gathering process. For example, a teacher may be able to spend far less time and gain much more useful and timely information about student learning through daily formative assessment activities, such as asking a student a good question or having a student explain a solution, than she or he might gain by administering, scoring, and interpreting the results of a mandated scheduled benchmark test to the whole class.

What Can We Do?

First and foremost, be a smart consumer of research. A smart consumer of research with a healthy degree of skepticism will look cautiously at statements that begin, "Research shows that...." Ask questions of any person making such a claim and see if accumulated evidence supports the claim. Think about:

- What is fact? Opinion?
- Is the conclusion justified by the study reported?
- Does the research inappropriately draw conclusions or make generalizations from a specific situation?
- Are there conflicting studies?
- If the report is about a single study, how closely does the setting and sample for the study match your setting and population?
- Is there an accumulation of evidence or an analysis of multiple studies that help the reader draw reasonable conclusions?
- What questions does the study or report raise for you? How can you address these questions?

Secondly, when dealing with your own classroom, school, or school system, if possible, gather data to inform decisions about potential changes in practice or adoption of new programs. Is the change you're contemplating justified by the data? As you implement a new program or practice, gather data throughout the implementation process to continually evaluate whether adjustments might be necessary or helpful.

Finally, let us call for more individuals and organizations to choose to focus their work in the zone where research and practice overlap. We need people in school systems who are intimately knowledgeable about research, both how to conduct it and how to interpret it. And we need researchers who will work with practitioners to identify the important questions practitioners want answered, conduct research that addresses those questions, and report results in a usable format and timely process to inform practice and programs.

As a community of education professionals, let us create new resources that put usable information into the hands of those directly responsible for students' learning. Let us create new mechanisms for asking the research community for help on the questions that we most need answered. Let us use research to guide policy decisions that support improvements in teaching and learning. Let us expand the group of individuals and institutions whose focus is not primarily on teaching or primarily on research but primarily on the link between the two. In the process, let us maximize the likelihood that we will accomplish our goal of helping students learn more mathematics better.

Reflections and Discussion

FOR TEACHERS

- What issues or challenges does this message raise for you? In what ways do you agree with or disagree with the main points of the message?

- How do you (or can you) find time to read and discuss with fellow teachers how current research can affect your teaching?

- Select a research article or report from resources listed under "More to Consider" at the end of this message or from your own work. As you read, ask the questions summarized in the "What Can We Do?" section of this message. What can you can take away from the article or report in terms of the reported generalizations or conclusion?

- What do you most need to know to improve teaching and learning in your classroom, school, or district? Are your questions addressed in existing research reports or summaries such as *Adding It Up* from the National Research Council, *Research Briefs* and other research resources published by the National Council of Teachers of Mathematics, or the

(continued)

research-informed position statements published by the National Council of Supervisors of Mathematics?

- What existing research resources are most valuable to you?
- When you look up research results online, how can you tell if they're true, reliable, and relevant to your situation?
- What are your biggest challenges in implementing research-informed practices?
- What kinds of data do you collect in your classroom? Can you think of other types of data that might be informative for supporting student learning and might not require too much extra time?

FOR FAMILIES

- What questions or issues does this message raise for you to discuss with your son or daughter, the teacher, or school leaders?
- When you hear a dramatic headline about research in education (test scores, a particular teaching approach, and so on), how can you follow up to determine whether the headline reflects important information and how it might relate to your own classroom or school?
- How can you use the *Research Briefs* published by the National Council of Teachers of Mathematics and the position statements published by the National Council of Supervisors of Mathematics? (See Appendix E for research resources, including URLs.)

FOR LEADERS AND POLICY MAKERS

- How does this message reinforce or challenge policies and decisions you have made or are considering?
- What do you most need to know to improve teaching and learning in your classroom, school, or district?
- What existing research resources are most valuable to you?
- Select a research article or report from the resources listed in "More to Consider," at the end of this message, or from your own work. As you read, ask the questions summarized in the "What Can We Do?" section of this message. What can you can take away from the article or report in terms of the reported generalizations or conclusion?
- What are your biggest challenges in implementing research-informed practices?

RELATED MESSAGES

Smarter Than We Think

- Message 21, "Moving from Where We Are to Where We Want to Be," considers what research and best practice tell us about leading change, especially related to implementing new programs.

- Message 23, "Common Sense and the Common Core," considers the difference between fads and lasting change, with a focus on the standards being widely adopted throughout the United States.

- Message 19, "How to Know What They Know," considers how to gather data or evidence through assessment to inform teaching practice and improve student learning.

- Message 1, "Smarter Than We Think," looks at notions of intelligence in very different ways from IQ testing.

Faster Isn't Smarter

- Message 4, "Good Old Days," reminds us that memories of times gone by may not reflect the reality of the data and evidence.

- Message 12, "Beyond Band-Aids and Bandwagons," encourages us to look beyond the glitz and headlines to evidence and substance of movements in education that too easily become fads.

MORE TO CONSIDER

- *Disrupting Tradition: Research and Practice Pathways in Mathematics Education* (Tate, King, and Anderson 2011) describes a variety of types of research and practice collaborations and connections that can inform both teaching practice and educational policy.

- *Teaching and Learning Mathematics: Translating Research for Elementary School Teachers* (Lambdin and Lester 2010) considers research related to improving elementary mathematics teaching and learning.

- *Teaching and Learning Mathematics: Translating Research for Secondary School Teachers* (Lobato and Lester 2010) considers research related to improving secondary mathematics teaching and learning.

- *Teaching and Learning Mathematics: Translating Research for School Administrators* (Charles and Lester 2010) considers research relevant to key policy and leadership issues related to teaching and learning in school mathematics.

- *Second Handbook of Research on Mathematics Teaching and Learning, Volumes 1 and 2* (Lester 2007) is a comprehensive summary of current research in mathematics education.

- *A Research Companion to Principles and Standards for School Mathematics* (Kilpatrick, Martin, and Schifter 2003) summarizes the research base behind NCTM's *Principles and Standards for School Mathematics.*
- *Adding It Up: Helping Children Learn Mathematics* (National Research Council 2001) considers and defines five dimensions of mathematical proficiency and discusses implications of research studies to inform and guide teaching practice toward student proficiency.
- Common Core State Standards for Mathematics (NGA Center and CCSSO 2010) includes a list of research and other resources consulted.
- *Classics in Mathematics Education Research* (Carpenter, Dossey, and Koehler 2004) offers insights into historically important research, focusing on themes surrounding involving learning and teaching.
- *Lessons Learned from Research* (Sowder and Schappelle 2002) is a collection of research articles from NCTM's *Journal for Research in Mathematics Education*, recast with commentary for teachers.
- *Teachers Engaged in Research: Inquiry in Mathematics Classrooms, Grades Pre-K–2* (Smith and Smith 2006) is part of a series focusing on teachers as researchers and as consumers of research, gathering and using data to improve classroom practice and student learning.
- *Teachers Engaged in Research: Inquiry in Mathematics Classrooms, Grades 3–5* (Langrall 2006) is part of a series focusing on teachers as researchers and as consumers of research, gathering and using data to improve classroom practice and student learning.
- *Teachers Engaged in Research: Inquiry in Mathematics Classrooms, Grades 6–8* (Masingila 2006) is part of a series focusing on teachers as researchers and as consumers of research, gathering and using data to improve classroom practice and student learning.
- *Teachers Engaged in Research: Inquiry in Mathematics Classrooms, Grades 9–12* (Van Zoest 2006) is part of a series focusing on teachers as researchers and as consumers of research, gathering and using data to improve classroom practice and student learning.
- *The Mismeasure of Man, revised and expanded edition* (Gould 2012) updates the author's earlier attack on the measurement of intelligence via IQ scores and on standardized tests in general, and presents a renewed argument questioning the merits of any view of intelligence as a fixed quantity.
- *Formative Assessment: The Bridge Between Teaching and Learning in High School Mathematics* (Wiliam 2013) is a PowerPoint presentation for NCTM High School Interactive

Institute in which the author discusses formative assessment, including a brief discussion of decision-driven data collection. www.dylanwiliam.org/Dylan_Wiliams_website/Presentations.html.

- "Class Size: What Research Says and What It Means for State Policy" (Chingos and Whitehurst 2011) looks at and summarizes research on class size over the past several decades.

- *Journal for Research in Mathematics Education* (National Council of Teachers of Mathematics 1970–present) is a longstanding, respected, peer-reviewed journal presenting research reports and articles related to mathematics education.

- *Research Briefs* (National Council of Teachers of Mathematics 2007–2013) is a series of concise summaries of research on several topics of interest to educators and parents. www.nctm.org /clipsandbriefs.aspx. (See Appendix E for a list of research briefs.)

- *Improving Student Achievement Series: Research-Informed Answers for Mathematics Education Leaders* (National Council of Supervisors of Mathematics 2007–present) is a series of research-informed position statements on key issues in school mathematics. www.mathedleadership.org/resources/position.html. (See Appendix E for a list of position statements.)

- See Appendix E for a list of research briefs from the National Council of Teachers of Mathematics and research-informed position statements from the National Council of Supervisors of Mathematics.

23

Common Sense and the Common Core

MAKING LASTING CHANGE OR TRYING OUT THE
LATEST FAD ...

> Reforms in the United States often are tied to particular theories
> of teaching or to educational fads instead of to specific learning
> outcomes.... Policymakers adopt a program, then wait to see if
> student achievement scores will rise. If scores do not go up—and
> this is most often what happens, especially in the short run—they
> begin hearing complaints that the policy isn't working. Momentum
> builds, experts meet, and soon there is a new recommendation,
> then a change of course, often in the opposite direction.
>
> —James Stigler and James Hiebert
> (*The Teaching Gap*, 1999)

The United States seems to be approaching something very close to
national mathematics standards for kindergarten through grade
12. In 1989, the National Council of Teachers of Mathematics
(NCTM) began this process by taking the bold step of describing the
mathematics that should be included in K–12 mathematics programs.
For more than two decades, much additional work has advanced consensus in this area, including further contributions from NCTM (1991,
2000, 2006), the National Research Council (2001), and several other
groups and individuals within the mathematics education community.
In 2010, we saw the completion of the Common Core State Standards
for Mathematics, initiated by the National Governors Association
and the Council of Chief State School Officers, with the vast majority of states agreeing to use the Common Core State Standards as the
basis for their states' mathematics programs. The question is: Are
the Common Core standards simply the next educational fad to catch
the attention of desperate policy makers and educational leaders, or
is this an opportunity for lasting change?

Looking at the Bigger Picture

Even the best idea can be trivialized if we fail to remember its purpose and if we don't pay attention to the system and the people involved. I've seen schools implement professional learning communities (PLCs) in very superficial ways; for example, the professional learning communities may convene and document meetings and maintain checklists of their activities in those meetings without ever tapping into the powerful potential of such communities to improve student learning. In effective professional learning communities, participants may consider and discuss nuances of student work within and across grades, or compare what's working and what's not working in different classrooms, or delve deeply into improving instructional practice by collegial study and professional learning. Examples like this of a good idea used superficially, and often abandoned when results fall short of the original model or when key people leave, remind us of how complex and fragile the improvement of teaching and learning actually is.

The problem comes when policymakers and leaders view the approach, theory, or program as *the answer*—as something that will finally solve the troublesome problem of students not learning as much mathematics as we might like. Then, not long after implementation begins, they may abandon the approach, theory, or program before giving teachers adequate time or support to implement it well. We may exacerbate the problem by not considering the system as a whole. The original success, even if based on sound research or theories, cannot be separated from the contexts and support that surround the program or approach. Every school context is at least slightly different from others, and the likelihood of matching the system that supported the original idea with the system within which we are trying to implement the same idea is extremely slim. Of even more importance, sometimes a program or approach was created to address one type of problem in a particular content area or situation, yet it may be adopted for a variety of other content areas or situations.

For example, one model sometimes advocated for teaching a skill is called "Gradual Release" (Pearson and Gallagher 1983). The teaching model used in this approach involves teacher modeling and presentation, followed by guided practice with the teacher, and, eventually, individual student work. This approach was designed for reading instruction, yet it is sometimes adopted more broadly, perhaps with the belief that an effective instructional approach in one discipline will surely work with other disciplines. In mathematics, there might be value in using this kind of I-We-You approach if the intention is to learn a specific procedural skill (*I'll* show/tell you something. *We'll* work on it together. *You'll* eventually work on your own). However, if our goal is problem solving or helping students develop mathematical habits of mind, this approach falls short by not offering students the opportunity to wrestle with an unknown problem and figure out how to approach it without first being told exactly what they should do to solve the problem. A far more effective approach to such an outcome might be characterized as You-We-I: *You'll* work on a problem for a while. *We'll* discuss and share how you approached it. *I'll* make sure you connect your work to the intended learning.

Fad Today, Gone Tomorrow?

Education is littered with the remains of good ideas that failed to fulfill their promise. Too often, as suggested in the quote at the beginning of this message from *The Teaching Gap* (Stigler and Hiebert 1999), reforms are based on fads or a novel technique used in a particular setting. If a school or teacher somewhere has succeeded with a particular approach or tool, others may find out about the success and want to replicate it. There's nothing wrong with that; we are all trying to find ways to improve student learning.

One example of a fad, perhaps even an epidemic, is the use of benchmark testing to monitor whether students are on track to perform well on the year-end high-stakes test. The rationale I hear is that we need to base decisions on data (data-driven decision making) and that we need to monitor student learning throughout the year (formative assessment). Both of these underlying ideas have merit, if used judiciously with professional judgment. But in some school systems, students are given benchmark tests as many as six times during the year. The loss of instructional time and time to process results and give feedback to students is significant, and both student and teacher morale suffers with weariness over yet another test. By the time the real test comes, students may not be motivated to do their best. Unfortunately, in some schools the notion of benchmark testing has become a symbol of a good idea gone bad; the minimal benefits simply do not justify the tremendous costs.

A good idea may become a fad if it isn't implemented the way originally envisioned nor supported for optimal results. So as we consider any new initiative, we must continually ask whether this is simply an example of the next fad likely to fall by the wayside or a new direction we commit to take, providing the time and support necessary for its success.

What's Different About the Common Core?

In the quote from *The Teaching Gap* that opened this message, Stigler and Hiebert (1999) imply that new programs based on a fad or on one particular teaching approach, rather than on what students learn, may be short-lived. In terms of a teaching approach, the Common Core standards initiative begins with defining student outcomes and leaves much room for flexibility in classroom practice. While the standards carry some apparent implications for instruction, the authors have noted that they tried to avoid calling for specific teaching strategies. But Stigler and Hiebert's other point—that policymakers often change direction dramatically if positive results of a new program aren't immediately evident—offers an even more important caution. We simply cannot take the first opportunity to abandon ship and try something new. Rather, we have a responsibility to stay the course and adjust what may not be working, rather than dramatically change direction one more time.

The most important aspect of the Common Core State Standards for Mathematics is that these standards are *common*. Nearly every state is using these standards or standards very close to them, so we have much to gain by supporting their implementation. As students move from one state to another, they're less likely to fall behind if the standards are the same. And as we look for excellent instructional resources, we're more likely to find them if developers have been focusing on the standards we use. One other important distinction of the Common Core standards is that they are supported by cutting-edge assessments. It may take a few years of refining the tests to see their full impact, but what the assessment developers have promised has the potential to influence classroom teaching in positive ways. Maybe we finally will see assessments that focus on what we have said for years we value—problem solving, reasoning, sense-making, communication, and other key mathematical habits of mind. With this kind of long-term potential, our greatest obligation to both students and teachers is to use common sense and commitment in implementing and supporting the standards.

Using Common Sense

As is true with opinions surrounding any document aimed at serving a large population, many of us might quibble about the details in the Common Core standards. We might have issues with specific wording, the grade placement of certain standards, or even the choices made in prioritizing topics within a level or across grades. We might be uncertain whether students should study Algebra 1 in grade 8 or grade 9; the Common Core standards show considerable attention to some typical Algebra 1 content in grade 8, but not a full course, with much of that content repeated in the high school standards. And we might have questions about how to structure high school courses, since the high school mathematics standards are not presented in a grade-by-grade or course-by-course fashion.

These issues can be addressed as we move forward. After all, the Common Core standards are not designed to represent consensus, arguably an impossible goal to achieve. Rather, they reflect thoughtful professional decisions by a group of writers and advisors, based on a set of beliefs and assumptions. Perhaps different professionals might have made slightly different decisions based on different beliefs or assumptions. However, the benefits to be gained from focusing national attention on common standards are so significant—both for individual students and for the nation—that we have a responsibility to deal with the issues and move ahead. The widespread adoption of these standards represents a great opportunity to advance the work of more than twenty years toward improving mathematics teaching and learning in the United States.

EMBRACING ADJUSTMENTS

As we reflect on how classroom mathematics teaching may change in light of widespread adoption of the Common Core State Standards for Mathematics, we may realize that we need to make some adjustments, especially in terms of when we teach certain topics. We will likely need to address some skills or concepts at different grade levels than we have done in the past, perhaps beginning or ending a few topics at an earlier or later grade level than in current practice. With these changes, there may also be necessary shifts in how we teach topics when they are addressed across fewer grade levels. These adjustments are necessary if we are to adequately prepare students to demonstrate their mathematical knowledge and proficiency on common assessments.

AVOIDING NARROW INTERPRETATIONS

We must exercise caution by not narrowly interpreting every standard or example and by not trying to pay attention to specific words or phrases at the expense of what we know to be excellent mathematics teaching practices. Over the past several years, especially since the increase of sanctions and consequences based on high-stakes state tests, some researchers and educators have lamented the abuses to excellent teaching perpetrated in the name of accountability. Diane Ravitch (2010), for example, offered the insight, "Accountability makes no sense when it undermines the larger goals of education." Much of this abuse to teaching has come about because of a narrow focus on testable skills and test specifications that often overlooked the richest (and least easily assessed) elements of state standards and lost sight of the broader goals of a strong mathematical education. As we take constructive steps forward, let us view every step we take through the lens of the professional mathematics educator to truly consider how to teach for understanding, proficiency, and the development of lifelong mathematical problem-solving tools.

STAYING THE COURSE

The most important thing policy makers and leaders can do with a change as massive as the Common Core State Standards for Mathematics is to stay the course, both in terms of policies and in terms of supporting teachers. The worst thing we've done to teachers over and over and over again across the years has been to yank the rug out from under them every year or two as we first adopt, then abandon, program after program. It should come as no surprise when teachers, students, and the community resist whatever the next new program may be, when they have seen so many come and go before.

Staying the course involves making course corrections along the way. We absolutely must commit within our profession and to the public that we will continue to revisit the standards and the related assessments on a regular basis, revising them or updating them as needed. Rather than discarding this comprehensive effort if test scores (predictably) do not immediately reach the heights we hope for, let us commit to students and teachers that we will adjust the standards if needed and continue to refine the tests, so that they can fulfill their promise of assessing mathematical thinking and reasoning. Within the framework of standards and assessment, let us focus our energy where it counts most—on instruction—and commit to the most engaging and effective classroom practice the standards and assessments can support.

What Can We Do?

The Common Core standards are much more than a fad; at least they can and should be. It's up to educators and the broader public to see if we can use this unprecedented opportunity to work together across institutions, communities, and even states in support of raising achievement in the nation one student at a time. If we keep focused on the big picture represented by common standards, rather than narrowly interpreting and arguing about details we can address in the classroom, we can accomplish this goal.

We often ignore common sense in education as we buy into a new program or approach, especially in terms of abandoning ship prematurely or dramatically changing course before a program has had a chance to work. Let's use our common sense this time more than ever before to follow through on implementation and commit to make adjustments as necessary, rather than dismissing the hard work and financial commitment of teachers and school systems at the local and state level.

Let's make smart, grounded, evidence-based decisions that will serve our students well. The introduction to the Common Core standards concludes with a charge to all of us:

> It is time for states to work together to build on lessons learned from two decades of standards based reforms. It is time to recognize that standards are not just promises to our children, but promises we intend to keep. (NGA Center and CCSSO 2010, 5)

It will take the commitment, dedication, and common sense of every player in every sector involved with education to keep the promises and help reach the goal of the standards—for every student to learn high-quality mathematics and become flexible mathematical thinkers who can use what they learn throughout their lives.

Reflections and Discussion

FOR TEACHERS

- What issues or challenges does this message raise for you? In what ways do you agree with or disagree with the main points of the message?
- What are the biggest changes you foresee in implementing new standards or whatever new program you may be facing?
- What kinds of support do you wish your administrators and leaders would provide? How can you let them know what you need?
- How can you use common sense to influence the implementation of the "next new idea" in constructive ways that support improved mathematics teaching and learning?

FOR FAMILIES

- What questions or issues does this message raise for you to discuss with your daughter or son, the teacher, or school leaders?
- How can you find our more information about any new program your school is implementing? How can you constructively add your voice—your own common sense—to interpreting the school's plans and supporting your son's or daughter's teacher in using his or her common sense and best professional judgment?
- Where can you go for help if you don't understand what's being asked of your daughter or son in school?

FOR LEADERS AND POLICY MAKERS

- How does this message reinforce or challenge policies and decisions you have made or are considering?

- How well does common sense infuse your implementation plans for new standards or a new program, especially the one you're currently implementing?

- How can you best find out what kinds of support teachers need in order to do their best with the new standards or new program? What will it take to provide them with the support they need?

RELATED MESSAGES

Smarter Than We Think

- Message 31, "Developing Mathematical Habits of Mind," focuses on the Common Core State Standards for Mathematical Practice (NGA Center and CCSSO 2010, 6) and is followed by eight messages addressing each of the eight practice standards.

- Message 21, "Moving from Where We Are to Where We Want to Be," offers thoughts about facilitating change and leading the implementation of new programs.

- Message 22, "Building Our Work on Evidence," reminds us of the importance of considering context when evaluating research on the effectiveness of programs.

Faster Isn't Smarter

- Message 12, "Beyond Band-Aids and Bandwagons," considers the difference between adopting a potentially short-lived fad versus implementing long-term systemic change.

MORE TO CONSIDER

- Common Core State Standards for Mathematics (NGA Center and CCSSO 2010) presents mathematics standards for kindergarten through high school as well as background information and a list of references consulted. www.corestandards.org.

- *The Common Core Mathematics Standards: Transforming Practice Through Team Leadership* (Hull, Miles, and Balka 2012) offers strategies for leaders to work with teams on implementing the Common Core State Standards for Mathematics.

- *Common Core Mathematics in a PLC at Work, Grades K–2* (Larson et al. 2012a) is part of a five-book series on using professional learning communities as a vehicle for teacher development around the implementation of the Common Core State Standards for Mathematics.

- *Common Core Mathematics in a PLC at Work, Grades 3–5* (Larson et al. 2012b) is part of a five-book series on using professional learning communities as a vehicle for teacher development around the implementation of the Common Core State Standards for Mathematics.

- *Common Core Mathematics in a PLC at Work, Grades 6–8* (Briars, Asturias, Foster, and Gale 2012) is part of a five-book series on using professional learning communities as a vehicle for teacher development around the implementation of the Common Core State Standards for Mathematics.

- *Common Core Mathematics in a PLC at Work, High School* (Zimmerman, Carter, Kanold, and Toncheff 2012) is part of a five-book series on using professional learning communities as a vehicle for teacher development around the implementation of the Common Core State Standards for Mathematics.

- *Common Core Mathematics in a PLC at Work, Leader's Guide* (Kanold and Larson 2012) is part of a five-book series on using professional learning communities as a vehicle for teacher development around the implementation of the Common Core State Standards for Mathematics.

- *The Death and Life of the Great American School System: How Testing and Choice Are Undermining Education* (Ravitch 2010) considers the unintended negative consequences of large-scale, high-stakes accountability testing.

- *The Teaching Gap: Best Ideas from the World's Teachers for Improving Education in the Classroom* (Stigler and Hiebert 1999) offers insights into how we can improve student learning, with an emphasis on targeting teaching as the primary lever for change.

- "The Instruction of Reading Comprehension" (Pearson and Gallagher 1983) discusses the gradual release approach to teaching reading.

24 Beyond PowerPoint

USING TECHNOLOGY TO SUPPORT STUDENTS' MATHEMATICAL LEARNING

> I think it's fair to say that personal computers have become the most empowering tool we've ever created. They're tools of communication, they're tools of creativity, and they can be shaped by their user.
>
> —Bill Gates (2004)

In the late 1960s, when I was a senior in high school, a computer scientist from MIT spoke at an assembly at my high school. I was captivated with what he had to share, including not only his work in computer science for business and scientific enterprises, but also the work of his colleague, Seymour Papert, in developing the Logo computer language for children. Their work, almost forty years before Bill Gates made his observations in the quote above, was focused on giving children a way to control and use technology to support their learning.

I was so inspired by what I heard that day that I decided I needed to learn as much as I could about this exciting new world. I set aside the three liberal arts colleges where I had already been accepted and decided to find an engineering school where I could study computers. I ended up at Virginia Polytechnic Institute, where I majored in mathematics and minored in statistics and computer science. The computer I used at Virginia Tech was housed in a large building it nearly fully occupied, even though that computer was far less powerful than the phone I carry today. Every evening, I submitted my computer programs as a stack of punch cards held together with a rubber band. That night, some unknown person would take all the stacks of cards and run each one through a card reader connected to the computer, generating page after page of printed verification of the steps the computer had taken (or tried to take) based on the instructions punched onto the cards. Every morning, like other students and faculty, I would pick up my print-out to see if I had accomplished what I thought I had and submit a revised stack of cards that evening if I had not.

Initially I thought I would become a computer scientist, working for some wonderful technology company and maybe contributing to the next new innovation. But this was an era when women were still not widely accepted into such a male-dominated field, and the job offers I received were primarily to wear short skirts and demonstrate computers at trade shows. Fortunately I had followed my mother's advice to get certified to teach as part of my college education, and that decision led me to pursue the field of education.

The inspiration I felt in my early experiences with the world of computers has stayed with me throughout my career, however. I've been fascinated by the potential of technology to reshape our thinking, our practice, and to influence how people learn.

The Challenge of Investing in Technology for Schools

Technology changes the world on a daily basis. Yet schools continue only to scratch the surface of what the increasingly diverse and expansive array of tools might offer in terms of dramatically rethinking what we do in schools and how we do it.

It's not surprising that educators and policy makers miss many opportunities to seize the full potential of technology to reform education, especially mathematics education. First, technology is changing so fast—with new tools appearing almost every day—that it's impossible to keep up with everything that is available. Second, once a new technology appears, we need to figure out how the new tool might be best exploited in the interest of student learning. Most of all, it's incredibly difficult to implement any kind of change in the prekindergarten through college educational arena, especially if that change might cause us to question long-standing practices or even challenge our most solidly held beliefs about what we want students to know and be able to do when they leave us. Keeping a few key ideas in mind as we consider new technologies can help us focus on how these versatile tools can support improved mathematics learning for all students.

MAKING MATHEMATICS THE FOCUS

Conventional wisdom and substantive expert opinion tell us that using a particular type of technology or particular instructional program, such as interactive whiteboards, networked clickers, individualized programmed software, and so on, should not be the focus of our mathematics classrooms (NCTM 2011). Rather, the driver should be the mathematics we want students to learn, with mathematics content guiding us in determining what technology we use and how we use it. It's only worth investing in technology if the tools and the people using them have the capacity to focus the investment on helping students learn the mathematical knowledge and habits of mind that will serve them in their future. Along the way, it's easy to be distracted by the glitz and extraordinary capabilities each new tool offers and lose sight of how much additional time a tool might cost the teacher. So keeping the goal of student learning in mind can help sort out the superficial from the substantive.

INVESTING IN CRITICAL MATHEMATICAL TOOLS

Many school systems are choosing to invest in interesting and appealing technologies such as Wi-Fi, interactive whiteboards, e-reading devices, and computer projectors. Some of these technologies can provide helpful resources to professional teachers focused on student learning or to students themselves. However even before considering such general types of technology, schools should ask the question whether they have adequately provided critical tools specific to mathematics teaching and learning, such as calculators appropriate to each grade level, graphing technology, dynamic geometry software, or computer algebra systems. These mathematics tools are essential in raising the bar for the nature of mathematical problems and tasks we can offer students, and they are just as essential to teaching mathematics today as maps and globes are to teaching social studies or lab equipment is to teaching science.

The availability of these powerful mathematics tools allows us to offer students opportunities to wrestle with different kinds of mathematics—to approach formerly unapproachable content, such as extended work with matrices or complex geometric constructions, and to deal with more challenging kinds of problems. Rather than hanging on to old notions of doing mathematics without technology, or maybe doing a few special technology-supported lessons, we should embrace the possibility that students can raise their level of mathematical expertise. Thus, it may make sense to advocate for the purchase of such fundamental tools before a school chooses to invest in other more trendy or widely known tools.

In choosing to invest in basic technological tools like computational, algebraic, and geometric technology, we want to help students learn to make good decisions about when and how to use these tools. Without guidance from teachers and experience using the tools, students are likely to use them inappropriately, reaching for a calculator to do a simple mental math calculation or trying to deal with a basic linear equation using a computer algebra system. Rather, we want students to understand the power of the technology and also the power of being able to use mathematics to represent problems, know what tool will be helpful and how and when to use it, and make sense of whatever information the tool provides within the context of the problem. Students need experience developing their judgment and common sense if they are to become productive users of technology in or outside of school.

BEING SELECTIVE

Using an interactive whiteboard or a PowerPoint presentation on a computer projector to show the same thing that could be shown on a chalkboard is not necessarily an improvement over the old way. I once visited a high school where the mathematics department was excited over their use of technology. In particular, one teacher had put into PowerPoint presentations all of his algebra lectures previously delivered via overhead projector (the kind with a rolling screen where the writing could be rewound and reused with the next class). In the class I visited right after lunch, the teacher turned down the lights so the screen was more clearly seen and proceeded to use the slideshow and accompanying audio to "present" the lesson. Neither I nor the students were the least bit engaged in what was being shown. It was all I could do to stay awake, and it was obvious that students had the same problem. Such a use of technology misses the point—it did not improve students' learning experience over the same thing done live by the teacher with the overhead. On the contrary, taking the real person out of the picture, especially in subdued lighting after lunch, made the presentation even more boring than it probably was originally. As I watched, I kept thinking that there could be so many more interesting ways to approach the content. In particular, I kept wondering if there might not be an engaging mathematical task (with or without technology) that students could be tackling to get them involved in the content. Such a problem might kick off a lesson where students had the opportunity to discuss their ideas and possible approaches to the problem before considering the content the teacher wanted them to learn.

Likewise, putting otherwise boring worksheets into an individualized computer system does not necessarily make the mathematics engaging. The cost of purchasing and developing technology for this purpose, and the consequential isolation of a student working alone on a computer, can make these types of uses of technology a waste of money and often a waste of students' time.

What Can We Do?

[The] education movement … seems remarkably reluctant to use computers for any purpose that fails to look very much like something that has been taught in schools for the past centuries. This is all the more remarkable since the computerists are custodians of a momentous intellectual and technological revolution.

—Seymour Papert and Cynthia Solomon
("Twenty Things to Do with a Computer," 1971)

As Papert and Solomon noted several decades ago, computers—and the various technological derivatives since then—have the potential to radically change our world and the world of school mathematics. Yet we sometimes seem reluctant to tap into technology, believing that it might replace the mathematics we've always taught or cause us to have to change long-standing instructional practices. This reluctance may reflect a healthy skepticism that can help us make sound instructional decisions, provided we are open to the possibility that the future of mathematics teaching and learning might not look like the past. It's quite reasonable to rethink instructional practices of the past, not only because of technology, but also because of what we now know about how to engage students in learning mathematics. With the aid of calculators, dynamic geometric construction software, computer algebra systems, and new tools that may arrive tomorrow, students can engage in challenging mathematics problems both from the real world and from the world of mathematics that they could not tackle readily without such tools.

Fortunately, there is a wealth of high-quality technologically-based tools and resources available at every grade level to help students learn mathematics, from essential mathematical tools to comprehensive instructional packages, to many other types of innovative resources, with more appearing all the time. The challenge is to first match the tool or resource to the need, then to evaluate the quality of the resource to meet the need, and finally to look at what is involved in implementing or using the resource. Ask these questions in deciding whether or how to use a particular technological resource:

- Does the tool allow access to higher-level or more challenging mathematics than might otherwise be available to students?
- Does the program or resource support, enhance, or promote student learning of important mathematics in ways that would not otherwise be possible or practical?
- Does the program support the teacher with useful resources that enhance teacher-student interactions or that help with managing student information?

- If the program is intended to be a complete, comprehensive program, does it provide students with opportunities for important exploration and discourse about mathematics?
- If the program is intended to be supplemental, does it support and coordinate with the primary instructional approach?
- Considering the time involved for students to participate in the program from day to day and for teachers to prepare and use the program, are the benefits to students and teachers worth the expense?

Perhaps the most important question is the one Seymour Papert wanted to address when he and his team invented the Logo computer language in the 1960s: How can humans be in control of technology rather than at the mercy of technology? Logo put students in control, allowing them to manipulate the technology and, in the process, learn mathematics. Further, they could use the mathematics they were learning to refine their interactions with the technology.

This idea—that technology can be used to serve students and their learning—leads us to consider how to best incorporate the deluge of technological resources available to us into our school settings. We must always consider how the tools can be used by humans and, even more important, how mathematics can be used to interact with the tools.

Reflections and Discussion

FOR TEACHERS

- What issues or challenges does this message raise for you? In what ways do you agree with or disagree with the main points of the message?
- What technology do you use in your classroom? Do your students use? How satisfied are you with the way you integrate this technology into your classroom?
- Do you allow students access to appropriate tools as they tackle challenging problems or do you withhold the use of technology, especially calculators, dynamic geometric construction software, or computer algebra systems, thinking that students need to strengthen their computational, geometric, or algebraic skills? If you withhold the use of such tools, how open are you to the possibility that perhaps the tools might allow them to learn more, rather than learn less (consider reading "Crystal's Calculator," listed under the "Related Messages" section of this message)?

- What kinds of professional learning would most help you in more effectively using technological tools to raise students' level of mathematical thinking?
- Do you have a technology wish list? What tools would you ideally like to have access to in your classroom and why? Use the questions suggested in the "What Can We Do?" section of this message to carefully consider the value of each tool in your classroom.

FOR FAMILIES

- What questions or issues does this message raise for you to discuss with your son or daughter, the teacher, or school leaders?
- How do you use technology in your daily life? In what ways can you use these tools to engage your daughter or son in conversation about each tool's appropriate uses?

FOR LEADERS AND POLICY MAKERS

- How does this message reinforce or challenge policies and decisions you have made or are considering?
- In making decisions about technology expenditures, have you prioritized essential mathematical tools? Think of some tools your school/program would like to invest in. Use the questions suggested in the "What Can We Do?" section of this message to carefully consider the value of each tool.
- What kinds of professional learning can you provide to teachers so that they can more effectively use technological tools to help students learn higher-level mathematics and develop mathematical habits of mind?

RELATED MESSAGES

Smarter Than We Think

- Message 36, "Building and Using a Mathematical Toolbox," offers thoughts about selecting and using appropriate tools in teaching mathematics, including technological tools.
- Message 28, "Bringing Testing into the 21st Century," discusses necessary shifts in high-stakes testing, including appropriate use of technology both for administering tests and as an accessible tool for allowing students to tackle more challenging questions.

Faster Isn't Smarter

- Message 5, "Technology Is a Tool," considers the benefits and appropriate uses of technology in the classroom and the role of the teacher in determining how and when to use it.
- Message 30, "Crystal's Calculator," tells the story of a ninth-grade student who used technology to access high-level algebraic problem solving and along the way mastered her long-standing deficiencies on fractions.

MORE TO CONSIDER

- *Stratosphere: Integrating Technology, Pedagogy, and Change Knowledge* (Fullan 2012b) offers a provocative look at the possibility (in the near future) and urgent necessity of transforming education in dramatic ways that raise the bar and open doors for all students, from one of the world's renowned thought leaders regarding the change process in educational settings.
- *Rethinking Education in the Age of Technology: The Digital Revolution and Schooling in America* (Collins and Halverson 2009) considers the need for and potential benefits of using computers and other technologies to reshape the forms and structures of how we educate.
- *Mindstorms: Children, Computers, and Powerful Ideas* (Papert 1993) discusses the benefits of connecting children directly with technology they control from the inventor of the computer language *Logo*.
- *The Children's Machine: Rethinking School in the Age of the Computer* (Papert 1994) shares thoughts on using technology to transform the classroom, supporting the teacher's role but putting control of technology into the hands of students.
- "Twenty Things to Do with a Computer" (Papert and Solomon 1971) is the source for the quote under "What Can We Do?" in this message. http://dailypapert.com/?p=1058.
- "Technology in Teaching and Learning Mathematics" (National Council of Teachers of Mathematics 2011) makes recommendations for the use of technology in support of mathematics teaching and learning.
- "Project Euler" by Colin Hughes provides a set of challenging problems for students or adults requiring mathematical insights and the use of technology to solve. http://projecteuler.net/.

25

We've Got the Village, Now What?

MAKING THE MOST OF PROFESSIONAL LEARNING COMMUNITIES

Coming together is the beginning. Keeping together is progress. Working together is success.

—Henry Ford

It takes a village to raise a child. In education, we recognize that it's especially critical for us to work together if we are going to be effective in helping students learn all they can—we need the whole village. So it should be no surprise that the notion of working in professional learning communities resonates with educators at all levels, from the classroom to the central office.

Certainly, the purpose of a professional learning community begins with coming together, but it cannot end there. If students are to benefit from our participation in professional learning communities, we need to look beyond the simple notion of getting people together; we need to address the fundamental ideas that make professional learning communities successful. By doing so, we can maximize our collective efforts and make a significant positive difference in student learning.

Rick DuFour is often acknowledged as a leading expert in professional learning communities. In "What Is a Professional Learning Community?," an article he wrote for *Educational Leadership*, DuFour (2004) reflected that the only way for professional learning communities to survive the inevitable dilution of other educational reform movements gone awry is for educators to critically consider the big ideas that underlie professional learning communities. While there are several important aspects of a comprehensive approach to organizing a school as a professional learning community (DuFour, Eaker, Harhanek, and

DuFour 2004; DuFour et al. 2010), he suggests three big ideas that capture the essence of this powerful structure:

- ensuring that students learn;
- focusing on results; and
- creating a culture of collaboration.

These big ideas are exactly what we mean when we use the three words that make up the acronym PLC: *professional, learning,* and *community.* Let's use these words as a lens to further understand DuFour's ideas (2004).

Professional: Ensuring That Students Learn

The most important aspect of an effective professional learning community in mathematics is that professional educators assume responsibility for ensuring that every student learns relevant and challenging mathematics. As professionals, we need to identify what mathematical knowledge, thinking, and skills are important for all students to learn; accept no excuses and make no excuses for students not to learn; and direct all of our work inside and outside the classroom toward ensuring that every student learns.

Professional educators need not worry so much about following steps, reading scripts, or completing lesson plans, but rather direct their attention to questions of practice like these:

- What can I/we do to help this/these student(s) learn this content?
- What is this student doing well and where is the student having difficulties?
- How can I restructure my lesson for tomorrow based on what happened today?
- What can I learn from my colleagues about how to help this student?
- What will it take to keep this student from falling behind?

In a professional learning community, participants pay attention to their practice and continually fine-tune it so that their students learn what is expected. It is this commitment to student learning that distinguishes educators as true professionals, in the same sense that doctors and lawyers are considered professionals by the nature, importance, and consequences of their work.

Learning: Focusing on Results

Focusing on results means focusing on what students are learning and not learning. Often, the word *results* makes us think of test scores. Certainly, this is one measure of results. The mistake educators can make, however, is acting as if test scores are always accurate and complete indicators of what students know. Any single test score, especially a score on a large-scale test, is simply a pointer to particular topics or skills the teacher might explore further in terms of what a student really knows. A teacher working within an effective professional learning community can use what he or she learns about teaching and learning to look more closely at other examples of a student's work—such as projects, daily work, or one-on-one conversations with the student— and work with colleagues in order to determine the student's next steps.

There are many other kinds of data and information that can help educators know how well students are learning. However, as DuFour (2004) notes, too many schools and teachers suffer from the DRIP syndrome—data rich/information poor. Data-driven instruction can be an important tool for professional learning communities, but, like some educational fads, data can also be misused to the point of hurting students if we make incorrect assumptions about what the data tell us.

Perhaps the most important thing teachers, in particular, can learn is how to use information well to inform and adjust their teaching in support of improving student learning. This leads directly to another question: Outside of test scores, what information should educators use to drive their instructional decisions? In a rich discussion of using data effectively to guide teaching and learning, Robert Marzano (2003) identifies eleven school, teacher, and student factors that influence student learning:

FACTORS THAT INFLUENCE STUDENT LEARNING

School-Level Factors
- guaranteed and viable curriculum
- challenging goals and effective feedback
- parent and community involvement
- safe and orderly environment
- staff collegiality and professionalism

Teacher-Level Factors
- instructional strategies
- classroom management
- classroom curriculum design

Student-Level Factors
- home atmosphere
- learned intelligence and background knowledge
- student motivation

Effectively gathering and using information is pivotal to identifying what students are learning and not learning and how we can improve their learning. Learning how to use appropriate information in useful ways is a great activity for professional learning communities. Even more effective is when members of a professional learning community actually engage in collaboratively looking at student work and other available information to decide what concrete steps a teacher might take next in order to advance the learning of that teacher's students.

As teachers focus on student learning, they monitor what students are doing, who's learning, what they are learning, and how well they are learning, and they take actions that guide that learning in positive ways. If we do this effectively, we can intervene with a student before the student gets so far behind that he or she requires remediation.

EDUCATOR LEARNING

The word *learning* here must also incorporate how the members of a professional learning community continue to learn and grow in support of their common goals. This notion of adult learning is a critical element of a professional learning community.

Learning in this sense might involve learning new ways of organizing the classroom, learning different ways of asking students questions that probe their thinking, learning what another teacher might do differently from what you do, or learning how the role and nature of mathematics might be evolving in our rapidly changing world. We can engage in this learning through classes, readings, discussions, and visits to other classrooms, or from many other sources. All of these activities emphasize not only the learning aspect of a professional learning community but also the professionalism involved. Strengthening knowledge in any of these areas is an invaluable learning experience for a community to undertake as a group. As participants engage in learning activities where they evaluate student work or share approaches to a challenging mathematics problem, they strengthen not only their professional knowledge but also their professional relationship with each other.

Community: Creating a Culture of Collaboration

The third critical word is *community*—our village. Professional learning communities might involve groups of teachers within or across grade levels or content areas, might involve teachers within a single department or school, might reach across school boundaries, or might involve people from various schools undertaking a common initiative together (such as implementing a new program within one grade or part of several schools). The community might involve teachers with administrators, families, community members, and even students.

Interacting within a culture of collaboration connects the work of individuals so that students continually benefit, building on consistent and coherent expectations from one year to the next. A collaborative culture can be a powerful tool to prevent or remedy teacher isolation. It can help us move beyond "pockets of wonderfulness" (Seeley 2009)—where great things happen in isolation from each other and therefore fail to achieve their full potential. Teachers who work together in meaningful ways don't waste valuable time each year reviewing what they consider priorities that might differ from those of last year's teacher. They don't have to spend weeks or months establishing guidelines for student behavior or written work, communicating operational rules, or setting group expectations. Coming together in a community can improve communication among teachers and streamline articulation across grade levels, increasing the likelihood that students will engage in a coherent learning process.

Simply coming together, however, is inadequate if there is not a clear, shared, relevant purpose for doing so. Effective collaboration within a professional learning community is focused on common goals for student learning—it is purposeful and professional. More than that, it is based on a shared responsibility for learning. In a true culture of collaboration, all participants work together toward their common goals with a collective commitment to achieve those goals for all students. It is this focused, purposeful, and committed work that can yield the most powerful outcomes for both teachers and their students.

What Can We Do?

Clearly there is more to working together in a professional learning community than we might initially realize. If we are to use professional learning communities to advance students' learning, then we have to go beyond a superficial implementation, perhaps not using the acronym but instead focusing on what the words mean in terms of professional activities:

- Is our professional learning community centered on a professional commitment to student learning of important mathematical content and processes?

- Is our professional learning community focused on results in terms of students learning mathematics and learning to think mathematically, as measured by multiple sources of information?

- Does our professional learning community take advantage of the power of our collaboration to mutually learn from each other and improve our practice toward greater student learning in mathematics?

The power of our village is to work together as professionals, embracing our commitment to every student's mathematics learning. When we center our work on learning (ours and our students'), we know what results to look for and how to continually improve them. And when we function as a community, within a culture of collaboration, we will accomplish the goals we share.

Reflections and Discussion

FOR TEACHERS

- What issues or challenges does this message raise for you? In what ways do you agree with or disagree with the main points of the message?
- Are you part of a professional learning community? If so, how effective is your community in supporting improved mathematics teaching and learning?
- If you are not part of a professional learning community, what steps might you take to start one?
- Regardless of whether you are part of an existing professional learning community, how can you use the ideas related to professional learning communities to improve your students' learning, ideally with colleagues or, if necessary, on your own?

FOR FAMILIES

- What questions or issues does this message raise for you to discuss with your son or daughter, the teacher, or school leaders?
- How can you find out what kinds of professional learning activities are part of the work of your daughter's or son's teacher?

FOR LEADERS AND POLICY MAKERS

- How does this message reinforce or challenge policies and decisions you have made or are considering?
- If your school(s) include professional learning communities, how well do they address the questions listed under "What Can We Do?" How effective are they in supporting substantive educator interaction and the improvement of mathematics teaching and learning?
- If you do not currently support professional learning communities, how might you implement this concept and how can you ensure that the community or communities operate in ways that go beyond superficial meetings to look deeply at student work and teacher practice? How can you help teachers find the time to do this kind of in-depth work?

RELATED MESSAGES

Smarter Than We Think

- Message 22, "Building Our Work on Evidence," considers the use of research and data in general and as a way to focus on results.
- Message 1, "Smarter Than We Think," offers insights into our notions of intelligence that can stimulate discussion within a professional learning community about student learning.
- Message 8, "Oops!," explores the nature of mistakes and the power of learning from them; looking at student work from this perspective can be a valuable activity for a professional learning community.
- Message 14, "Effectiveness and Efficiency," considers what it takes to teach well the first time including the importance of working with peers.

Faster Isn't Smarter

- Message 26, "Beyond Pockets of Wonderfulness," encourages us to collaborate professionally within and across grade levels.
- Message 12, "Beyond Band-Aids and Bandwagons," cautions us against adopting educational fads without considering the substance of the potentially good idea(s) underlying the fad.
- Message 34, "Forgetting Isn't Forever," reminds us that sometimes students forget what they have learned, but that they can get back what they forgot without extensive review; looking at student work through this lens can be a powerful tool for a professional learning community.

MORE TO CONSIDER

- *Learning by Doing: A Handbook for Professional Learning Communities at Work, 2nd edition* (DuFour, DuFour, Eaker, and Many 2010) provides updated guidance from recognized experts on effectively using professional learning communities in support of improved teaching and learning.
- *Cultures Built to Last: Systemic PLCs at Work* (DuFour and Fullan 2013) offers insights and practical suggestions for creating and improving the effectiveness of professional learning communities from DuFour, the leader in work on professional learning communities, and Fullan, an internationally known expert on facilitating change and educational improvement.
- *Common Core Mathematics in a PLC at Work, Grades K–2* (Larson et al. 2012a) is part of a series organized around grade bands for using professional learning communities as an effective strategy in the implementation of the Common Core State Standards.

- *Common Core Mathematics in a PLC at Work, Grades 3–5* (Larson et al. 2012b) is part of a series organized around grade bands for using professional learning communities as an effective strategy in the implementation of the Common Core State Standards.

- *Common Core Mathematics in a PLC at Work, Grades 6–8* (Briars, Asturias, Foster, and Gale 2012) is part of a series organized around grade bands for using professional learning communities as an effective strategy in the implementation of the Common Core State Standards.

- *Common Core Mathematics in a PLC at Work, High School* (Zimmerman, Carter, Kanold, and Toncheff 2012) is part of a series organized around grade bands for using professional learning communities as an effective strategy in the implementation of the Common Core State Standards.

- *Common Core Mathematics in a PLC at Work, Leader's Guide* (Kanold and Larson 2012) accompanies the related grade-band books to help leaders facilitate the use of professional learning communities as an effective tool for implementing the Common Core State Standards.

- *The Common Core Mathematics Standards: Transforming Practice through Team Leadership* (Hull, Miles, and Balka 2012) focuses on using leadership teams to implement the Common Core State Standards for Mathematical Practice.

- *Leaders of Learning: How District, School, and Classroom Leaders Improve Student Achievement* (DuFour and Marzano 2011) is a handbook for leaders on using professional learning communities to improve student learning.

- "What Is a Professional Learning Community?" (DuFour 2004) gives a basic overview of the key features of an effective professional learning community.

- *Whatever It Takes: How Professional Learning Communities Respond When Kids Don't Learn* (DuFour, Eaker, Harhanek, and DuFour 2004) looks at how to use professional learning communities to help students who are not learning what we intend.

- "Using Data: Two Wrongs and a Right" (Marzano 2003) details the use of appropriate kinds of data to inform instructional decisions; this article is the source for the eleven factors noted in this message.

26 Leading Change

LESSONS FROM STUDYING SPANISH IN COSTA RICA

> If you just keep bringing in a new agenda every two or three years, you will be mediocre, period.
>
> —Jim Collins (*Good to Great: Why Some Companies Make the Leap … and Others Don't,* 2001)

I have wanted to learn Spanish for several years. I wanted to be able to communicate with the many students whose first language is Spanish, and I wanted to be able to relate to the teachers who work with those students. Despite this longstanding interest, however, I never seemed to get beyond learning to count to ten.

In 2006, I had the privilege of being invited to serve on an international committee putting together the program for the 2008 International Congress on Mathematics Education (ICME) to be held in Monterrey, Mexico. I went to a planning meeting in Mexico City, where we were working together in English most of the day and accompanied by our Mexican hosts for the evening events. It was possible to get by without speaking Spanish. But being with Spanish speakers from many Latin American countries reminded me how much I wanted to be able to speak the language. While I was there, with the help of my dictionary, I figured out how to ask for a towel in my hotel—and that was the extent of improving my Spanish beyond my counting-to-ten prowess.

When I got home, I was fired up about learning Spanish. Soon after my return, I bought Spanish tapes, thinking it would be great to be able to communicate in Spanish when I went to the next year's committee meeting. Unfortunately, as the months passed, I realized that purchasing the technology does not ensure learning.

The next committee meeting was held in 2007 in Monterrey. Shortly before the meeting, I pulled out the tapes and managed to work through the first cassette a couple of times. By the time I traveled to Monterrey,

I could more or less count to twenty, and I was very proud that I learned how to say, "Estoy aprendiendo español" ("I am learning Spanish"). That sentence served me well during my brief trip to northern Mexico. Any time I encountered someone speaking Spanish, I would say my one sentence, and the person would immediately smile with encouragement and say something to me in Spanish that I couldn't understand. I promised my new friends from Latin America that I was going to be able to speak some level of Spanish by the time I would see them at ICME the following year. This time I went home and decided to try a Spanish textbook. But apparently, having a nice new textbook on the shelf doesn't ensure learning, either.

ICME arrived in 2008, and I was off for a week to Monterrey. My small hotel was far from the site of the conference and far from the other committee members I had come to know. I was essentially on my own to figure out how to eat and how to get from one place to another. When I saw my Spanish-speaking friends from the committee, I was embarrassed to share that I hadn't quite mastered the language—yet. During that week, I added to my repertoire the words *derecha* and *izquierda* (*right* and *left*).

After returning home, this time I tried to pick up Spanish words and phrases by listening and watching whenever I heard or saw Spanish. I still wasn't learning much. My friends then suggested that I spend time in a Spanish-speaking country.

I finally did just that. For two weeks, I studied Spanish for four hours a day at a language school in Nicoya, Costa Rica. I lived with a family who didn't speak English and spent my afternoons using my emerging Spanish to get around town. I came home to visit with the family and to study—a lot. On the television in the background were talk shows and novellas, with Spanish spoken so fast I could only pick up a word or two at a time. But by the time I left, not only had I made several new friends, I was also communicating without relying on English. Wow.

By now, you are probably thinking that my message is about not relying on just having tapes (or other tools) or buying new textbooks to make change. Or that we will never learn if we always have someone right next to us who tells us what to do or does the work for us. These are important ideas for any educator, however my reflections on learning Spanish and the journey that led me to the town of Nicoya made me think of three key principles about facilitating change. I believe these principles may be useful for leaders as we tackle the challenges of improving mathematics teaching and learning.

Principle 1: Making Change Means Making a Commitment

Making change means making a commitment. It's fine to take small steps toward a vaguely defined ideal for a while. You can make a little progress, as I did in the early stages of my journey to speak Spanish. But you are not likely to see big gains until you make a serious commitment. A real commitment calls for investing resources—including money, but especially time. Leaders need to provide time for people to work together—to identify issues and to generate a pathway to move through those issues toward a shared vision. It is also critical to commit to enough time for a reasonable transition that has a possibility of getting to that vision.

Commitment may involve investing in new tools or programs, but we know that simply buying things will never get us where we need to go. We also need a significant investment in the teachers and students who will use those resources. The most important part of a serious commitment is the individual and collective will of leaders and policy makers to sustain the effort and support the teachers and students doing the work.

Principle 2: Take into Account All Parts of the System

An effective leader takes into account all the parts of the system when developing a plan. Thinking systemically means targeting the most strategic point in the system for your attention at this point in time, while keeping an eye on the rest of the system as you try to anticipate what may be coming.

It isn't enough to focus solely on books, or technological tools, or even on a new program. In teaching, the resource is only the starting point for a professional teacher's judgment and custom tailoring to meet students' needs. In my case, while attending classes was a critical component of my learning, being able to practice at home and being required to communicate in Spanish every day were invaluable—I needed to immerse myself in all parts of "the system."

For the work we do to improve mathematics in schools and school systems, we know that we must target not only the standards that define our goals, but also the materials we provide students, the tools we use to assess learning, the pedagogical approaches we employ, the professional development we provide, the administrative support we seek, and the engagement of the community. Thinking systemically involves working as a team with those who have ideas about and connections with other elements of the system from those you may have.

Principle 3: Provide Ongoing Support

We know that the work is not over when one benchmark is reached. Obviously I am not yet a fluent Spanish speaker. My two weeks of immersion and intensive study provided a great start. But if I don't nurture my new learning, practice what I know with people who speak

Spanish, and continue to learn, I will lose the progress I have made thus far. In the case of improving mathematics teaching and learning, the goals are numerous and complex, and they require sustained support over long periods of time.

With any improvement effort in a school or school system, the key to long-term success is ongoing support, and then more support, and then more. Teachers need encouragement, discussion, and sometimes guidance, as they move from step to step. They need opportunities to work together and learn from their colleagues and from others as they continue to improve their practice. As leaders, we need to pay attention to teachers' needs and ideas to determine where to invest our resources next. Do they need a workshop? An open discussion session? Opportunities to network with colleagues they may not see everyday about common issues? A professional learning community focused on how to improve student learning? Paying attention to what teachers need and providing them time and resources to address those needs are important ways for leaders to support their teachers in improving student mathematics learning.

What Can We Do?

In your work, the stakes are higher than in my relatively straightforward journey toward Spanish fluency. The job of teaching mathematics is more complex and the results affect the lives of students. Both leaders and teachers are being asked to take a risk—a calculated leap of faith—that you are moving in the right direction. Taking this risk is worthwhile only if you listen to your inner voice about what is best for students. Any steps you take must be guided by your professional judgment. Sometimes you may need to raise a voice of reason in an arena where there may seem to be little. Ask yourself if you are paying attention to the three principles described in this message—commitment, systemic thinking, and ongoing support:

- Have you committed not only philosophically to the change you want to lead, but also committed the necessary resources to support students and teachers as they undergo the change?
- Have you considered all the aspects of the system within which the change will happen?
- What kinds of support have you planned to help teachers and others as they move through the change process? Are you committed to providing support for the long haul?

As you travel on your journey, there will be bumps in the road. You are likely to stumble. (Space doesn't allow for analogies based on my salsa dancing lessons in Costa Rica. ...) When the bumps and stumbles come, remember to pause and take a look at your work from a broader perspective as you focus on your destination, and remember the benefits for students as you succeed.

Reflections and Discussion

FOR TEACHERS

- What issues or challenges does this message raise for you? In what ways do you agree with or disagree with the main points of the message?
- How do the three principles—commitment, systemic thinking, and ongoing support—relate to your work with helping students learn mathematics?
- What insights can you gain from your own efforts to learn something new that might help you help students learn mathematics? What kinds of things helped you? What got in the way?

FOR FAMILIES

- What questions or issues does this message raise for you to discuss with your daughter or son, the teacher, or school leaders?
- What insights can you gain from your own efforts to learn something new that might help you support your son or daughter to persevere and succeed in her or his study of mathematics?

FOR LEADERS AND POLICY MAKERS

- How does this message reinforce or challenge policies and decisions you have made or are considering?
- What insights can you gain from your own efforts to learn something new that might help you in your work with teachers to improve mathematics teaching and learning?
- For any new program or initiative you implement, consider the questions suggested in the "What Can We Do?" section of this message. How well have you addressed (or can you address) the three principles in your planning?

RELATED MESSAGES

Smarter Than We Think

- Message 21, "Moving from Where We Are to Where We Want to Be," offers thoughts about facilitating educational change and improvement, including the need to stay the course and make adjustments as needed.
- Message 25, "We've Got the Village, Now What?," considers the power of effective professional learning communities in supporting teachers to improve student mathematics learning.
- Message 13, "Clueless," includes another lesson from the trip to Costa Rica, this time in relation to helping students learn to think mathematically and solve problems.

Faster Isn't Smarter

- Message 12, "Beyond Band-Aids and Bandwagons," cautions against adopting educational fads and reminds us that substantive change takes time and commitment.
- Message 13, "Seek First to Understand," advocates using collaboration to improve teaching and learning.

MORE TO CONSIDER

- *Cultures Built to Last: Systemic PLCs at Work* (DuFour and Fullan 2013) offers insights and practical suggestions for creating and improving the effectiveness of professional learning communities from Fullan, an internationally known expert on facilitating change and educational improvement, and DuFour, the recognized leader in work on PLCs.
- *Leading in a Culture of Change* (Fullan 2007) provides insights into facilitating change for leaders.
- "Mathematics Professional Development" (Doerr, Goldsmith, and Lewis 2010) considers the nature of professional development most likely to support improved teaching and learning.
- *The Common Core Mathematics Standards: Transforming Practice Through Team Leadership* (Hull, Miles, and Balka 2012) discusses issues related to improving mathematics teaching, including a discussion of how to deal with resistance to change.
- *Sensible Mathematics: A Guide for School Leaders in the Era of Common Core State Standards, 2nd edition* (Leinwand 2012) offers practical suggestions for school leaders dealing with implementing new standards and assessments.
- "Scaling Up Instructional Improvement Through Teacher Professional Development: Insights from the Local Systemic

Change Initiative" (Weiss and Pasley 2006) provides research insights for leaders about using professional development and high-quality instructional materials for effective improvement of mathematics teaching and learning.

- *Taking Charge of Change* (Hord, Rutherford, Huling-Austin, and Hall 1987) remains the frontline educators' manual on leading change, with a good overview of the Concerns-Based Adoption Model (CBAM).

- *Implementing Change: Patterns, Principles and Potholes, 2nd edition* (Hall and Hord 2005) considers the change process, including the CBAM model.

- *Good to Great: Why Some Companies Make the Leap ... And Others Don't* (Collins 2001) proposes now–widely accepted principles for business leaders; the accompanying monograph, *Good to Great and the Social Sectors: Why Business Thinking Is Not the Answer* (Collins 2005), relates and adapts principles to the social sector, including education.

27 Fixing High School

IS IT FINALLY TIME TO RESTRUCTURE 9–12
MATHEMATICS?

If you stop at general math, you're only going to make general
math money.

—Calvin Cordozar Broadus, Jr.
(rapper stage name Snoop Dogg)

One day when my middle daughter was a high school junior, I
asked her how school was going. She remarked, "Well, I really
love chemistry. But I *hate* Algebra 2!" I swallowed hard at
the negative opinion about the discipline in which I work every day,
then asked her what it was about chemistry that she loved so much.
She quickly answered, "It's all those cool equations we get to solve!" I
laughed and noted that Algebra 2 was all about exactly that. She immediately replied, "Oh no it's not!"

The State of High School Mathematics

How sad that such a direct application of algebra can escape students.
But, unfortunately, much of high school mathematics seems elusive, boring, and irrelevant to many students today. Even more sadly, in today's
world of rising expectations and increased high school graduation
requirements, more students are being placed into or counseled toward
higher-level mathematics courses, especially Algebra 2, in the belief that
passing this particular course will open doors for them and give them a
bright future. Thus, an increasing number of students—in some schools
all students—are by default placed in retrofitted Algebra 2 courses originally designed to prepare a relatively small, specific segment of the student
population for advanced mathematics, including calculus. This purpose
is fine for students preparing for careers in science, engineering, or other
calculus-dependent fields, but the precalculus/calculus pinnacle may

not be the best preparation for students in many fields, especially non-mathematics-dependent fields. Even some STEM (Science-Technology-Engineering-Mathematics) fields might rely more on statistics or discrete mathematics than on calculus. For many years, mathematician Lynn Steen and others have called for an alternative goal to calculus (Mathematical Sciences Education Board and National Research Council 1989; Steen 2006). The content of most Algebra 2 programs—arguably the launch pad for precalculus and calculus—is simply not useful today to many students enrolled in the course. Maybe the time has finally come to change not only the goal, but also the path to reach the goal.

Algebra 2—the Crux of the Problem

Many states have raised graduation requirements, reflecting the call for stronger college and career readiness skills for all students and reflecting the expectations for all students in the Common Core State Standards for Mathematics (NGA Center and CCSSO 2010). A few states that previously raised requirements are now going through a predictable pendulum swing away from the more rigorous standards and high graduation requirements adopted over the past decade or more. Algebra 2 seems to lie at the center of the discussion for both proponents and opponents of the course. Advocates call for Algebra 2 as a way to prepare students for calculus and for greater options after high school. Opponents call for removing Algebra 2 as a requirement for all students because they worry that some students may drop out of school or simply not pass the course, thereby not graduating. Both points of view have merit. However, maybe it's not that Algebra 2 is too hard, but that it's the wrong course. It may well be that changing the nature of the course—or replacing it with an equivalently rigorous level of mathematics addressing somewhat different topics—may be preferable to lowering standards or sentencing some students to limited future options.

Algebra 2 as it stands in most U.S. high schools today is not the course all students need. It addresses valuable skills for those students going into precalculus and calculus programs. But not all students need the advanced symbol-manipulation skills that form the backbone of the course, nor its emphasis on abstraction and symbol manipulation rather than on applications and modeling, which might be both more engaging and more relevant. While some individuals and organizations are doing great work to create more relevant and engaging Algebra 2 programs, there remains a built-in ceiling as to how much we can accomplish within the prevailing course structure. There is a fundamental problem with a program that includes Algebra 2 as a third course in a three-year sequence built on Algebra 1 and Geometry, in that mathematics content for all three of these courses has become relatively well established, even entrenched, leaving little room for increasingly important content like statistics and finance or topics from discrete mathematics such as matrices, networks, graph theory, and the use of algorithms.

This doesn't mean that students should enroll in lower-level courses covering arithmetic, such as the remedial courses that represented the basis of a high school diploma just a few decades ago. What we need is a revamped high school mathematics program that prepares every student in the twenty-first century with the kind of rigorous mathematics that applies to their lives and prepares them for a promising future. Such a goal might be accomplished with two to three years of a rich, multifaceted common foundation in mathematics followed by a couple of optional paths that might prepare STEM-focused students, business-focused students, or students heading in a variety of other directions.

Creating a New Vision

Fixing high school involves rethinking the mathematics we teach, how we organize it, and how we assess it. But it also involves looking closely at the nature of day-to-day teaching and learning—how we can teach in ways that help students make sense of what they learn and apply it to solve relevant, in-depth problems. Consider these central elements of a new vision for high school mathematics.

RECONSIDERING WHAT WE TEACH

The study of data and statistics has become indispensable in the mathematics curriculum throughout the grades, but especially at high school. Understanding quantitative situations, interpreting data, and using statistical processes to make sense of numerical information are now among the most useful and important life skills for every adult. By the time they graduate from high school, students are beginning to encounter increasingly sophisticated types of data, and they regularly see statistical conclusions—often misleading or inaccurate—on the Internet and in the media. Many careers involve interpreting data or using statistics, and consumers come in contact with quantitative information every day in the media and in their lives. Students need opportunities to wrestle with data-rich situations, using increasingly sophisticated statistical tools.

Likewise, topics in discrete mathematics have become important for a variety of fields. Applied mathematicians who can develop algorithms to identify consumer patterns, for example, or who can interpret the floods of data gathered technologically in the retail sector are in high demand in today's business world. Various aspects of discrete mathematics are applied in scheduling plane routes, organizing supply chain processes, designing robots and other complex machines, developing video games ... the list goes on.

Of course, to make room for these critical twenty-first-century topics, we may need to reconsider how much of the content in traditional courses like Algebra 1, Geometry, and Algebra 2 is still relevant for all students today (see Usiskin 2006a and 2006b for discussions of what might be left out). In a world where technology offers tools to perform many of the algebraic procedures and geometric constructions traditionally addressed in such courses, perhaps there is room for at least modifying how much we include of such procedures and constructions in light of the demand to include new topics and in light of increased emphasis on reasoning, thinking, and modeling. And some current course content still relevant today might be reorganized into later specialized courses like precalculus.

REORGANIZING COURSES

One lesson we can learn from innovative programs in the United States and from widespread practice abroad is that we can gain a lot in terms of student learning by focusing our teaching around fewer expectations in greater depth than we currently address in traditional programs. This notion of increased focus is evident in the kindergarten through grade 8 portion of the Common Core State Standards for Mathematics (NGA Center and CCSSO 2010), but it is not nearly as evident in the Common Core listing of high school standards. Thus, describing high school courses with a tighter focus on fewer topics should be an important goal of any restructuring of secondary mathematics.

Arguably, the best way for high school mathematics to serve the needs of a broader audience is for United States schools to adopt an integrated high school mathematics program, as is used in essentially every other country in the world. In such a model, topics from algebra, geometry, statistics, and other strands of mathematics are organized into a coherent sequence of units blending content from multiple topics as appropriate, based on the mathematics called for in particular tasks. A few innovative programs developed in the past twenty years in the United States organize high school mathematics in this way. (See the list of high school programs in Appendix A.) Much has been done to center agreement on general topics addressed in every grade through grade 8, but no such broad agreement exists for high school courses other than for the traditional Algebra 1, Geometry, and Algebra 2 sequence. Ideally, we might agree nationally on what topics would be addressed in each of two to three years of common mathematics expectations for all students, regardless of whether they are aiming at a STEM field or not. Beyond generating the national will to move in this direction, agreeing on content to be addressed in each year is likely to be the most difficult part of making such a shift, and would probably call for some kind of national convening and/or extensive input-gathering process. In addition to the almost universal use of an integrated approach outside of the U.S., a few states and some school systems have adopted this integrated structure for high school mathematics courses, either using one of the high school programs identified in Appendix A or developing their own unique standards and materials. But implementing this approach on a large scale represents a dramatic change from most current programs, even though the Common Core State Standards for Mathematics (NGA Center and CCSSO 2010) include a suggested integrated high school pathway. Because this notion is so different from current practice, moving to an integrated program can raise concerns from parents and teachers about what might be involved in such a big change, so it's not surprising that such a change has not been implemented on a broad scale in the United States yet.

CREATING CAPSTONE MATHEMATICS COURSES

Every student needs to experience a rich, relevant, rigorous mathematics course before graduating from high school, preferably in the last year. Such a capstone course might be tailored to the particular direction a student has chosen or it might be more general, but it would likely follow two or three years of a common mathematics curriculum designed for all students. While few sixteen- to eighteen-year-olds are likely to have identified the career path they will eventually follow, some students are fairly certain that they would like to pursue mathematics-dependent fields. These students should definitely select strong courses in precalculus and, if possible, calculus. Other students, however, may benefit from one of a new generation of courses we might classify as *quantitative reasoning* or *decision making*, with strong statistics and finance components (see the following box for some examples). Whatever the type of capstone course, including precalculus or calculus, the course should involve a heavy expectation that every student will research questions, support a point of view with evidence, construct strong arguments, and communicate findings through written reports and oral presentations. Assuming this kind of personal responsibility for learning is rarely expected of high school students, even in high-level rigorous courses, and yet it is exactly this kind of experience that employers seek and that can prepare both college-intending and workforce-bound students for success.

A SAMPLING OF CAPSTONE MATHEMATICS COURSES

- Advanced Mathematical Decision Making, utdanacenter.org/amdm
- Mathematics INstruction using Decision Science and Engineering Tools (MINDSET), www.mindsetproject.org
- Transition to College Mathematics and Statistics (TCMS; related to Core-Plus), www.wmich.edu/cpmp/tcms
- Functions, Statistics, and Trigonometry and Precalculus and Discrete Mathematics, UCSMP, http://ucsmp.uchicago.edu
- COMAP, www.comap.com

Some states and other organizations have developed or are developing additional options in quantitative reasoning, statistics, finance, discrete mathematics, or other applied fields.

Seizing the Opportunity

If we ever accept the challenge of widely changing our model for high school mathematics courses, we will have an unprecedented opportunity to make related changes many have advocated for some time. In particular, we may finally be able to incorporate the modeling, applications, reasoning, sense-making, and mathematical habits of mind every student needs in order to actually use the mathematics they learn once they leave high school.

INCORPORATING MATHEMATICAL MODELING

In restructuring high school mathematics, we have a perfect opportunity to incorporate mathematical modeling as called for in the Common Core State Standards for Mathematics (NGA Center and CCSSO 2010, 7). Modeling has become increasingly important as a goal for school mathematics, so that students are able to use what they learn to solve the real, often ill-defined, problems that arise in everyday life and in the workplace. At its most basic level, modeling involves giving students opportunities to apply mathematics to problems in real contexts, often going beyond simple application of a single recently learned procedure. Deeper levels of modeling call for students to make sense of and mathematize situations unlike those they may have experienced before. Problems may not be clear or well defined, and students would be expected to select from their mathematical tool box possible strategies or combinations of strategies that might work. From a practical standpoint it's far easier to incorporate modeling into a program when creating courses anew than it is to retroactively impose modeling into an existing curriculum.

FOCUSING ON THINKING AND REASONING

Beyond modeling and applications, we must also consider how students can learn to think and reason with mathematics. As Thomas Friedman (2007), the National Center on Education and the Economy (2008), and others have suggested in recent years, workers of the future will need to be able to apply skills they learn in school today to tomorrow's jobs, many of which don't yet exist. In particular, there is nearly universal agreement that adults need to be able to reason, communicate—including communicating about quantitative ideas—and solve all kinds of problems. Recognizing the importance of general, transferable skills in life and work, we must recommit to the National Council of Teachers of Mathematics's process standards (1989, 2000) and the Council's focus on reasoning (2009) and the mathematical practices in the Common Core State Standards for Mathematics (NGA Center and CCSSO 2010). Each of these resources offers specific ways to incorporate reasoning and thinking into day-to-day mathematics instruction. These critical skills cross all educational levels and job categories and they help adults manage their everyday lives. Thus, they should be a focus of every high school mathematics program claiming to prepare students for their future, not just for their next mathematics course or their high-stakes test. As with modeling, shifting what we do in teaching for thinking is more likely to be possible in creating new courses than in trying to impose another dimension onto an existing course.

Tests to Support the Vision

Finally, in order to allow and encourage high schools to address this significant restructuring, the tests that drive instruction need to be brought into the twenty-first century. This includes, at a minimum, college entrance tests, college placement tests, tests for awarding college credit in high school, and state accountability tests. We are seeing the birth of a few new and innovative types of mathematics tests that call for students to reason and communicate their thinking, using powerful technological tools as appropriate to solve complex problems. Only when the promise of this kind of test is realized and when we see such tests in widespread use will educators feel they have permission to implement the kind of mathematics programs and teaching approaches we know will help students learn what they need to learn in order to thrive after high school.

What Can We Do?

We don't have to sacrifice rigor and high standards in order to fix high school mathematics. We don't have to lower standards for some students, just because Algebra 2 isn't what they need. Perhaps this is finally the moment in time when mathematics educators, mathematicians, employers, and community members can join forces to create the kind of goals for high school mathematics that will really serve students and society. We must work together, across states and using the full power of the mathematics education community, first and foremost, to agree on what outcomes we want as a common experience for all students when they graduate from high school and how we can organize courses to address those outcomes. We must tackle the hard questions no one has yet been able to resolve nationally:

- How many years of a common high school mathematics program should all students share; at what point might they diverge paths and choose options based on their possible future path?
- Which traditional high school topics from algebra and geometry can or should be retained, and which should be left out of the common curriculum in order to make room for more current, relevant topics?

- What topics in current algebra and geometry courses might be bumped up to be addressed in precalculus or other courses designed for STEM-intending students?
- When should students begin the high school sequence so that some have time to get to calculus? How do we address the needs of students who are well ahead of or far behind their peers?

These questions call for thoughtful, extended discussion across various audiences.

We must continue the good work of recent years in helping high school teachers focus their teaching on reasoning and sense making and seize the opportunity to incorporate mathematical modeling as a fundamental part of the high school mathematics program. We need to work with higher education institutions to recognize the rigor and benefits of the new generation of capstone courses students will bring to college. And we need to vigorously advocate the realization of the potential of a new generation of tests that actually measure what we say we value.

Let us make a collaborative effort to define the next iteration of common mathematics standards and finally take on the challenge we seem to have been avoiding for decades to restructure high school mathematics so that all students receive a solid, rigorous mathematical foundation with a few rigorous and flexible options for those going in different directions after that foundation. Let us give policy makers a real option for the future—take on the hard challenge of fixing high school mathematics instead of going back and forth raising and lowering the bar when it isn't even the right bar.

Reflections and Discussion

FOR TEACHERS

- What issues or challenges does this message raise for you? In what ways do you agree with or disagree with the main points of the message?
- How relevant do you feel your mathematics program is to students' lives after high school? How might you make it more relevant?
- How well does your current mathematics program support students in learning to think, reason, make sense of, and apply mathematics?
- How well do you think you address mathematical modeling in your classroom? Where can you go to learn more about how to incorporate modeling in your teaching?

(*continued*)

- How well do you think you address thinking and reasoning in mathematics in your classroom? How can you learn more about incorporating thinking and reasoning into your teaching?
- If your school is not using an integrated approach to high school mathematics, how open are you to the possibility that such a structure might be a good approach for students? What kind of support would you need and might your colleagues need in order to move toward an integrated mathematics program?
- What kinds of shifts in large-scale testing would support the changes you want to make in your classroom? How can you prepare students in meaningful ways for tests that address thinking and reasoning?

FOR FAMILIES

- What questions or issues does this message raise for you to discuss with your son or daughter, the teacher, or school leaders?
- If you learned mathematics in high school courses called Algebra or Geometry, how open are you to the possibility that your daughter or son might learn mathematics well using a more integrated approach, with courses titled something like Math 1, Math 2, and so on?
- How can you help your son or daughter learn to think and reason about the mathematics problems they solve, rather than trying to memorize rules and procedures for certain types of problems?
- Look for how mathematical models are used in predicting weather, tracking economic trends, predicting the impact of certain events on the economy, looking at housing trends, and so on. How can you involve your daughter or son in looking for such examples of modeling and how can you discuss them together?

FOR LEADERS AND POLICY MAKERS

- How does this message reinforce or challenge policies and decisions you have made or are considering?
- If your high school still offers courses called Algebra 1, Geometry, and Algebra 2, how open are you to the possibility that offering integrated courses, such as Math 1, Math 2, and so on, might be a better option?
- How can you influence the direction of college placement tests and procedures to reflect improvements and changes in your high school program?
- In planning for reorganizing high school mathematics, what kinds of support will your teachers need and how can you provide that support?
- How can you help parents understand the need for reorganizing high school mathematics?

RELATED MESSAGES

Smarter Than We Think

- Message 21, "Moving from Where We Are to Where We Want to Be," considers issues related to facilitating change, such as the changes advocated in this message.

- Message 28, "Bringing Testing into the 21st Century," addresses the need to reflect problem solving and the use of technology in our high-stakes tests, as called for in this message.

- Message 35, "Math in the *Real* Real World," discusses Mathematical Practice 4 on modeling from the Common Core Standards and supports the inclusion of modeling in school, as called for in this message.

- Message 16, "Let It Go," reminds us that, in order to focus our curriculum, and include emerging topics, we may need to let go of some topics or activities previously taught.

Faster Isn't Smarter

- Message 7, "Not Your Grandpa's Algebra," advocates restructuring algebra to make it more relevant to more students.

- Message 8, "More Math, More Dropouts?," looks at the impact of increasing expectations and graduation requirements in mathematics.

MORE TO CONSIDER

- *Common Core Mathematics in a PLC at Work, High School* (Zimmerman, Carter, Kanold, and Toncheff 2012) looks at schoolwide work to facilitate implementation of the Common Core State Standards for Mathematics in a high school.

- "Math Will Rock Your World" (Baker 2006) discusses the uses of mathematics in business. www.businessweek.com/printer /articles/201872-math-will-rock-your-world?type=old_article.

- Math Is More includes information on rethinking high school mathematics. www.mathismore.net.

- *Closing the Expectations Gap 2013* (Achieve, Inc.) annually reports states' efforts to raise requirements for high school graduation. www.achieve.org/closing-expectations-gap-report.

- *Shop Class as Soulcraft: An Inquiry into the Value of Work* (Crawford 2009) is a fascinating look at the world of work that can influence our thinking about how to educate all students.

- *The World Is Flat 3.0: A Brief History of the Twenty-First Century* (Friedman 2007) predicts demands on the worker of the future.

- Focus in High School Mathematics Series: Reasoning and Sense Making (Martin, Carter, Forster, Howe, Kader, Kepner, Quander, et al. 2009) lays out recommendations for improving high school mathematics instruction. www.nctm.org/standards/content .aspx?id=23749.

- *Everybody Counts: A Report to the Nation on the Future of Mathematics Education* (Mathematical Sciences Education Board and National Research Council 1989) is one of the early calls to aim school mathematics at a goal other than calculus.

- "Facing Facts: Achieving Balance in High School Mathematics" (Steen 2006) considers refocusing high school mathematics to better meet the needs of more students.

- "From the 1980s: What Should Not Be in the Algebra and Geometry Curricula of Average College-Bound Students?" (Usiskin 2006a) suggests what to delete from high school mathematics courses in order to provide time to include new topics.

- "Reconsidering the 1980s: A Retrospective After a Quarter Century" (Usiskin 2006b) updates the author's recommendations for what not to include in courses.

- "Pattern" (Steen 1990) discusses the need to rethink high school mathematics.

WWW This message is also available in printable format
at mathsolutions.com/smarterthanwethink.

Thanks to Eric Robinson and Zalman Usiskin for their comments on an earlier version of this message.

28

Bringing Testing into the 21st Century

SUPPORTING HIGH-LEVEL TEACHING AND LEARNING WITH HIGH-STAKES TESTS

> Standardized testing has swelled and mutated, like a creature in one of those old horror movies, to the point that it now threatens to swallow our schools whole. Of course, on the late, late show no one ever insists that the monster is really doing us a favor by making its victims more "accountable."
>
> —Alfie Kohn (*The Case Against Standardized Testing*, 2000)

Alfie Kohn's scathing indictment of high-stakes standardized tests reflects growing concerns among policy makers, educators, parents and students about the negative impact of standardized tests on high quality learning. In mathematics, especially, such tests can hold back large-scale systemic improvement aimed at raising the nature of mathematical thinking, potentially destining students to an education consisting of primarily—or only—easily tested, superficial facts and procedural skills. This devastating trend comes at a time when twenty-first-century employers desperately seek workers who can think, reason, analyze, create, communicate, and work together in teams to solve problems we don't know how to answer. When employers can't find workers in the United States with these skills, they are forced to look elsewhere.

We know how to teach mathematics well and how to help students learn to think and reason mathematically from their entrance into kindergarten, or even preschool, through their graduation from high school. Implementing on a wide scale what we know about high-quality mathematics teaching, however, comes with many challenges. Perhaps the most significant of these challenges is the pressure for students to perform well on *the test*. Unfortunately, *the test* almost universally fails to assess the kind of learning and thinking we say we value and essentially ignores the existence of powerful technology that can be used to allow students to tackle and solve complex problems. Worse, most of the tests used to make high-stakes decisions today tend to push mathematics teaching and learning down to the lowest levels of recall and guesswork. The sad reality is that we will never accomplish our goal of raising the bar on high-level mathematics teaching and learning until we put testing in its proper perspective with respect to teaching and learning and until the tests we use actually measure what we say we value.

Tests Drive Instruction

Many types of tests influence mathematics teaching and learning in the classroom: college-entrance tests like the ACT or SAT, state accountability tests, tests given to award college credit in high school, college placement tests used to determine entry-level course placement, national and international comparative tests, and even tests used for contests and competitions. The influence of such tests can be positive or negative, but overwhelmingly these tests tend to solidify the status quo and inhibit even small improvement efforts at the school level. They especially inhibit any efforts to restructure, rethink, or redesign school mathematics. The amount of inhibiting tends to increase from elementary through middle school to high school and comes to a peak for students ending their high school experience and preparing for what will follow high school. Consider the following illustrative examples:

- A school decides to reorganize their high school mathematics courses to integrate mathematics topics from algebra and geometry and to include more statistics, but their state gives end-of-course tests for Algebra 1, Geometry, and Algebra 2.

- A curriculum developer creates an exciting new grade 12 capstone mathematics course focused on high-level quantitative reasoning, statistics, and discrete mathematics. But the developers determine that they have to include algebraic skills, even though this takes away time for other topics, because college placement tests are focused on algebra.

- A school district wants to have students use calculators in middle school mathematics, graphing calculators in algebra classes, and computer algebra systems (CAS) in advanced mathematics classes, but these tools are not allowed on state accountability tests, college entrance tests, or college placement tests. So they struggle with whether to allow students to use the tools in the classroom.

- A school implements a program to focus on collaborative learning, where students gather evidence to answer questions and communicate results orally or in written reports, but the state accountability test consists of multiple-choice questions on specific facts and procedures, so they have to take time away from their instruction to prepare students for the test—or they decide they cannot implement the program because of the test.

- A mathematics contest organization decides to allow the use of calculators, so they write problems where the calculator won't interfere with the question being asked. Eventually they run out of such questions, so they go back to not allowing calculators on the contest. Schools preparing students for the contest therefore also decide to do so without students having access to calculators.

All of these examples demonstrate how strong the influence of testing can be on teaching and learning. Perhaps the most powerful influence comes in the form of college entrance and college placement tests, where bright, capable, hard-working students continue to be disadvantaged in high school by restrictions pushed onto their high school mathematics programs from their higher education neighbors.

Three Shifts We Need to See in Tests

High-stakes test can have a positive impact on mathematics teaching and learning, but if this is going to happen, we need tests that model three fundamental shifts:

- an innovative, in-depth *theoretical framework* that allows for new ways of thinking about measuring and reporting student knowledge and skills,
- *testing formats* that reflect the nature of mathematical thinking, reasoning, and overarching habits of mind that we see as the best outcomes of school mathematics, and
- the incorporation of appropriate *technologies*, not only for administering and scoring a test, but as *fundamental mathematical tools*, the availability of which shapes the nature of questions being asked of students.

SHIFT 1: A NEW THEORETICAL FRAMEWORK

The arrival of the first criterion-referenced tests several years ago heralded a major improvement in the alignment of large-scale tests with the school curriculum. Rather than reporting how a student's score compared to that of other students (locating a student on a bell curve), tests were now designed with a certain number of questions targeted at a particular learning objective or standard. Thus, educators (and parents) could largely predict the kind of questions that would be asked and students could prepare accordingly to perform on the test. The release of sample test questions made such a test even more predictable. On the down side, classroom teaching reflected this kind of fragmented, targeted attention to individual objectives rather than attention to connections among standards or overall mathematical thinking or reasoning. Scores were reported for every student on every objective or standard. Criterion-referenced tests thus seemed quite reasonable and, allegedly, it was possible for all students to do well, instead of previous norm-referenced models where half the students would always be "below average." Unfortunately, the theoretical measures used in test development and scoring were often based on beliefs about the fixed nature of intelligence and the predictability of student performance. Thus, the potential of criterion-referenced tests to accurately reflect student learning remained somewhat debatable.

Today, testing experts are questioning the limitations of previous evaluation models, including criterion-referenced testing. A new model, *evidence-centered design*, allows for testing connections across standards and takes into account the nature of student thinking the test aims to elicit. Within evidence-centered design, there is room for new formats for test questions, new ways to incorporate technology, and new ways of reporting results. Most of all, this new model allows us to assess multiple standards within a single task on a test, requiring students to draw on their knowledge not only of individual standards, but also of connections between standards. The resulting questions can be structured to call for students to demonstrate mathematical thinking and reasoning in unprecedented ways. Evidence-centered design also allows test developers to create large banks of such tasks, including extended-response performance tasks where students have to solve complex problems and write their explanations of solutions. We can be cautiously optimistic that evidence-centered design can yield better tests even as we watch carefully to monitor and adjust how well the testing industry implements the theory in both state accountability tests and across the broader testing field.

SHIFT 2: NEW TEST FORMATS

We can also be optimistic that the rhetoric associated with new tests being developed today will result in tests that do what they say they will do. In particular, the two major national assessments being developed to accompany the Common Core State Standards have taken into account an evidence-centered design model in their planning and test development (see PARCC and the Smarter Balanced Assessment Consortium in the "More to Consider" section of this message). Moreover, both tests promise to include a variety of machine-scored test item formats that go beyond single-answer multiple-choice items, such as multiple-answer multiple-choice, fill-in-the-blank, drag-and-drop, and interactive graphing on both the number line and a coordinate plane. Beyond these more challenging and more interesting types of short-answer items, both assessments promise in-depth, extended performance tasks that may require students to spend ten or twenty minutes or more on a single problem and may involve multiple questions about a single problem situation. We can hope that such advances, if they fulfill their potential, will also take hold for other types of tests, including national college-entrance tests and college and university placement tests.

SHIFT 3: MATHEMATICAL TECHNOLOGY AND THE TEST

While developers of most new tests are exploring delivery, administration, and scoring with the aid of computers, the use of mathematical technology as a problem-solving tool is still new territory. Existing tests have barely scratched the surface in terms of how to incorporate the use of computational, scientific, or graphing calculators; dynamic geometric construction software; or computer algebra systems. And new tools are being invented every day. The question cannot be whether to incorporate such tools; it's critical that tests send a message to schools equivalent to a permission slip, saying that using technology tools in teaching and learning *must* be part of what they do. More than that, the use of these tools allows us to ask more challenging, interesting questions and offer more engaging tasks than we could do without the tools. Schools need examples of the kind of higher-order tasks supported by mathematical tools like graphing calculators or computer algebra systems. We have to get past old thinking that calculators or even CAS are just there to help with hard computations. On a test involving performance tasks, students should be able to simulate situations, explore relationships, try out different approaches, represent ideas in multiple ways, visualize the symbolic representation of a problem in graphical form, and more. It's irresponsible, even foolhardy, not to take advantage of well-researched technology to raise the bar and increase the level of students' mathematical thinking and reasoning in the classroom. But it's even more irresponsible and foolhardy to try to raise the bar in this way within the classroom without likewise changing the nature of the tests that determine major decisions for students and their teachers—tests that too often today block the use of technology tools that could make higher-level thinking possible.

Interpreting Test Scores and Making Instructional Decisions

Regardless of how much we improve large-scale, high-stakes testing, we must always be careful in not over-interpreting the results of any single evaluation. Every test comes with limitations in terms of content, format, scope, and testing conditions. Combine these factors with student factors such as health, sleep, attitude, and anxiety, among others, and we have a recipe for a snapshot of student performance that may or may not accurately or fully reflect what a student knows or doesn't know. The best we can hope for is that a well-designed, well-aligned test will point toward areas the student may know or not know well. Beyond such broad summative observations, a teacher needs more formative, day-to-day, information about a student's thinking and learning in order to make sound instructional decisions. Despite any rhetoric to the contrary, large-scale tests simply do not provide the kind of focused, detailed information a teacher needs in order to make appropriate instructional decisions. To make such decisions, teachers instead, or additionally, need to formatively assess students by paying attention to student work, asking questions about students' thinking, and using a variety of informal and formal methods to monitor student learning on a day-to-day basis and adjust teaching as needed.

What Can We Do?

Over the years, we've seen a few cases where tests have driven positive change. One state's end-of-course Algebra 1 test several years ago required the use of graphing calculators to address certain problems on the test. Within a couple of years, algebra teachers across the state were learning how to use graphing calculators to ask different kinds of questions and to have students solve more complex problems than had been possible without the calculators. Teachers realized that their students would be disadvantaged if they did not know how to use these tools when it came time for the test. And a handful of states experimented in the 1990s and early 2000s with incorporating open-ended problem-solving questions on their state accountability tests, thus encouraging teachers to teach in deeper ways than they might have taught when preparing for tests consisting solely of multiple-choice questions. Unfortunately, the additional cost of administering and scoring such tests led to the elimination of most or all of them over time. What we lacked then perhaps we can put into action now—a coming-together of minds and the mobilization of a national will to finally bring testing into alignment with the kind of mathematics teaching and learning we say we value.

The critics of high-stakes tests are undoubtedly right that these tests occupy too much of our time and attention in schools. Yet, for the foreseeable future, we are likely to continue to see the effect of such tests on what and how we teach. As professional educators and responsible adults who care about young people, our first responsibility is to raise our voices when necessary to ask for reasoned use of tests and test scores and to call for decisions to be based on more than one piece of data. Numerous organizations have joined forces in calling for just that; see, for example, FairTest's posting of a multiorganizational national resolution on high-stakes testing (http://fairtest.org/national-resolution-highstakes-testing). As we raise our voices in support of high-quality mathematics teaching and learning, let us also demand tests that better reflect our priorities in schools and the outcomes we want for students.

Meanwhile, there is no shortage of criterion-referenced standards-based tests claiming to be aligned with the Common Core standards. The question is whether they will be any different from the prevailing model of economical, often superficial, standardized tests being used today to make high-stakes decisions. And, if they are different, will states and other administering agencies be willing to invest in the inevitable additional costs of tests that do more than offer single-answer, multiple-choice questions?

More locally, let us exercise caution in interpreting test scores and recognizing their limitations. Teachers can help students put tests and test scores in perspective by focusing teaching and learning around mathematical thinking, reasoning, and problem solving, even when these may not be the focus of the test students will face in the spring, and by relying on ongoing formative assessment to guide instructional decisions. If given the choice between teaching the kinds of low-level questions likely to appear on a test and teaching students mathematical habits of mind like thinking and reasoning, I would argue for the latter. I'm convinced that students who can think mathematically and figure things out will do at least as well on almost any test, even on a multiple-choice test of superficial knowledge and skills, as students who have practiced test-like items. And I'm even more convinced that students who can think mathematically and figure things out are far better prepared for their future than students who have spent most of their time preparing for the test.

We may not be able to change the well-established reality that tests drive instruction. But we can push for tests that drive instruction in appropriate ways. And we can advocate responsible and reasonable use of test results, realizing that a test score is merely a snapshot of student learning, not a single, absolute measure of what a student knows or how the student thinks. Let us continue to raise the bar in our mathematics classrooms to help every student learn the kinds of mathematical habits of mind that will serve them throughout their lives. And let us not accept any test that interferes with that goal.

By confusing means and ends, by making testing more important than learning, today's students are held hostage to yesterday's mistakes.

—Lynn Steen (as quoted in *Everybody Counts, MSB and NRC* 1989)

Reflections and Discussion

FOR TEACHERS

- What issues or challenges does this message raise for you? In what ways do you agree with or disagree with the main points of the message?
- How can you reasonably prepare students to succeed on your high-stakes accountability test without disrupting teaching and learning?
- How open are you to the possibility that teaching students mathematical habits of mind, rather than practicing test-like items, might help them both perform better on the test and learn more for their future?
- How confident are you that the high-stakes test(s) your students will face are appropriate measures of what you think is important for them to learn? If this is not the case, how can you advocate for more appropriate tests?
- What kinds of formative assessment tools help you gather evidence about student learning to inform instructional decisions?
- How can you help students and families make sense of test scores within a broader perspective of mathematical learning?

FOR FAMILIES

- What questions or issues does this message raise for you to discuss with your son or daughter, the teacher, or school leaders?
- How can you get the most accurate information on how well your daughter or son is doing in math without relying on annual test scores?
- Where can you find out more information about what the test scores mean that your son or daughter brings home?
- How can you advocate for more appropriate tests that support higher-level thinking and for reasonable use of test scores as only one piece of data about student learning?

FOR LEADERS AND POLICY MAKERS

- How does this message reinforce or challenge policies and decisions you have made or are considering?
- How confident are you that the high-stakes test(s) your students will face are appropriate measures of what you think is important? If this is not the case, how can you advocate for more appropriate tests?

- How can you support teachers in giving them permission to teach mathematical habits of mind rather than low-level skills, even if the latter is what students will see on a high-stakes test?

RELATED MESSAGES

Smarter Than We Think

- Message 14, "Effectiveness and Efficiency," looks at investing instructional time wisely, including not being interrupted with extensive test preparation.
- Message 19, "How to Know What They Know," discusses assessment in general and encourages us to consider the costs and benefits in determining how much time to spend preparing for and administering various types of tests.

Faster Isn't Smarter

- Message 19, "Embracing Accountability," calls for us to welcome accountability and also raise our voices as professionals about reasonable ways to demonstrate student learning without interfering with it.
- Message 11, "Weighing Hens," cautions against excessive benchmark testing and looks at the benefits and limitations of various types of testing.
- Message 35, "Putting Testing in Perspective," considers the nature of tests and how we can best prepare students for accountability tests and make sense of the results.

MORE TO CONSIDER

- *The Death and Life of the Great American School System: How Testing and Choice Are Undermining Education* (Ravitch 2010) considers the unintended negative consequences of large-scale, high-stakes accountability testing.
- *The Case Against Standardized Testing: Raising the Scores, Ruining the Schools* (Kohn 2000) raises serious questions about the dangers to students of relying on high-stakes standardized testing to make critical decisions about students, teachers, and programs.
- "Ten 'Must-Know' Facts About Educational Testing" (Popham) provides useful information to parents and the lay public about

high-stakes tests and how to make sense of test scores. www.pta.org/programs/content.cfm?ItemNumber=1724.

- "Testing Our Schools: A Guide for Parents" (Beaupré, Nathan, and Kaplan) offers tips for making sense of high-stakes testing issues, including how test scores are interpreted and used. www.pbs.org/wgbh/pages/frontline/shows/schools/etc/guide.html.
- "The Case Against High Stakes Testing" (FairTest) advocates changing the practice of using high-stakes tests to make critical decisions for students, teachers, and programs. www.fairtest.org/arn/caseagainst.html.
- "A Brief Introduction to Evidence-Centered Design: CSE Report 632" (Mislevy, Almond, and Lukas 2004) is an introduction to evidence-centered design, a new model for large-scale tests.
- "The Ethics of Using Computer Algebra Systems (CAS) in High School Mathematics" (Usiskin 2012) considers issues related to the use of CAS, including implications from and for testing. http://ucsmp.uchicago.edu/resources/conferences/2012-03-01.
- "Computer Algebra Systems and the Challenge of Assessment" (Brown 2001) considers the use of Computer Algebra Systems (CAS) in high-stakes assessments, including current practice and future recommendations regarding the use of CAS to support increased mathematical thinking of students.
- Charles A. Dana Center at The University of Texas at Austin, "PARCC Prototyping Project," provides prototypes of assessment tasks associated with the PARCC assessment, including some designed to address mathematical practice standards. ccsstoolbox.org
- *This Is Only a Test: Teaching for Mathematical Understanding in an Age of Standardized Testing* (Litton and Wickett 2008) offers strategies for helping students succeed on large-scale tests while teaching for understanding.
- *The Mismeasure of Man, revised and expanded* (Gould 2012) presents the author's attack on the measurement of intelligence via IQ scores and on standardized tests in general, and argues in support of a growth model of intelligence.
- "Smarter Balanced Assessment Consortia" (Washington State Office of Superintendent of Public Instruction); the Smarter Balanced Assessment Consortia is one of two national assessment consortia developing large-scale accountability tests to accompany the Common Core State Standards. www.smarterbalanced.org.
- PARCC, the Partnership for Assessment of Readiness for College and Careers, is one of two national assessment consortia developing large-scale accountability tests to accompany the Common Core State Standards. http://parcconline.org.

29

Finding Great Teachers

HELPING TEACHERS GROW THEIR MINDS
(PAUL'S STORY)

> Not everyone can be a great artist, but a great artist can come from anywhere.
>
> —Anton Ego, from *Ratatouille*

When I was a new mathematics coordinator for kindergarten through grade 12, my job included supporting mathematics teachers in any way I could. Although I didn't really intend to do so, over time I formed opinions through my visits and conversations with both teachers and principals about who the best teachers were. I recall thinking that one particular junior high school mathematics teacher seemed to be somewhat weak based on what I thought a good teacher acted like or did in the classroom. Paul was very quiet and not at all outgoing. However, my thinking changed as I passed by Paul's classroom just after the end of school one day. I noticed a group of students hanging around talking with Paul. I paused (lurked, actually) in the hallway for a while to see what might transpire. I watched as one of the students walked to a shelf and pulled out a couple of chess boards. Paul and the students started playing chess. I overheard parts of their conversation and realized that Paul was connecting with these students, most of whom were fairly shy, probably like he was at that age. As they played, Paul and the group talked a bit about the chess move someone had just made, a bit about school in general, a bit about mathematics and the work they were doing in class, and a bit about the students themselves.

I came away from that brief experience seeing Paul in a whole new light. He had conscientiously formed personal relationships with these students, and likely other students, who may well have not experienced such relationships with other teachers, and he did it over a game based on logic and reasoning. What he gave them may have seemed small, even to Paul, but I saw it as significant. During the traumatic years of adolescence, connecting with someone at school and feeling genuine caring from an adult can be a powerful motivator for coming to school and doing the work of learning.

The best educators know that the talents and intelligence of many students get overlooked and that every student has the potential to become smarter. The best leaders also know that the gifts of many teachers may get overlooked and that every teacher has the potential to become more effective.

Every Teacher Can Grow

We now know that any student's intelligence can grow and that effective teachers can choose tasks and structure classrooms in ways that support that growth for every student. For a leader working with teachers, we need to recognize that every teacher can also grow—in intelligence and in becoming more effective. Not long ago, I attended a presentation to a group of teacher educators at a state conference. A speaker from a major university in the state reported on a study that looked at how long new teachers stayed in teaching and how many left for different careers. She challenged the group to guess what was the primary reason teachers gave for choosing to leave. Contrary to the group's belief that teachers probably left the field because the work was too hard or the pay wasn't enough, the study had found that the major reason teachers gave for their decision to leave the profession was a lack of efficacy. They were frustrated in not being able to make the difference in students' lives that they hoped they would. Teachers generally are motivated to teach well and to help students learn. A teacher's drive to improve students' lives offers a good starting point for leaders in determining how to best support all teachers, not only those new to the field.

In his book, *What the Best College Teachers Do*, Ken Bain suggests that teaching involves "anything we might do that helps and encourages students to learn—without doing them any major harm" (2004, 173). His observation certainly holds for prekindergarten through grade 12 teaching as well. He notes the limitations of teaching as telling and suggests that "teaching occurs only when learning takes place" (173). Learning takes place in all kinds of classrooms with all kinds of teachers. We can identify factors or behaviors effective teachers seem to have in common, but good teachers don't all teach in the same way. Rather than a leader looking for a particular kind of teaching, we can start by believing that every teacher can be effective and can continue to grow and improve.

Thinking Positively (Acting as If)

The idea of *acting as if* has been used in self-improvement work for some time, including helping people improve athletic performance, overcome adversity, deal with a challenging time in a relationship, and other areas where someone might want to achieve a goal or make a change. *Act as if* we are the better team. *Act as if* I like my adolescent child, even when she drives me crazy. *Act as if* I will find a job. Often we can *act as if* long enough and effectively enough to work through a situation or internalize our behavior.

We want teachers to raise their expectations of all students—to believe in their intelligence and ability—to *act as if* these are smart students and offer them all appropriate learning opportunities to develop their intelligence. Maybe as leaders, we can also *act as if* we are working with smart, effective professional teachers who want to improve student learning. From that basis, we can continue to offer constructive ways for each teacher to improve, letting teachers know we believe they can do it.

Build on Strengths

Every teacher, including Paul, has strengths. One of Paul's strengths was his ability to connect with students who might not be outgoing or willing to speak up in class. He also prepared his lessons thoroughly and was well organized. He carefully attended to students' errors on quizzes and tests and tried to help them understand whatever they had misunderstood or done incorrectly. And he was willing to do whatever he could, during the school day or after, to let students know that he wanted them to learn and cared about them as people.

Every teacher brings certain strengths to his or her work. Some have engaging personalities and relate well with students. Some bring creativity to planning engaging lessons. Some are good at generating questions that push students' thinking. Some are effective facilitators of classroom discourse, drawing out participation from all students in thinking about a problem. Some are informal leaders or mentors, to whom other teachers turn for advice or guidance. Some are voracious consumers of research and professional readings, staying on top of what might improve their students' learning. Whatever strengths a teacher brings to teaching, those strengths and talents offer a foundation to build improved skills in other areas. Educators have learned that when we tap into students' strengths in learning mathematics, rather than focusing primarily on remediating deficiencies, students will have more confidence and become more open to learning. We can apply this same philosophy to helping teachers grow.

What Can We Do?

Paul participated in every professional development opportunity offered to him. He was open to having me visit his classroom and give him feedback. But, frankly, I was inexperienced at the time and I'm not sure I gave him the support I now know can help a teacher grow. In the decades since my start as a new mathematics supervisor working with Paul and other teachers, we have learned much about effective coaching, mentoring, peer teaching, and professional learning. Andy Hargreaves and Michael Fullan note that "getting good teaching for all learners requires teachers to be highly committed, thoroughly prepared, continuously developed, properly paid, well networked with each other to maximize their own improvement, and able to make effective judgments using all their capabilities and experience" (2013, 3). Several of these factors are well addressed in collegial teams or within professional learning communities. By collaborating with colleagues, often around student work, committed teachers learn from each other's strengths, support each other in improving areas that aren't as strong, and together extend their capabilities to improve their practice more significantly than any of them could do alone.

It may well be that some teachers will determine that they need to find another place or a path other than teaching. By focusing on teachers' strengths, giving them opportunities, and working to help them learn and grow, these individuals may decide that becoming an excellent teacher is not a goal they choose. For other teachers, the same approach will indeed help them become continually more effective at their chosen field.

As leaders, we can accept the task of helping every teacher identify his or her strengths and build on those strengths, no matter whether the teacher is highly proficient or still learning the ropes. By supporting professional learning communities, recognizing the hard work of all teachers, advocating for appropriate compensation and working conditions, and ensuring the availability of high-quality, job-embedded professional development, we can make sure teachers continue to grow and improve, while at the same time helping them stay excited about the profession they've chosen. We can offer the help of coaches and mentors for teachers without strong mathematics backgrounds.

Good teachers, just like smart students, can be found wearing many disguises. Some brilliant (or potentially brilliant) mathematics students may be tuning out in the back row of a seventh-grade classroom. Some outstanding (or potentially outstanding) teachers may be quietly slipping away in a school where they aren't supported. Strong leaders will find ways to bring out the best in those teachers so that they, in turn, bring out the best in their students.

Reflections and Discussion

FOR TEACHERS

- What issues or challenges does this message raise for you? In what ways do you agree with or disagree with the main points of the message?
- How do you build on the strengths of your students to help them learn mathematics?
- What do you see as your strengths as a teacher? How can you use those strengths to continue to grow and improve in areas you don't see yourself as strong?
- What would you most like in terms of support from the leaders, coaches, supervisors, or others you work with? How can you increase the likelihood that you will receive what you need from them?

FOR FAMILIES

- What questions or issues does this message raise for you to discuss with your daughter or son, the teacher, or school leaders?
- What do you see as your son's or daughter's strength(s) as a mathematics student? Does he or she agree with you? How can you reinforce his or her strength(s) to help when facing a challenging problem?
- What else can you share with the teacher about your daughter or son to support the teacher in helping your daughter or son learn mathematics?
- What does your son or daughter like most about his or her mathematics teacher? When you talk with the teacher, how can you acknowledge what your son or daughter sees positively in class?

FOR LEADERS AND POLICY MAKERS

- How does this message reinforce or challenge policies and decisions you have made or are considering?
- How can you help teachers identify their strengths? How can you help teachers build on their strengths as they work on areas where they can become stronger?
- How can you support the positive use of professional learning communities to help all teachers learn to improve their practice so that their students improve their mathematics learning?

RELATED MESSAGES

Smarter Than We Think

- Message 1, "Smarter Than We Think," provides background on mindsets and a growth model of intelligence applicable to both students and adults.
- Message 30, "Walking the Walk," asks leaders to work with teachers in the same ways we ask teachers to work with students.

Faster Isn't Smarter

- Message 41, "Thank You, Mr. Bender," acknowledges great teachers who make a difference.
- Message 39, "Standing on the Shoulders …," reminds us of the power of collaboration and learning from others.
- Message 40, "Seven Steps Toward Being a Better Math Teacher," offers thoughts about how to continue to learn and grow.

MORE TO CONSIDER

- *Professional Capital: Transforming Teaching in Every School* (Hargreaves and Fullan 2013) suggests that making lasting improvement in schools depends on investing in professional capital—primarily in high-quality teachers and teaching.
- *David and Goliath: Underdogs, Misfits, and the Art of Battling Giants* (Gladwell 2013) describes how people who may seem weak or deficient might have hidden talents that come out under the right circumstances.
- *Outliers: The Story of Success* (Gladwell 2008) explores the lives of prodigies and geniuses to consider what it takes to succeed at high levels.
- *Common Core Mathematics in a PLC at Work, Leader's Guide* (Kanold and Larson 2012) accompanies a series of related grade-band books to help leaders facilitate the use of professional learning communities as an effective tool for implementing the Common Core standards.
- "Mathematics Professional Development" (Doerr, Goldsmith, and Lewis 2010) summarizes research about effective professional development to help teachers improve their practice and help students learn mathematics.
- *Focus on the Good Stuff: The Power of Appreciation* (Robbins 2007) reminds us of the power of paying attention to what is good about people and ourselves.

- *What the Best College Teachers Do* (Bain 2004) offers insights into excellent teaching relevant for prekindergarten through grade 12 as well as higher education.
- *Frames of Mind: The Theory of Multiple Intelligences* (Gardner 2011) describes different types of intelligence and notes that people may be stronger in some areas than others.
- *Strengths Based Leadership: Great Leaders, Teams, and Why People Follow* (Rath and Conchie 2008) offers insights we might learn from the business world about leading and supporting teams based on identifying and building on people's strengths.

30 Walking the Walk

CAN WE DO WITH TEACHERS WHAT WE ASK TEACHERS TO DO WITH STUDENTS?

> It is not fair to ask of others what you are not willing to do yourself.
>
> —Eleanor Roosevelt

In my first role as a district mathematics coordinator, I had the privilege of working for one of the best leaders I've ever known. Every day he demonstrated how to nurture and support those of us working with him. He taught his small team, all of us recently from the classroom, how to become leaders ourselves, overseeing the district's instructional program and, most of all, working with teachers. Since that excellent beginning, I've had the opportunity to work in various roles with school systems and organizations, including doing what I could to support teachers in improving mathematics teaching and learning. I've returned to the classroom twice to remind myself how challenging it is to help students learn to think and reason mathematically. I've reflected on what we ask teachers to do to help students. More recently, I've been reflecting on whether, as leaders, in addition to just *talking the talk*, we can also *walk the walk....* Can we do with teachers what we ask teachers to do with students?

Don't Tell Everything

We ask teachers not to tell students everything we think they need to know. We have learned that this kind of one-way delivery of information does not always stick for students. Rather, we know that engaging students in tackling a hard problem can be valuable, even if students don't know in advance all the rules and procedures they may want to know in order to solve the problem (see the teaching model described in Message 12, "Upside-Down Teaching"). Students are more likely to be open to learning mathematics when they have spent some time and energy working on a problem first.

Likewise, can we agree that it may not always be desirable to tell teachers everything we think they need to know, especially when dealing with a problem? For example, if a school or district is facing a need to make some kind of change to address a problem of student performance or unexpected implementation of a new mandate, rather than telling teachers what the administration has decided to do to address the problem, consider engaging teachers in the problem. Invite teachers' participation in solving the problem. In addition to raising their interest in the solution, they may even arrive at strategies the administration didn't consider.

Develop Reasoning and Thinking Skills

We ask teachers to help students develop reasoning and thinking skills (see, for example, Message 31, "Developing Mathematical Habits of Mind" and Message 32, "Problems Worth Solving"). In order for students to be prepared with the most important mathematical habits of mind that will equip them for twenty-first-century study and employment, they need to be able to figure things out, think mathematically, and reason through increasingly complex situations.

Likewise, can we agree that teachers need the opportunity to further hone their thinking and reasoning skills? For example, an excellent activity for a professional learning community is to examine student work and engage in collaborative discussion about the work. Sometimes teachers can discover patterns in misconceptions that a teacher can use to redirect a student. Sometimes they may discover that one teacher's students seem to understand a concept better than students in other classes. By thinking together and discussing the differences, teachers may uncover ways to adjust their teaching approach to help all students. In the process, teachers not only become better thinkers, they also can become more effective teachers.

Listen

We ask teachers to engage in conversation with students and listen to what they have to say, especially as an informal formative assessment tool (see the kinds of formative assessment described in Message 19, "How to Know What They Know" and Message 13, "Clueless"). When a teacher lets a student explain his or her thinking and really listens to the explanation, the student sometimes catches his or her own mistake or may reinforce a correct approach. Or the teacher may be able to notice an emerging or existing misconception that the teacher can address.

Likewise, can we agree that it's important to give teachers opportunities to talk with their supervisors, coaches, mentors, or administrators? More than that, can we agree that it's important for leaders to *listen* to teachers? Teachers are the professionals on the front line, expected to implement school or district plans and essentially responsible for the mathematics learning of their students. Their perspective is critical for any leader in making decisions about current challenges or future directions. Teachers will not only feel valued when their input is considered, they will often be able to help shape a solution to a problem or help forge a future direction.

Reflect

We ask teachers to help students learn to reflect on their work—to zoom out and look metacognitively at their thinking. We ask teachers to help students construct explanations and analyze and critique the thinking of others (see, for example, the importance of constructing and critiquing in Message 34, "Who's Doing the Talking?"). Having students routinely look deeply at their own thinking and the thinking of others and discuss what they notice helps them strengthen their grasp of mathematical ideas and gives them opportunities to communicate with and about mathematics.

Likewise, can we agree that it's valuable to give teachers opportunities to reflect on their teaching practice and to participate collaboratively in helping colleagues reflect on their practice? The practice of lesson study, where peers collaboratively plan, observe, and debrief a mathematics lesson offers a beautiful opportunity for this kind of reflection and collaborative professional critiquing. Likewise, an effective professional learning community provides an excellent venue for routinely reflecting and critiquing collaboratively. Setting norms and expectations for this kind of ongoing professional activity reinforces teachers' view of themselves as true professionals and also contributes in concrete ways to shifts in teaching practice and positive benefits in terms of student learning. Reflecting isn't a bad practice for leaders, as well, in considering how to improve their effectiveness.

Encourage, Nurture, and Support

We ask teachers to encourage, nurture, and support their students (see, for example, several related messages in Part I, including Message 1, "Smarter Than We Think" and Message 8, "Oops!"). Many students have not had the opportunity to develop their mathematics potential or become as smart as they can be. We now know that we can support students' mathematical learning and continuing growth by giving them opportunities to work with challenging problems and by recognizing their hard work as they tackle problems. We can pay attention to their learning on a daily basis to help them focus on what is most important and to help them redirect their attention before they solidify potential misconceptions. Perhaps most important, we can establish personal relationships with students to help them feel connected to school and take ownership in their own learning.

Likewise, can we agree that we can encourage, nurture, and support our teachers? In today's world of high-stakes accountability and seemingly endless mandates, it's easy for any teacher to become discouraged. Sometimes it seems that no matter how hard they try, something gets in the way of a teacher being able to do his or her best work to help students learn the kind of mathematics and mathematical thinking they need. As leaders, we can find ways to recognize teachers' hard work in meaningful ways, provide opportunities for them to continue to learn and grow, and support them in as many ways as possible. We can pay attention to the number of mandates, forms, requirements, and restrictions teachers receive, perhaps even serving as a filter or central source to coordinate such communications with teachers. During that first year of my job as a new mathematics coordinator, a wise principal took me aside for a conversation. In a warm and nurturing way, he suggested that perhaps I could spend less time being formal with what I perceived as my job responsibilities and spend more time "hanging out" in the teachers' lounge' getting to know his teachers personally. It was some of the best advice I ever received. The relationships I formed helped us all work through the inevitable challenges of implementing a comprehensive new mathematics program, among other things. These relationships also made my work more rewarding for me and, I hope, for the teachers doing the hard work every day with students.

What Can We Do?

Just as we ask teachers to believe in the ability of every student to learn, can you as a leader believe that every teacher has the talent and ability to be an excellent teacher? We can surely invest in teachers in ways we ask them to invest in their students. We can give them opportunities to engage in the real problems of schools and to think with colleagues about professional issues. We can talk with them and really pay attention to what they have to say. We can help them grow as professionals by leading them in learning how to reflect on their work and the work of their colleagues. And we can encourage, nurture, and support them as human beings on our team.

Essentially we want to treat teachers as professionals. These teacher-focused behaviors reinforce our commitment to the professional educators who are part of our team—the part of our team that makes the biggest difference in the lives of the students we all serve.

Reflections and Discussion

FOR TEACHERS

- What issues or challenges does this message raise for you? In what ways do you agree with or disagree with the main points of the message?
- How well do you think you accomplish the five goals described in this message for teachers to accomplish with students?
- How well do you think your leaders (administrators, supervisors, coordinators, coaches, and so on) accomplish these goals with you and your colleagues? What can you do to increase the likelihood that you will have opportunities to grow in the ways described here?

FOR FAMILIES

- What questions or issues does this message raise for you to discuss with your son or daughter, the teacher, or school leaders?
- How can you get to know your daughter's or son's teacher and let them know that you support their efforts to help your daughter or son learn?
- How can you apply the five goals in this message to supporting your son or daughter?

FOR LEADERS AND POLICY MAKERS

- How does this message reinforce or challenge policies and decisions you have made or are considering?
- Which of the five goals described in this message do you think you accomplish the best with the teachers you serve? Which of the goals do you think you can improve, and how might you proceed to do so?

RELATED MESSAGES

Smarter Than We Think

- Message 21, "Moving from Where We Are to Where We Want to Be," provides guidance on facilitating educational change.
- Message 29, "Finding Great Teachers," discusses supporting every teacher in becoming more effective.
- Message 1, "Smarter Than We Think," provides background on mindsets and a growth model of intelligence that applies to both students and adults.
- Message 31, "Developing Mathematical Habits of Mind," discusses the kinds of thinking, reasoning, and mathematical habits of mind advocated in this message for both students and teachers. Several other messages noted throughout this message address particular aspects of teachers' work with students.

Faster Isn't Smarter

- Message 29, "The Evolution of a Mathematics Teacher," looks at how mathematics teachers grow over time.
- Message 13, "Seek First to Understand," reminds us of the power of working together.

MORE TO CONSIDER

- *The Moral Imperative of School Leadership* (Fullan 2003) focuses on the importance of personal leadership, especially in the role of school principal.
- *Professional Capital: Transforming Teaching in Every School* (Hargreaves and Fullan 2013) suggests that making lasting improvement in schools depends on investing in professional capital—primarily in high-quality teachers and teaching.
- *Good to Great: Why Some Companies Make the Leap ... And Others Don't* (Collins 2001) proposes now–widely accepted principles for business leaders; the accompanying monograph, *Good to Great and the Social Sectors: Why Business Thinking Is Not the Answer* (Collins 2005), relates and adapts principles to the social sector, including education.
- *Lesson Study Research and Practice in Mathematics Education: Learning Together* (Hart, Alston, and Murata 2011) summarizes research and describes current practice about *lesson study*, an important professional tool for improving mathematics teaching and learning.
- "Walking the Walk" (Seeley 2010b) is my five-minute talk that inspired this message. www.youtube.com/watch?v=RwCc8po4Vkc.

IV

Messages About Thinking Mathematically in a Common Core World

MATHEMATICAL PRACTICES AND MORE

31

Developing Mathematical Habits of Mind

LOOKING AT THE BACKGROUND, CONTEXT, AND CONTENT OF THE COMMON CORE STANDARDS FOR MATHEMATICAL PRACTICE

The only way to know mathematics is to do mathematics.

—Paul Halmos (*I Want to Be a Mathematician*, 1985)

In a fascinating glimpse into his mind, Paul Halmos (1985) describes his journey as a learner, teacher, and practitioner of mathematics. His description reflects a career of sometimes elegant and often messy exploration with mathematical ideas, properties, problems, and theorems, including wrong turns, dead ends, and unproductive thinking. His depiction of his life as a mathematician reflects the insights, *Aha!* moments, epiphanies, and serendipities that resulted from analyzing what worked and what didn't. He shares the value of interacting with colleagues, persevering through difficult paths that often required considerable amounts of time, and sometimes stepping away from a problem for a while. In looking at this image of what a "real" mathematician does, we can see the power of learning from mistakes and of backing up to reflect, consider, analyze, regroup, redirect, and move forward, building on curiosity and a desire to find solutions. What he describes in a very personal way are the *mathematical habits of mind* we want every student to develop.

What Are Mathematical Habits of Mind?

For years, mathematicians, educators, and other experts have tried to describe the heart of what it means to do mathematics and think mathematically, often using terms like *mathematical habits of mind, mathematical processes,* or *mathematical practices.* Students who learn only mathematical facts, definitions, rules, and procedures may do fine on large-scale tests that address these relatively easy-to-score elements of mathematics. But many of these same students later find that they cannot use what they know when they encounter any problem or situation they haven't specifically learned how to solve. On the one hand, we lament the poor preparation of students who can't apply what they've learned, but on the other hand, too often we continue to cling to the old notion that mathematics consists primarily of a checklist of knowledge and skills.

There is no one correct or complete list of mathematical habits of mind. Many descriptions overlap or address similar aspects of the nature of mathematics. (See the "More to Consider" section of this message for several descriptions and lists of mathematical habits of mind as conceptualized by various experts.) Almost all descriptions of mathematical habits of mind, mathematical thinking, practices, or processes center on a person's ability to solve mathematical problems, especially those that go beyond simple word problems related to a recently learned procedure.

Closely connected to solving problems is the ability to explain one's thinking and engage in productive discourse with others about the problem or observations about the mathematics in the problem. Thus, almost all discussions of mathematical habits of mind involve dimensions of thinking and reasoning. Some descriptions of mathematical habits of mind build from general intellectual habits of mind, such as perseverance, persistence, listening and communication skills, or metacognitive skills like reflection and analysis. Others may be uniquely associated with mathematics, such as considering multiple ways of representing mathematical ideas, zooming in and zooming out on particular aspects of a problem and on the problem as a whole, the ability to connect ideas within and outside of mathematics, making conjectures and generalizations, understanding the structure of mathematics, considering mathematical relationships, justifying and explaining mathematical solutions, and so on. These habits of mind span grade levels and ages; students can develop and demonstrate them in appropriate ways from their earliest experiences with mathematics. Given the right kinds of opportunities, a student's level of expertise in using mathematical habits of mind will increase year after year, ideally with students graduating from high school having developed a powerful set of mental abilities.

Connecting Mathematical Habits of Mind and the Common Core State Standards

The design of the Common Core State Standards includes both Standards for Mathematics Content and Standards for Mathematical Practice in acknowledgment of what mathematicians and mathematics educators have recognized for years—that it is not possible to be knowledgeable about mathematics if all a person knows is mathematical content. The essential partner to mathematical content is a set of mathematical ways of thinking and reasoning that can equip a person to navigate through hard or unknown mathematical territory.

The Common Core's descriptions of the Standards for Mathematical Practice address many, if not most, of the dimensions of mathematical thinking and habits of mind articulated in previously published discussions. Thus, considering these practices can give us a good, broad overview of the nature of mathematical habits of mind essential for today's students. In considering these practices, we should also keep in mind excellent recommendations from other sources in recent years, most notably the process standards from the National Council of Teachers and the mathematical proficiencies described in *Adding It Up* (National Research Council 2001). Both of these sources are acknowledged in the Common Core State Standards documents, and the writers have also considered other important discussions of mathematical habits of mind listed in the "More to Consider" section of this message, especially the work done by Al Cuoco, E. Paul Goldenberg, and June Mark (2010).

PRACTICES

The Common Core State Standards' explicit attention to mathematical habits of mind is represented by the Standards for Mathematical Practice (NGA Center and CCSSO 2010, 6). Increasingly these critical practices are being recognized as central to the goals of mathematics teaching and learning and many consider the practices the most central component of the standards. One vision of the Standards for Mathematical Practice groups the eight standards into four related pairs (see Figure 31.1).

In this organization of the eight Standards for Mathematical Practice, the four major types of mathematical practices are:

- reasoning and explaining,
- modeling and using tools,
- seeing structure and generalizing, and
- overarching habits of mind of a productive mathematical thinker, including problem solving and communication.

One of the important messages in this graphic is a reminder that the eight practices may not be as discrete as they initially appear. Rather, they function together, not only as pairs of standards, but as a cohesive set of descriptors contributing to our notions of what mathematical habits of mind we hope to help every student develop.

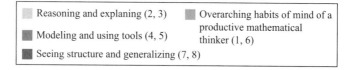

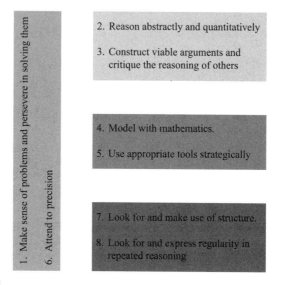

Figure 31.1

Common Core Eight Standards of Mathematical Practice

Source: http://commoncoretools.me/2011/03/10/structuring-the-mathematical-practices

PROCESSES

As we explore these practices, consider NCTM's standards for mathematical processes:

- Problem Solving
- Reasoning and Proof
- Communication
- Connections
- Representations

In thinking about how the eight CCSSM Practices relate to NCTM's five process standards, groups of experts and practitioners are likely to arrive at different ways of cross-matching the two sets of standards, demonstrating how overlapping and nondiscrete any list of mathematical habits of mind is likely to be. One model from *Connecting the*

NCTM *Process Standards and the CCSSM Practices* (Koestler, Felton, Bieda, and Otten 2013) shows the following matches:

NCTM Process Standards	CCSSM Practices
Problem Solving	1, 2, 4, 5
Reasoning and Proof	1, 3, 8
Communication	1, 2, 4, 6
Connections	1, 2, 4, 7, 8
Representations	1, 2, 4, 5, 6, 7

It's far less important to identify which particular standard(s) a given problem or practice addresses than it is to look for opportunities to focus on and help students develop one or more of the practices within the context of the problem. In fact, a valuable professional learning experience, especially among colleagues or within a professional learning community, can be to do a matching among these two sets of standards for mathematical habits of mind, considering the intent of each standard and what each standard seems to address most directly. My own matching differs somewhat from the one above, but agreeing on a list is much less important than the discussions that can arise as individuals and colleagues consider the standards in depth as related to their own work.

PROFICIENCY

The National Research Council's *Adding It Up* (2001) offers a vision of *mathematical proficiency* that echoes many of the same notions as the practices and processes described earlier. The NRC identifies five components describing what is necessary for a person to learn mathematics successfully:

- *conceptual understanding*—comprehension of mathematical concepts, operations, and relations;
- *procedural fluency*—skill in carrying out procedures flexibly, accurately, efficiently, and appropriately;
- *strategic competence*—ability to formulate, represent, and solve mathematical problems;
- *adaptive reasoning*—capacity for logical thought, reflection, explanation, and justification; and
- *productive disposition*—habitual inclination to see mathematics as sensible, useful, and worthwhile, coupled with a belief in diligence and one's own efficacy.

Looking at these strands of proficiency, the first two—conceptual understanding and procedural fluency—seem to address the kind of mathematics knowledge and skills most often represented in state mathematics standards. The last three strands—strategic competence, adaptive reasoning, and productive disposition—seem to reflect mathematical habits of mind related to solving problems, reasoning, justifying, and persistence and willingness to tackle mathematical problems, as well as confidence. As in other discussions of mathematical habits of mind, the NRC notes that these five strands are interwoven and interdependent, a notion they try to represent graphically using a rope metaphor (see www.nap.edu/openbook.php?record_id=9822&page=117).

Incorporating Mathematical Habits of Mind

Through in-depth consideration of the Common Core Standards for Mathematical Practice and these processes and strands of proficiency, educators can begin to determine how mathematical habits of mind might be developed in their particular schools and classrooms. Ideally, we can learn how to seamlessly incorporate these habits into our mathematics programs so that both teachers and their students come to routinely view mathematics as a rich, powerful, and useful set of thinking and analytical tools they can use to make sense of and tackle a wide variety of problems both in and outside of mathematics.

Many good mathematics tasks offer opportunities for students to use or develop mathematical habits of mind. Even some routine or procedural problems can offer opportunities for students to call on mathematical habits of mind if we ask appropriate questions to push students' thinking beyond an obvious or superficial response. Consider the following problem:

> *A store is advertising a sale with 10% off all items in the store. Sales tax is 5%. A 32-inch television is regularly priced at $295.00. What is the total price of the television, including sales tax, if it was purchased on sale?* (Charles A. Dana Center at The University of Texas at Austin 2012b)

This is a good applied problem—a fairly traditional word problem. To solve it requires a couple of steps, but the solution path is clear if a student understands percent, and the answer will either be right or wrong. Now consider the following extension of the same problem:

> *Adam and Brandi are customers discussing how the discount and tax will be calculated. Adam says that to find the total cost for any item in the store, you can take 10% off the original price, then add the sales tax to the discounted price. Brandi says that to find the total cost for any item in the store, you can determine the original price of the item, including sales tax, and*

then take 10% off. Are both Adam and Brandi correct? Justify your answer. (Charles A. Dana Center at The University of Texas at Austin 2012b)

The extended problem requires students to apply further mathematical reasoning, not just aiming at a numerical answer, but comparing two different procedures with subtle but important differences. Further, simply by asking students to justify their answer, we ramp up the thinking and reasoning involved.

Messages 32–39 in this section look more closely at each of the eight mathematical practices and urge us to consider how each might be addressed. Appendix A includes several sources for problems and tasks that allow for deep thinking, reflection, analysis, explanation, and reasoning, among other mathematical habits of mind.

Assessing Mathematical Habits of Mind

The fundamental idea of building mathematics programs grounded in mathematical habits of mind has been advocated for decades, but its actualization has sometimes eluded teachers, textbook authors, curriculum developers, standards writers, and test developers. Even though we can see the value in helping students develop ways of using mathematical thinking to make sense of their world and solve the many types of problems they will encounter, sometimes it simply seems too time consuming, expensive, or difficult to make real. In particular, we have not seen widespread use of appropriate assessments to support the teaching of mathematical thinking and habits of mind.

While the NCTM standards from both 1989 and 2000 included specific standards on mathematical processes, most state standards in the late 1990s and well into the twenty-first century consisted primarily of lists of mathematical content. If mathematical processes were addressed at all, they may have appeared in relatively invisible introductory paragraphs or accompanying narratives describing how important it was to incorporate problem solving, reasoning, and so on. Since state tests tended to focus on the standards themselves, rather than on the invisible paragraphs and narratives, rarely, if ever, did these dimensions of mathematical thinking appear on such tests.

Now, however, we see indications that the Common Core standards may bring with them the promise of a new era in assessment to support the mathematical thinking and habits of mind we value. In contrast to the widespread lack of attention on assessments in the past to mathematical habits of mind and mathematical processes, the two primary large-scale tests designed to accompany the standards have indicated a commitment to focus primarily on these practices, at least in rhetoric and intention (PARCC/SBAC). It may take several years for these tests and other state and local assessments to realize the full power of inclusion of these elements of mathematical thinking. While there are noteworthy examples of assessing mathematical thinking in the classroom and in some curriculum programs, this is new ground for large-scale testing. If high-stakes tests can measure and reward deep aspects of mathematical thinking, reasoning, and problem solving, perhaps teachers will feel that they're allowed to teach toward these habits of mind.

What Can We Do?

Being a mathematician is no more definable as "knowing" a set of mathematical facts than being a poet is definable as knowing a set of linguistic facts.... Being a mathematician, again like being a poet, or a composer or an engineer, means doing, rather than knowing or understanding.

—Seymour Papert ("Teaching Children to Be
Mathematicians Versus Teaching
About Mathematics," 1972)

The work of mathematicians who *do* mathematics as Seymour Papert describes involves thinking, reasoning, looking for patterns, noticing and connecting elements of structure, and solving complex problems using mathematical tools, among other things. It's only when our students engage in actually doing mathematics—working on hard problems, engaging in discussion, arguing, explaining, interacting with mathematical ideas and paying attention to their thinking as they do so—that they come to know mathematics well and develop a positive disposition toward the subject. Becoming proficient at particular habits is not so much an end goal as a lifelong journey. Even professional mathematicians continue to hone and refine these habits throughout their career. And there is no profession where this kind of lifelong learning and growth is a higher priority than for teachers of mathematics, regardless of the level or age of students they teach. Teachers not only become better at helping students learn mathematics; they serve as powerful role models. If students are to develop mathematical habits of mind in ways that will serve them in the future, then we need to examine our curricula, assessments, and instruction in light of such habits of mind. We can also help students themselves become aware that the purpose of their mathematics learning is much more than the skills, facts, procedures, and even concepts they learn. When we do so, we not only improve students' understanding and proficiency, we also improve their attitudes toward mathematics and their interest in doing more of it.

The Common Core State Standards provide us with a unifying vehicle to help students develop the crucial habits of mind they need in order to learn mathematics well and, especially, in order to use what they learn after they leave school. The Standards for Mathematical Practice offer a new structure for understanding mathematical habits of mind. As we work to implement these eight practices, we need to be patient with ourselves and remember that the practices involve mathematically sophisticated ideas that are not always easy to understand, even for those with a mathematics background. It will take learning and effort to implement the practices well. Mathematics is a science of patterns, and we can look for patterns ourselves as we make sense of the Standards for Mathematical Practice. The patterns we find might take the form of common themes such as:

- looking for, articulating, and using patterns in mathematics (to make generalizations, to recognize mathematical structure, to solve problems, and so on);
- learning to reason and make sense of mathematics (reasoning takes many forms and crosses many practices);
- zooming out and zooming in (backing up to look at the big picture of a concept, problem, or connecting topic, and focusing back in on the specifics); and
- representing mathematical situations and ideas in many ways and moving back and forth between representations.

Mathematics is also held together by a web of connections. As we continue to learn about the practices and collaborate on how best to help students internalize them, we will discover that the practices are neither separate nor sequential. The practices blend together and overlap in beautiful and messy ways, sometimes confusing us about which practice we're seeing or using. We need to remember that such distinctions are contrary to the vision of the Standards for Mathematical Practice—the vision of every student possessing a unified and useful set of mathematical habits of mind. It's much better to keep our eye on the overall picture painted by the set of eight practices together, rather than keeping our eye on a checklist of which practices a student may or may not have mastered.

In our fast-paced, technology-driven world and competitive global workplace environment, today's students—tomorrow's workers—must be able to reason, think, and figure out how to approach and solve problems they've never specifically learned how to solve. If students leave school having learned mathematical content alone, without having learned these twenty-first-century survival skills, we will have woefully underprepared them for their future. Perhaps the time has come when enough people realize the importance of these powerful habits and when we have learned enough about how to help students develop them. Perhaps the time has come when we can finally garner the national will to actualize the goal of helping every student develop mathematical habits of mind that can serve them throughout their lives.

Reflections and Discussion

FOR TEACHERS

- What issues or challenges does this message raise for you? In what ways do you agree with or disagree with the main points of the message?
- When do you use mathematical habits of mind in your everyday life?
- How do you demonstrate mathematical habits of mind in your work with students?
- How familiar are you with the Common Core Standards for Mathematical Practice? Which practices do you find most challenging to understand or implement?

FOR FAMILIES

- What questions or issues does this message raise for you to discuss with your son or daughter, the teacher, or school leaders?
- How can you help your daughter or son understand that succeeding in mathematics involves more than learning facts and procedures—that it involves learning how to think and reason?
- How familiar are you with the Common Core State Standards for Mathematics (available online at corestandards.org), especially the set of eight Standards for Mathematical Practice? Which practices do you find most challenging to understand? Where can you go for help to make sense of any standards you may not fully understand?

FOR LEADERS AND POLICY MAKERS

- How does this message reinforce or challenge policies and decisions you have made or are considering?
- How familiar are you with the Common Core Standards for Mathematical Practice? Which practices do you find most challenging to understand or implement?
- How well does your curriculum address mathematical habits of mind?
- How can you support your teachers in balancing the teaching of mathematical content and the development of mathematical practices?

RELATED MESSAGES

Smarter Than We Think

- Messages 32 through 39 explore each of the Standards for Mathematical Practice from the Common Core State Standards, incorporating important ideas from NCTM's process standards and other sources.

- Message 40, "Mathematical Habits of Instruction," pulls together ideas from this message and the other messages in Part IV, as well as drawing from messages throughout the book, to suggest how we can implement what we know to help students develop their abilities to think mathematically.

- Message 1, "Smarter Than We Think," reminds us of the importance of mathematical thinking for all students and, based on a growth mindset, emphasizes the role that challenges can play in helping students improve their intelligence, develop mathematically, and learn to think, reason, and make sense of mathematics.

Faster Isn't Smarter

- Message 14, "Balance Is Basic," makes a case for teaching a balanced program of knowledge, skills, understanding, and, most of all, mathematical thinking.

- Message 1, "Math for a Flattening World," considers the rapidly changing workplace and world around us in terms of the need to help individuals learn how to reason, think creatively, and solve problems we don't know the answers to.

- Message 3, "Making the Case for Creativity," emphasizes the importance of teaching creativity as part of a broader vision of mathematical thinking and reasoning.

MORE TO CONSIDER

- *I Want to Be a Mathematician* (Halmos 1985), a description of Halmos's life's work in mathematics, includes his wrong turns and approaches that didn't work out and offers an intimate view of what it means to be a mathematician, to do mathematics, and to think mathematically.

- "Implementing the Mathematical Practice Standards" (Education Development Center) gives background and overview of the CCSS practices and resources for classroom lessons, including both online resources and professional development. http://mathpractices.edc.org.

- "Mathematical Practice Institute" (Education Development Center) is a professional development institute for high school teachers focused on the mathematical practices. https://mpi.edc.org.

- Common Core State Standards: A New Foundation for Student Success: *The Importance of Mathematical Practices* (McCallum and Zimba 2011) is a four-minute video by Bill McCallum and Jason Zimba on the Standards for Mathematical Practice of the Common Core State Standards.
- "From the Inside Out" (Fillingim and Barlow 2010) describes the kind of mathematical thinking involved in helping children become "doers of mathematics" in and outside of school.
- "Mathematics, Mathematicians, and Mathematics Education" (Bass 2005) shares insights from a mathematician about priorities in school mathematics teaching, including the importance of mathematical thinking, and describes the role of mathematicians in working collaboratively to support those priorities.
- *Thinking Mathematically: Integrating Arithmetic and Algebra in Elementary School* (Carpenter, Franke, and Levi 2003) offers background and strategies on how to focus elementary mathematics instruction on mathematical habits of mind that support the transition from numbers to symbols.
- *Learning and Leading with Habits of Mind: 16 Essential Characteristics for Success* (Costa and Kallick 2008) looks at productive habits of mind in general (not specifically related to mathematics), including personal behaviors and intellectual habits, and offers steps for educators on helping students develop habits of mind.
- "Contemporary Curriculum Issues: Organizing a Curriculum Around Mathematical Habits of Mind" (Cuoco, Goldenberg, and Mark 2010) suggests using mathematical habits of mind, rather than content topics, as a way to organize a mathematics program.
- "A Collection of Lists of Mathematical Habits of Mind" (Lim 2013) is a list of bullet points from several sources addressing aspects of mathematical habits of mind or general habits of mind.
- *Adding It Up: Helping Children Learn Mathematics* (National Research Council 2001) reports research around a conceptual definition of mathematical proficiency (and accompanying "rope" model), incorporating mathematical habits of mind, as cited in this message.
- *Connecting the NCTM Process Standards and the CCSSM Practices* (Koestler, Felton, Bieda, and Otten 2013) unpacks each practice and relates it to NCTM's five process standards, including sample classroom vignettes for elementary, middle, and high school.
- "Growth Mindset and the Common Core Math Standards" (Bryant 2013) looks at a growth mindset as it relates to students developing mathematical habits of mind described in the Common

Core Standards for Mathematical Practice. www.edutopia.org/blog
/growth-mindset-common-core-math-cindy-bryant.

- "What Is Mathematics? A Pedagogical Answer to a Philosophical
  Question" (Harel in Gold and Simons 2008) discusses mathematical
  habits of mind as a central part of the discipline of mathematics and
  is influential as background for the Common Core Standards for
  Mathematical Practice (NGA Center and CCSSO 2010).

- *Mathematics and Plausible Reasoning, Volume II: Patterns of
  Plausible Inference* (Polya 2009) discusses multiple dimensions of
  mathematical thinking and reasoning, generally at the high school
  level, from the widely respected expert on problem solving.

- *Curriculum and Evaluation Standards for School Mathematics*
  (NCTM 1989) offered a description of mathematical process
  standards as part of the first set of mathematics standards offered
  from the profession.

- *Principles and Standards for School Mathematics* (National Council
  of Teachers of Mathematics 2000) describes mathematical content
  and process standards.

- Common Core State Standards for Mathematics (NGA Center
  and CCSSO 2010) includes descriptions of the eight Standards for
  Mathematical Practice.

WWW This message is also available in printable format
at mathsolutions.com/smarterthanwethink.

Standard for Mathematical Practice 1

Make sense of problems and persevere in solving them.
(Solving problems)

Mathematically proficient students start by explaining to themselves the meaning of a problem and looking for entry points to its solution. They analyze givens, constraints, relationships, and goals. They make conjectures about the form and meaning of the solution and plan a solution pathway rather than simply jumping into a solution attempt. They consider analogous problems, and try special cases and simpler forms of the original problem in order to gain insight into its solution. They monitor and evaluate their progress and change course if necessary. Older students might, depending on the context of the problem, transform algebraic expressions or change the viewing window on their graphing calculator to get the information they need. Mathematically proficient students can explain correspondences between equations, verbal descriptions, tables, and graphs or draw diagrams of important features and relationships, graph data, and search for regularity or trends. Younger students might rely on using concrete objects or pictures to help conceptualize and solve a problem. Mathematically proficient students check their answers to problems using a different method, and they continually ask themselves, "Does this make sense?" They can understand the approaches of others to solving complex problems and identify correspondences between different approaches.

—Common Core State Standards for Mathematics (NGA Office and CCSSO 2010, 6)

CONNECTIONS TO STANDARD FOR MATHEMATICAL PRACTICE 1

The Common Core standards have renewed interest in problem solving, listing it as the first standard for mathematical practice. Standard for Mathematical Practice 1 on problem solving may not add a lot of new elements to what's been called for in problem solving for several years, but it's a reminder of the importance of problem solving as a central pillar of mathematics. Problem solving has connections to every other Standard for Mathematical Practice, but especially Practice 3 on communicating reasoning, Practice 4 on modeling, Practice 5 on tools, and Practice 6 on communication. These connections are reflected in the emphasis on problem solving and related discourse in messages related to these practices. In addition to the obvious connection to NCTM's problem-solving standard, the description of Practice 1 in the Common Core standards emphasizes the need for students to be able to move back and forth among different representations for the problem situation, thus also connecting this standard to NCTM's representation standard. (See Appendix F for a description of the five NCTM process standards [2000].)

MESSAGE

32

Problems Worth Solving—And Students Who Can Solve Them

STANDARD FOR MATHEMATICAL PRACTICE 1:
MAKE SENSE OF PROBLEMS AND PERSEVERE
IN SOLVING THEM

An expert problem solver must have two incompatible quantities:
a restless imagination and a patient pertinacity.

—Howard W. Eves, mathematician

In 1980, the National Council of Teachers of Mathematics (NCTM) published a thirty-page booklet called *An Agenda for Action*. The *Agenda* identified eight priorities for the decade, beginning with their highest priority: "Problem solving must be the focus of school mathematics programs." That recommendation followed from the results of surveys of various audiences of educators, mathematicians, and interested non-educators (school boards and PTA/PTOs) in the late 1970s called the Priorities in School Mathematics Project (PRISM). The number-one priority of every group surveyed was to focus school mathematics on problem solving. Yet, the meaning of *problem solving* varies not only among groups surveyed but also among individuals within those groups. Some teachers and parents—and students—consider anything hard in math as being a "problem." However, not everything hard in mathematics is a good problem, and a good problem is defined by more than its level of difficulty. Finding good problems for learning mathematics requires care and thoughtfulness.

A Closer Look at Problem Solving

Much has been written in recent years about problem solving, and several resources are listed under "More to Consider" at the end of this message. There are many resources also identified within the "Related Messages" section of this message, including a discussion of worthwhile tasks from NCTM's *Professional Standards for Teaching Mathematics* (1991), which is excerpted in Message 12, "Upside-Down Teaching." Additionally, Appendices A, B, and C suggest useful resources on problem solving, including sources for finding and evaluating tasks, and Appendix D suggests a teacher's essential library.

Let's briefly explore a few important aspects of problem solving addressed in the CCSS description of Practice 1, drawing in what we know from the sources mentioned earlier and others about helping students become problem solvers.

WHAT MAKES A GOOD PROBLEM?

When we hear as much about problem solving as has been discussed since NCTM's *An Agenda for Action* (1980), it's easy to automatically think, "Oh yes, we do that." But problem solving is a central and fundamental aspect for every grade and level of school mathematics, and it's worth reminding ourselves what makes a problem worthwhile for students' and teachers' time.

NCTM's *Principles and Standards for School Mathematics* describes a problem as "a task for which the solution method is not known in advance" (2000, 52). Thus what is a worthwhile problem for one student might be a trivial exercise for another, even in the same classroom. Choosing a task worthy of all students' time might not always be possible, but an effective teacher will work to find a variety of tasks that engage and challenge most students, perhaps varying the questions asked for a student who may already know a way to approach the problem or perhaps offering that student or a group of students an extension to the problem. NCTM's process standard on problem solving notes that, in the course of solving a problem, a student might generate new knowledge, thus making problem solving both a goal of learning mathematics and a means to learn mathematics beyond the problem. When a student engages his or her mind on a problem, wrestling with the mathematical ideas related to the situation, the student also becomes more open to learning mathematics that may build on the mathematics involved in the problem. Engagement is the key to both persevering through the problem to a solution and learning important mathematics.

PROBLEMS VERSUS MODELING

Good mathematics problems—worthwhile tasks—may be presented in a purely mathematical context, such as exploring patterns in numbers or shapes or finding the vertex of a figure on a coordinate plane meeting certain conditions. Or they may be presented in applied contexts from outside of mathematics, such as computing the total cost of a school project or making a scale drawing of a building. In considering the Standards for Mathematical Practice, this may begin to sound like Practice 4 on mathematical modeling, but there are important distinctions between applied problems and modeling. As noted in Message 35, "Math in the *Real* Real World," in Practice 4 on modeling, applied problems generally arise by considering an idea, concept or skill from *within* mathematics and finding ways to use that idea *outside* of mathematics, whereas modeling involves looking at a situation *outside* of mathematics and searching for ideas, concepts, relationships, or skills from *within* mathematics to help resolve questions about the situation.

THE ROLE OF CONTEXT

When looking for a good applied problem in a nonmathematical context, it's important not to let the context overwhelm the problem or the solver. Contrived contexts or overly complicated contexts can distract from the potential mathematical benefits of a problem, and a context so far outside the experience of most students that they cannot relate to the problem might inhibit students from even trying to solve it. But it's worth searching for appropriate contexts to use selectively so that students come to see mathematics as something useful not only in math class, but also outside of school in real life.

THE DANGERS OF TEACHING KEY OR CLUE WORDS

Many lists of key words or clue words are available that might appear to be a helpful problem-solving tool for students. The words allegedly flag a particular operation to be used to solve the problem (for example, *altogether* means *to add*; *fewer than* means *to subtract*; *out of* means *to divide*). But such lists can be more a distraction than support, especially since any word on any list can be used in a problem where the identified operation would not be helpful. The focus of understanding a problem should be on getting to the mathematical heart of the situation, not on looking at the words in the problem to find a special word from the list. Spending students' time looking for such clues without understanding the related mathematics takes their attention away from identifying what is really important in a problem and can dangerously plant the seeds for possible misconceptions. (See Message 13, "Clueless," for a discussion of the unintended consequences of focusing on key words or clue words.)

GETTING INTO THE PROBLEM

Once students understand what a problem says and what it's asking for, they can begin to explore the underlying mathematical concepts and relationships related to the problem. What kinds of constraints or considerations might affect the solution? What patterns can we see when we represent the information from the problem mathematically? Using different representations—pictures, physical objects, diagrams, tables, graphs, equations—can help students uncover more information or make clearer what they think they see in a problem.

Even young children can learn to find a way into a problem, talking about the problem with each other and explaining to themselves or other students what the problem states and what it's asking for. They look for ways to represent the problem mathematically. For young children, this might mean using concrete objects or drawing a picture to visualize the situation. Older students might make a table or a graph as a tool to help make sense of a relationship in the problem.

Students should be able to move among representations and explain how the representations relate to each other and what these representations reveal about the relationships in the problem. Students should be able to recall similar problems they may have seen before or generate an analogous problem that can help make sense of the problem at hand.

THINKING AND THINKING SOME MORE

We want students to learn how to make conjectures—to think about what might be possible approaches to a solution based on what they see in the problem. As students wrestle with the mathematical ideas related to the problem, they can try out different approaches, adjusting or changing course based on whether their approach is productive. Rather than trying to remember what rule they are supposed to use for which type of problem, we want students to constantly be asking themselves whether what they are doing or trying makes sense—common sense and mathematical sense—and whether the answer(s) they find make sense with respect to the question or problem they're trying to address. Ideally, as students dig deeper and deeper into a good problem, they will push and expand their own thinking and analysis skills.

ENGAGING IN DISCOURSE TO MAKE SENSE OF THE SOLUTION

One of the most important aspects of developing students' abilities to persevere and solve problems is the use of student conversation about the problem and the underlying mathematics. Structuring class discussions around students' approaches to a problem and giving students opportunities to explain their thinking and discuss each other's thinking helps everyone advance their understanding of the problem and the mathematics, also addressing Practice 3 in the process. This kind of rich mathematical discourse engages students in thinking about mathematics and about the problem and keeps them interested in persevering—sticking with the problem through to a solution.

A Classroom View of Problem Solving

In order to develop the kind of problem solvers NCTM envisioned and the Common Core standards call for, the most important factor is a classroom environment where students spend considerable time solving problems and having opportunities to talk about what they're doing. Many teachers find that an upside-down teaching model—one focused on worthwhile problems and considerable student discussion about these problems—not only helps students become successful problem solvers, but also serves as a springboard from which to engage them in thinking about the important mathematical content they need to learn (see the teaching model described in Message 12, "Upside-Down Teaching").

The teacher's role in this kind of problem-centered classroom is crucial as a facilitator of student discourse—the interactions of students around the problem and the related discussion as they share their thinking and talk about the thinking shared by their peers. It's in this back-and-forth interaction that students begin to solidify their thinking and sharpen their ability to make sense of the problem and the processes that can be used to solve it. The nature of the questions a teacher poses can make a big difference in the richness of the discussion and in the amount of mathematical learning that takes place. How to ask questions that push students' thinking is one of the most important skills a teacher can learn (*Why do you think so? Can you say more about that? How did you decide to ...? Would it work the same way if ...? What's the same about these two problems? What's different? Where have you seen this idea before?*)

What Can We Do?

Never never never give up.

—Winston Churchill

Teaching problem solving can and should play a central role in school mathematics just as much today—or more—as when NCTM first identified problem solving as the top priority for school mathematics. Organizing instruction around good problems provides an excellent platform from which to develop not only the mathematical habits of mind described in the Standards for Mathematical Practice, but also a platform from which to develop the mathematical knowledge and skills described in every state's standards. When students learn mathematics from their experience and engagement with problems, they're more likely to make sense of what they're learning and to remember it. Most of all, they're more likely to be able to take on a hard problem when they encounter one, even if they don't already know a rule to use to solve the problem, and they're far more likely to persevere in dealing with the problem through to a solution. Helping students focus on the importance of effort and helping them understand a growth model of intelligence—that they can become smarter by working hard on challenging tasks—are important aspects of helping them develop as problem solvers (see also Message 1, "Smarter Than We Think").

By choosing engaging tasks that elicit thinking, we can focus instruction on helping students develop that thinking. We can give them opportunities and support for noticing patterns, making conjectures and predictions, representing mathematical ideas and problem situations in different forms, and analyzing the relationships in a problem. By engaging students in interactive discourse with each other and the teacher, we can help them communicate their thinking even as they clarify and illuminate their ideas, and encourage them to stick with a problem even if it's hard. The tasks we choose, the classroom environment we create, and the way we orchestrate student discussions are the keys to helping students become problem solvers willing and able to take on the many problems they will encounter in future mathematics classes and in their lives outside of school.

Reflections and Discussion

FOR TEACHERS

- What issues or challenges does this message raise for you? In what ways do you agree with or disagree with the main points of the message?
- How much of your teaching focuses on giving students opportunities to solve engaging, challenging problems?
- What have been some of your most successful strategies for helping students learn to persevere through a problem to a solution?
- What tips might you offer colleagues about selecting engaging tasks for problem solving?
- What kinds of questions have you found most helpful to ask students in order to advance their thinking when they're stuck—without taking away the challenge of the problem?
- How well do you orchestrate classroom discussions around problems? How can you continue to learn and grow in facilitating productive student discourse about mathematical problems and mathematical ideas?
- What barriers do you see in implementing this practice? How can you address those barriers?

FOR FAMILIES

- What questions or issues does this message raise for you to discuss with your son or daughter, the teacher, or school leaders?
- How well are you able to identify problems in everyday life that you can ask your daughter or son to notice and solve, either alone or with your participation? (See "More to Consider" for a few suggested resources for families in finding problems.)

FOR LEADERS AND POLICY MAKERS

- How does this message reinforce or challenge policies and decisions you have made or are considering?
- How can you reinforce teachers in centering their mathematics teaching around meaningful problem solving?
- What kinds of professional learning experiences do you (or can you) offer teachers to help them orchestrate classroom discussions around problems, where students share their thinking and reflect on and critique their thinking and that of their classmates?
- What resources can you make available to teachers for the kind of deep, engaging, worthwhile mathematical tasks that will help students grow as mathematical thinkers and problem solvers?

RELATED MESSAGES

Smarter Than We Think

- Message 31, "Developing Mathematical Habits of Mind," and Message 40, "Mathematical Habits of Instruction," address overall mathematical thinking and reasoning, much of which is supported through the use of worthwhile problems and productive classroom discourse.
- Message 35, "Math in the *Real* Real World," looks at the particular kind of problem solving used in addressing real-world problems that may not be well formed.
- Message 12, "Upside-Down Teaching," describes a problem-centered approach to teaching around worthwhile tasks, including a discussion of such tasks from NCTM's *Professional Teaching Standards* (1991).
- Message 13, "Clueless," provides a caution that teaching students key words or clue words as a problem-solving strategy can be both unhelpful and dangerous.

- Message 1, "Smarter Than We Think," looks at a growth model of intelligence and its impact on helping students learn to (and be willing to) persevere when facing a hard problem.

Faster Isn't Smarter

- Message 14, "Balance Is Basic," advocates a balanced mathematics program incorporating concepts, computation and operations, and problem solving.
- Message 17, "Constructive Struggling," recommends giving students opportunities to wrestle with challenging problems.

MORE TO CONSIDER

Several of the resources listed for Message 31, "Developing Mathematical Habits of Mind," and Message 40, "Mathematical Habits of Instruction," address problem solving, as well as the resources listed in Appendix D, "Essential Library." Also consider the following resources.

- *Connecting the NCTM Process Standards and the CCSSM Practices* (Koestler, Felton, Bieda, and Otten 2013) unpacks each practice and relates it to NCTM's five process standards, including sample problems and related classroom vignettes for elementary, middle, and high school.
- "Figure This!" (National Council of Teachers of Mathematics 2004) offers families mathematical problems to solve with their middle schoolers. www.figurethis.org/index.html.
- "Family Math (Various Resources)" (Lawrence Hall of Science/ EQUALS) is a long-standing program offering resources, games, and activities to engage families in solving problems with their children in mathematics, including sample games and activities. http://sv.berkeley.edu/showcase/pages/fm_act.html.
- "Learning and the Adolescent Mind" (Charles A. Dana Center at the University of Texas at Austin) presents background and resources related to a growth model of intelligence, as well as resources for helping students learn to solve problems; see especially "Resources/Classroom Tools" for two downloadable instruments to help students organize their thinking on a problem and reflect on their problem-solving processes. www.learningandtheadolescentmind.org.
- *Powerful Problem Solving: Activities for Sense-Making with the Mathematical Practices* (Ray 2013) provides guidelines and activities for teaching problem solving as a tool for developing mathematical habits of mind as described in the Common Core Standards for Mathematical Practice.

- *What's Your Math Problem? Getting to the Heart of Teaching Problem Solving* (Gojak 2011) offers practical strategies for finding good problems and organizing instruction to elicit the kind of thinking that helps all students become strong problem solvers.
- *Teaching Mathematics Through Problem Solving: Prekindergarten–Grade 6* (Lester and Charles 2003) is a collection of resources on teaching around mathematical problem solving.
- *Teaching Mathematics Through Problem Solving: Grades 6–12* (Schoen and Charles 2003) is a collection of resources on teaching around mathematical problem solving.
- "Delving Deeper: In-Depth Mathematical Analysis of Ordinary High School Problems" (Stanley and Walukiewicz 2004) considers the benefits of having students go deeply into the mathematics of apparently ordinary problems.
- *Mathematical Discovery: On Understanding, Learning and Teaching Problem Solving, Combined Edition, Volumes I and II* (Polya 1981) provides problems focused at the secondary level and related commentary, including insights and suggestions for thinking and solving problems, from the widely respected expert on problem solving.
- "Cornered by the Real World: A Defense of Mathematics" (Otten 2011) is a helpful look at the benefits and cautions related to using real-world contexts in problems we present to students.
- *Using Multiple Representations in Algebra (6–12), E-Seminar ANYTIME* (Nolan) is a professional development recorded seminar for teachers to learn how to help students use multiple representations as a problem-solving tool. www.nctm.org/catalog /product.aspx?id=14116.
- "Why Is Teaching with Problem Solving Important to Student Learning?" (Lester and Cai 2010) summarizes research and recommendations related to the importance of problem solving for developing students as mathematical thinkers.
- *An Agenda for Action* (National Council of Teachers of Mathematics 1980) offers eight recommendations for improving school mathematics teaching and learning, with the number one priority to focus programs on problem solving.
- *Principles and Standards for School Mathematics* (National Council of Teachers of Mathematics 2000) describes mathematical content and process standards, including the Problem Solving standard noted in this message.
- Common Core State Standards for Mathematics (NGA Office and CCSSO 2010) includes descriptions of the eight Standards for Mathematical Practice.

Standard for Mathematical Practice 2

Reason abstractly and quantitatively. (Reasoning)

Mathematically proficient students make sense of quantities and their relationships in problem situations. They bring two complementary abilities to bear on problems involving quantitative relationships: the ability to *decontextualize*—to abstract a given situation and represent it symbolically and manipulate the representing symbols as if they have a life of their own, without necessarily attending to their referents—and the ability to *contextualize*, to pause as needed during the manipulation process in order to probe into the referents for the symbols involved. Quantitative reasoning entails habits of creating a coherent representation of the problem at hand; considering the units involved; attending to the meaning of quantities, not just how to compute them; and knowing and flexibly using different properties of operations and objects.

—Common Core State Standards for Mathematics (NGA Center and CCSSO 2010, 6)

CONNECTIONS TO STANDARD FOR MATHEMATICAL PRACTICE 2

Practice 2, "Reason abstractly and quantitatively" goes hand-in-hand with Practice 3 on constructing arguments; together they echo what the National Council of Teachers of Mathematics called for in their reasoning and proof standard (NCTM 2000). Practice 2 also relates closely to Practice 1 on problem solving and connects with Practice 7 on mathematical structure, likewise linking to NCTM process standards on problem solving and representation. The focus of Practice 2 is on using reasoning to understand relationships and quantities, especially in problem situations, thus distinguishing it somewhat from Practice 3, which focuses more on making and evaluating reasoned arguments. (See Appendix F for a description of the five NCTM process standards [2000].)

MESSAGE

33

Making Sense of Mathematics—The Multiple Facets of Reasoning

STANDARD FOR MATHEMATICAL PRACTICE 2:
REASON ABSTRACTLY AND QUANTITATIVELY

I had come to an entirely erroneous conclusion which shows, my dear Watson, how dangerous it always is to reason from insufficient data.

—*Sherlock Holmes* (Doyle 1891)

Sherlock Holmes was a master of deduction—applying logic, reasoning, and scientific principles to the work of "detecting," as author Sir Arthur Conan Doyle called it. Today's personifications of Sherlock Holmes appear on television and in real life as forensic scientists, computer analysts, and applied mathematicians, surrounded by the latest high-tech tools and scientific equipment, but sharing with Sherlock Holmes the notion that mathematics, science, and, most of all, logic and reasoning can help us figure out mysteries.

Mathematics makes sense because of the logical reasoning that undergirds and is so essential to all parts of the discipline, across all grade levels and into higher education and advanced mathematics. Reasoning is not limited to abstract proofs or symbolic arguments, although these are central to mathematics. Of increasing importance in the twenty-first century is quantitative or numerical reasoning—the ability to reason through problems involving real quantities represented by numbers in increasingly complex situations—not unlike the reasoning used by Sherlock Holmes or today's scientists and mathematicians.

A Closer Look at Reasoning

Looking at the first words in the title of Standard for Mathematical Practice 2 (Reason abstractly) can allow a reader to focus on a traditional view of abstract mathematical reasoning, a critical element in mathematics. This kind of reasoning involves making reasoned arguments and applying logic to solving problems, covered as well in Practice 3. Students also apply abstract reasoning as they make sense of the steps in an algorithm or make a guess about the relationship between the angles of a triangle and justify or prove their guess. But Practice 2 involves more than abstract reasoning.

QUANTITATIVE REASONING

Standard for Mathematical Practice 2 calls for not only abstract reasoning, but actually emphasizes quantitative reasoning, a broad area of mathematics receiving increasing attention in all levels of school mathematics. Problems in this field of study are also sometimes labeled as *quantitative literacy* or *numerical reasoning*. Quantitative reasoning begins with elementary students exploring and understanding the relationships among numbers and operations as they build their understanding of the number system and use that understanding to solve problems such as determining how many cases of soft drinks to order for a school event. As students get older, they can explore and reason about a wide range of situations they see in their everyday lives or on television or the Internet, eventually expanding their skills with numbers to algebraic representations. Middle school students, for example, might be asked to deal with a problem about professional baseball or basketball players' salaries generated by a headline about a strike over money. *How typical is the "average" salary reported in the headline for most players? How much money would it take to raise the average by a certain amount?*

Quantitative reasoning shares many characteristics with mathematical modeling, the topic of Message 35, "Math in the *Real* Real World." However, in quantitative reasoning, problems are more likely to be posed in a way that can be framed directly with a mathematical representation, whereas a mathematical modeling problem is more likely to be offered with some (or many) parts of the problem undefined or even unclear, thus calling for the additional step of mathematizing the situation and posing a well-formed problem before attempting a solution. In tackling problems calling for quantitative reasoning (problems involving quantities and the relationships between them), students need to pay attention to the particular units for each quantity as they represent the situation and deal with the relationships among the quantities.

BACK AND FORTH AND IN AND OUT

One of the most interesting aspects of Standard for Mathematical Practice 2 is the emphasis on students being able to move back and forth while solving a contextual problem between a situation and the mathematical representation of the situation. The standard talks about *decontextualizing*—translating the problem situation into a mathematical representation without the context—and *contextualizing*—checking back with the situation along the path of solving the problem or considering the situation to make sure the mathematics makes sense with the original context. As students move back and forth between the problem and their mathematical representation of the problem, they refine their understanding of both the problem and the mathematics and begin to make sense of their work. This sense-making aspect is critical in helping students learn to reason and it applies even to our very youngest students.

Likewise, it's useful for students to be able to *zoom in* and *zoom out* on a problem, sometimes focusing on the overall problem and sometimes focusing on certain aspects of the problem. This ability to focus on the big picture or overall process as well as focusing on particular details is an important mathematical habit of mind, and it appears again in Practice 8 on regularity in reasoning.

A Classroom View of Reasoning

In developing students' mathematical habits of mind, the structure of the classroom environment and plans for daily activities need to allow for extensive opportunities for students to wrestle with challenging problems and engage in discourse around those problems. As students are routinely expected to figure out mathematical processes and ideas, communicate their own thinking, and interact with other students about theirs, they hone their reasoning ability.

Standard for Mathematical Practice 2 emphasizes the need for students to pay attention to quantities and relationships among quantities in problems. Students need specific opportunities to develop this kind of quantitative reasoning skills. Selecting problems from the real world can help students realize how much numerical information surrounds them, giving them a platform for developing their reasoning in quantitative situations and solving quantitative problems. As they translate from the problem to a mathematical representation and vice-versa, they begin to make sense of the quantities and units in a problem, even as they work with appropriate mathematical operations and processes to arrive at a solution that makes sense.

What Can We Do?

Standard for Mathematical Practice 2 on abstract and quantitative reasoning reminds us of the growing importance of considering a broad view of reasoning. Today more than ever, every person needs to be able to reason about situations represented numerically and apply their reasoning to solve problems, including problems that can involve somewhat complicated situations, especially if problems come from real stories in the news. Quantitative information bombards us daily, and much of it can be useful to us if we know how to deal with it. We can help students develop their quantitative reasoning skills by incorporating opportunities to reason with numbers and quantities from the earliest grades, helping students learn to represent quantities and explore relationships among the quantities using their growing understanding of mathematical properties, and paying attention to the units associated with the quantities.

We can encourage students to continually move back and forth and in and out as they deal with various aspects of a problem and check the consistency between the context and the related mathematics. As they advance through the grades, they can refine their ability to pull back and look at the problem and move closer to focus on parts of the problem. Central to developing both their abstract reasoning skills and their quantitative reasoning skills is the opportunity to continually be asked to explain and communicate their thinking as part of regular classroom discourse. None of this happens in isolation of other practices or the broader vision of mathematical habits of mind. Together these practices and processes equip students to deal with increasingly complex and relevant problems as they become the kind of thinkers and reasoners Sherlock Holmes would be proud to call colleagues.

Reflections and Discussion

FOR TEACHERS

- What issues or challenges does this message raise for you? In what ways do you agree with or disagree with the main points of the message?
- How effectively do you incorporate the development of reasoning skills, including quantitative reasoning, into your teaching?
- What can you do to increase attention to quantitative reasoning?
- What barriers do you see in implementing this practice? How can you address those barriers?

FOR FAMILIES

- What questions or issues does this message raise for you to discuss with your son or daughter, the teacher, or school leaders?
- Can you look for opportunities in your daily life to identify problems involving numbers and relationships? How can you find time to talk about such problems with your daughter or son?

FOR LEADERS AND POLICY MAKERS

- How does this message reinforce or challenge policies and decisions you have made or are considering?
- Is quantitative reasoning part of your mathematics program? How can you increase awareness of reasoning, especially quantitative reasoning, among teachers, parents, and students?

RELATED MESSAGES

Smarter Than We Think

- Message 31, "Developing Habits of Mind," and Message 40, "Mathematical Habits of Instruction," address overall mathematical thinking and reasoning.
- Message 34, "Who's Doing the Talking?," discusses Practice 3 on constructing logical arguments, closely related to Practice 2.
- Message 27, "Fixing High School," describes several aspects of needed change in the high school mathematics program, including calling for grade 12 courses that strengthen quantitative reasoning.

- Message 12, "Upside-Down Teaching," proposes a problem-centered teaching model ideal for developing both abstract and, especially, quantitative reasoning skills.

Faster Isn't Smarter
- Message 1, "Math for a Flattening World," calls for new skills for workers in the twenty-first century, including a range of mathematical thinking and reasoning skills and topics like quantitative reasoning.
- Message 17, "Constructive Struggling," advocates giving students challenging problems so that they develop their thinking and reasoning skills.

MORE TO CONSIDER

- *Connecting the NCTM Process Standards and the CCSSM Practices* (Koestler, Felton, Bieda, and Otten 2013) unpacks each practice and relates it to NCTM's five process standards, including sample classroom vignettes for elementary, middle, and high school.
- *Thinking Mathematically: Integrating Arithmetic and Algebra in Elementary School* (Carpenter, Franke, and Levi 2003) is a particularly helpful resource, since much of Practice 2 addresses particular content related to numbers, operations, and algebra. The book shows how to help students develop reasoning and thinking skills as they explore the transition from numbers and operations to algebraic thinking, including practical classroom strategies, research findings, and sample problems.
- *Focus in High School Mathematics: Reasoning and Sense Making* (Martin et al. 2009) offers guidelines and recommendations for incorporating reasoning and sense-making as a fundamental focus of high school mathematics; NCTM has developed a suite of related print, virtual, and interactive professional development resources around this foundational description of the vision of reasoning and sense making. www.nctm.org/standards/content.aspx?id=23749.
- *Mathematics and Plausible Reasoning, Volume II: Patterns of Plausible Inference* (Polya 2009) discusses multiple dimensions of mathematical thinking, argumentation, and reasoning, generally at the high school level, from a widely respected expert on problem solving.
- *Advanced Mathematical Decision Making* (also called *Advanced Quantitative Reasoning* in Texas) (Charles A. Dana Center at the University of Texas at Austin 2010) is a grade 12 course focused on quantitative reasoning, including statistics, finance, and topics from discrete mathematics.

- National Numeracy Network is a national organization focused on resources and networking to increase awareness of and educational attention to quantitative reasoning and related topics at the secondary and post-secondary levels, including publishing an online journal. http://serc.carleton.edu/nnn/index.html.

- *Principles and Standards for School Mathematics* (National Council of Teachers of Mathematics 2000) describes mathematical content and process standards, including the process standards on reasoning, problem solving, and representation discussed in this message.

- Common Core State Standards for Mathematics (NGA Office and CCSSO 2010) includes descriptions of the eight standards for mathematical practice.

Standard for Mathematical Practice 3

Construct viable arguments and critique the reasoning of others.
(Make and evaluate arguments)

Mathematically proficient students understand and use stated assumptions, definitions, and previously established results in constructing arguments. They make conjectures and build a logical progression of statements to explore the truth of their conjectures. They are able to analyze situations by breaking them into cases, and can recognize and use counter-examples. They justify their conclusions, communicate them to others, and respond to the arguments of others. They reason inductively about data, making plausible arguments that take into account the context from which the data arose. Mathematically proficient students are also able to compare the effectiveness of two plausible arguments, distinguish correct logic or reasoning from that which is flawed, and—if there is a flaw in an argument—explain what it is. Elementary students can construct arguments using concrete referents such as objects, drawings, diagrams, and actions. Such arguments can make sense and be correct, even though they are not generalized or made formal until later grades. Later, students learn to determine domains to which an argument applies. Students at all grades can listen or read the arguments of others, decide whether they make sense, and ask useful questions to clarify or improve the arguments.

—Common Core State Standards for Mathematics (NGA Center and CCSSO 2010, 6)

CONNECTIONS TO STANDARD FOR MATHEMATICAL PRACTICE 3

The center of Practice 3 is on generating reasoned arguments in age-appropriate ways and evaluating arguments. Practice 3 is closely connected to Practice 2 on abstract and quantitative reasoning, together reflecting NCTM's reasoning and proof process standard (2000). Practice 3 reflects the heart of the logical reasoning and proof part of NCTM's standard. Here, the emphasis is on students learning to explain their thinking and developing the habit of making a case that can convince others. Because of the emphasis on communicating logical reasoning, this practice is also closely related to Practice 6 involving precision in communication and to NCTM's communication standard (2000).

Also, the description of the adaptive reasoning mathematical proficiency strand in *Adding It Up* (National Research Council 2001) connects with this practice, noting that supporting the ability to make and evaluate logical arguments helps students develop personal responsibility for what they learn and believe about mathematics. The description of adaptive reasoning tells us that when disputes and disagreements arise in a classroom, if students can make or consider a reasoned argument, it's not necessary for them to check with the teacher or gather outside data to back up their point of view. Students can discuss whether the argument—explanation, justification, or proof—is valid as a tool for resolving the difference of opinion. (See Appendix F for a description of the five NCTM process standards [2000].)

34

Who's Doing the Talking? When Students Communicate Their Thinking

STANDARD FOR MATHEMATICAL PRACTICE 3:
CONSTRUCT VIABLE ARGUMENTS AND CRITIQUE
THE REASONING OF OTHERS

From my students' perspective, convincing me was pointless; I already knew whether or not their solutions were correct. If I really wanted to change students' dependence on my authority, I needed to shift the audience and get students talking to someone whom they could convince—one another.

—Corey Webel ("Shifting Mathematical Authority from
Teacher to Community," 2010)

The way we structure classrooms and the kinds of tasks we give students can help them learn to argue, and also help them learn to understand and question arguments, all in support of their broader mathematics learning. Consider a teacher-centered classroom where the teacher essentially tells students the important mathematics they need to know and is the primary source for validating whether a student's work is right or wrong. The teacher may call on students who raise their hand to answer questions, often a small number of the same students. Many classrooms organized in this way have traditionally been considered good classrooms. But in such a classroom, students come to rely on the teacher to tell them whether they're right or wrong, rather than reflecting on their work and interacting with each other to determine whether their thinking makes sense. Now imagine a classroom where students' explanations and discussions about each other's explanations take center stage. The teacher is still very much responsible for structuring the classroom, but the attention is on students and their thinking. Corey Webel (2010) notes a variety of benefits in terms of the learning community when student-developed *arguments*—explanations and justifications—take a central place in the classroom. It may well be that whoever's doing the talking in the classroom is who's doing the learning.

Much of the work of mathematicians involves arguing to prove hypotheses true or false. The kind of arguments mathematicians make consist of reasoned steps that lead logically to a conclusion in a mathematically valid way. For school mathematics, learning how to argue whether an assertion is true or false in age-appropriate mathematical language is a central part of learning to do mathematics and think mathematically (Koestler et al. 2013). And the notion of more formal proof based on sound principles of logic—sometimes a center of debate in schools and homes—is receiving renewed attention in our high-tech world where the ability to communicate logically with computers has become a high-demand career field.

A Closer Look at Making and Evaluating Arguments

The basis of Standard for Mathematical Practice 3 involves students being responsible for making their case—stating what they notice or believe to be true about a mathematical idea or offering a solution to a problem—and then backing up their case with an explanation or justification. Let's take a closer look at how students develop the ability to make and evaluate arguments.

CONJECTURES

The development of this practice can begin early when students are expected to make *conjectures*—informed guesses—about a mathematical situation or problem. NCTM (2000) describes systematic reasoning as a defining feature of mathematics, beginning with students making conjectures, either orally or in writing, based on their explorations and reflections about the mathematics in a task. Students of all ages can notice patterns and make age-appropriate conjectures with suitable rigor for their age. Young children might be able to notice and describe a pattern with even and odd numbers; older elementary children might make a conjecture about the perpendicularity of diagonals in a square after exploring the relationship informally with paper folding or technology; middle school students might predict the likelihood of whether the product of two unknown numbers will be even or odd; and high school students might consider why a number trick always produces a predictable outcome. The level of sophistication of a conjecture and the nature of an argument justifying it should increase across the grades as students become more mathematically sophisticated.

EXPLANATIONS AND JUSTIFICATIONS

After a student makes and communicates a conjecture, or articulates a possible generalization (a more informed hypothesis about an observed pattern), the student needs to learn to logically justify the conjecture or generalization, either informally or formally. Students should learn to explain their thinking and defend their point of view mathematically from their earliest school experience. These processes are both a means and ends to mathematics learning. Explanations and justifications themselves are important, as described in this practice, in NCTM's standards (2000), and in *Adding It Up* (National Research Council 2001). But even beyond developing the ability to explain and justify orally or in writing, as students practice these skills, they clarify and refine their thinking about important mathematical ideas or a mathematical problem they are addressing, thus strengthening their understanding. Justification represents the heart of this practice—constructing arguments—from explanations at the elementary level to more formal arguments and proofs by the time students reach advanced mathematics.

FORMAL ARGUMENTS

The most formal kind of argument is a mathematical proof, typically appearing in high school mathematics courses. But some forms of proof may be appropriate for young children. NCTM suggests that certain types of proof may be possible at a young age and that proof by contradiction, in particular, is a legitimate form of reasoning for children in the early grades (2000).

Proof need not be reserved for a high school geometry course, but can be a regular part of any mathematics course, especially a course integrating topics from algebra, geometry, and other strands of mathematics. For students at any grade level, creating a classroom environment where productive discourse, active discussion, and justification are routine parts of day-to-day activities can support the transition to formal proof, rather than reserving proof to a special unit or an occasional side trip.

A Classroom View of Making and Evaluating Arguments

When we help students learn to argue mathematically, both teachers and students reap benefits. Students learn new ways of thinking and teachers gain insights into their understanding of mathematics. Students cannot learn these skills sitting in a classroom listening to a teacher explain; they need opportunities to think out loud, explain their answer, convince their classmates, and back up their claim with evidence. The most effective way to provide such opportunities is in a classroom focused on rich, interactive, productive student discourse around mathematical tasks, problems, concepts, or skills. A classroom structured around this kind of student discussion and interaction helps students learn how to talk through their thinking and communicate with other students and the teacher. The processes of explaining and justifying, and of evaluating and critiquing explanations of others, serve as both means and ends to mathematics learning. They help students learn mathematics, and they also give students powerful thinking tools they can use throughout their lives.

MAKING ARGUMENTS

If students have an opportunity to engage in some kind of mathematical activity, they can be asked to offer their results, followed always by *Explain how you got that result*, or something similar (*Can you explain your thinking to the class? How did you decide to do ...?*). Simple conjectures and explanations when students are first asked to present their thinking can evolve over time to increasingly complete reasoned arguments with mathematical validity appropriate to the grade level. Constructing an argument can help the student clarify his or her own thinking. At the same time, listening or reading a student's argument can give a teacher insight into the student's mind—when the student explains, justifies, and proves something, we can determine what the student understands about the mathematics being discussed and decide how we can help the student correct a misconception or clarify or extend what they know.

EVALUATING ARGUMENTS

Critiquing the arguments of others is an important ability in itself, specifically called for in this practice. Being able to understand, question, and offer suggestions about the argument of another person can help a student not only make sense of the mathematics being discussed, but can also help the student improve his or her own ability to make an argument. Routinely incorporating back-and-forth conversations around explanations and justifications, and questioning or clarifying those explanations, or the presentation of a proof and evaluation by peers, should be part of a classroom environment that stimulates students to think mathematically. This kind of student-to-student conversation about mathematics may not come easily to students who have not experienced it. However over time, especially if teachers create an environment of mutual respect amongst the teacher and students, and students are taught to be respectful to each other, students will become more willing to participate in commenting on each other's work. In this kind of environment, students feel safe to take the risk of participating. They can become accustomed to thinking about and listening to different ways other students might have approached a problem or situation, engaging in increasingly thoughtful interactions with each other that help advance everyone's thinking. If teachers within a school or school system work together across grades to create this kind of safe and nurturing classroom environment, students will become more and more comfortable from year to year in routinely explaining their thinking and discussing the thinking of others.

PREPARING FOR THE TEST

One of the most exciting and important aspects of the expectation for students to construct arguments and critique others' arguments is that this practice is likely to receive more and more emphasis on high-stakes tests. Thanks in part to advances in technology and new types of testing models, tests being developed to assess the Common Core standards promise to call for students to write explanations and justifications. These tests also ask students to determine whether given explanations or arguments are valid and to make suggestions as to how to improve those explanations or arguments (see both the PARCC and Smarter Balanced Assessment Consortia). Such assessment tasks are likely to be hand-scored, although a few samples of the latter type have been released as examples of computer-scorable items. When tests reflect what we say is important in the mathematical practices, teachers are far more likely to be willing to include this central notion of reasoning into their instruction on a routine basis.

What Can We Do?

Corey Webel's (2010) quote at the beginning of this message is about fostering a sense of community in the classroom. Looking more closely, it's obvious that this sense of community is part of a rich classroom environment that also allows students to not only feel more in control of their learning, but also to hone and advance their reasoning skills. When students are routinely expected to explain and communicate their thinking, make conjectures, defend their point of view, and discuss others' arguments, they have to engage in both the problem situation and the underlying mathematical ideas. The resulting learning helps them reason more effectively, in and outside of mathematics class, and also helps them learn the mathematical content we want them to learn.

Reflections and Discussion

FOR TEACHERS

- What issues or challenges does this message raise for you? In what ways do you agree with or disagree with the main points of the message?
- How often do you give students opportunities to explain or justify their thinking?
- How often do you give students opportunities to discuss others' arguments, including asking questions and possibly making suggestions for improving those arguments? How can you help students become more comfortable critiquing each others' work constructively?
- What do you see as the biggest barrier to helping students learn to reason through arguments? How might you address that barrier?

FOR FAMILIES

- What questions or issues does this message raise for you to discuss with your daughter or son, the teacher, or school leaders?
- What opportunities do you offer (or can you offer) your son or daughter to explain the mathematics he or she does, whether on homework or on problems you may discuss together outside of homework?

FOR LEADERS AND POLICY MAKERS

- How does this message reinforce or challenge policies and decisions you have made or are considering?

- How well does your instructional program support teachers with time and resources to help students learn how to explain and justify their thinking and critique others' arguments?

- How can you find out whether teachers are giving students adequate opportunities to develop their reasoning abilities and construct and evaluate arguments? How can you help them improve in this area?

RELATED MESSAGES

Smarter Than We Think

- Message 31, "Developing Mathematical Habits of Mind," and Message 40, "Mathematical Habits of Instruction," address overall mathematical thinking and reasoning.

- Message 33, "Making Sense of Mathematics," discusses Practice 2 on reasoning, closely related to Practice 3.

- Message 12, "Upside-Down Teaching," advocates a problem-centered approach to teaching that fosters students' productive discourse, including explanations, justifications, and critiques of others' arguments.

- Message 28, "Bringing Testing into the 21st Century," discusses the nature of tests that call for students to explain their thinking.

Faster Isn't Smarter

- Message 37, "Boring!," looks at the difference between teacher-delivered instruction and a more student-centered approach where students engage in productive discourse as they are routinely expected to explain their thinking and discuss the thinking of others.

- Message 33, "Engaged in What?," reminds us that the point of student engagement and discourse cannot be just about activity, but must be focused on purposeful learning of mathematics content and mathematical habits of mind, especially mathematical thinking and reasoning.

MORE TO CONSIDER

- *Connecting the NCTM Process Standards and the CCSSM Practices* (Koestler, Felton, Bieda, and Otten 2013) unpacks each practice and relates it to NCTM's five process standards, including sample classroom vignettes for elementary, middle, and high school.

- "Connecting Research to Teaching: Shifting Mathematical Authority from Teacher to Community" (Webel 2010) offers lessons from two classrooms about having students assume increasing levels of responsibility for their learning in a problem-centered classroom with an emphasis on high-quality student discourse that includes student explanations and class discussion of explanations.

- *Adding It Up: Helping Children Learn Mathematics* (National Research Council 2001) reports research around a conceptual definition of mathematical proficiency, including a strand called "Adaptive Reasoning," related to this message.

- "PARCC Prototyping Project" (Charles A. Dana Center at The University of Texas at Austin 2012) provides prototypes of a variety of assessment tasks associated with the PARCC assessment, including some that are designed to address Practice 3 as well as other practices. www.ccsstoolbox.org.

- Smarter Balanced Assessment Consortia is one of two national assessment consortia developing large-scale accountability tests to accompany the Common Core State Standards. www.smarterbalanced.org.

- Partnership for Assessment of Readiness for College and Careers (PARCC) is one of two national assessment consortia developing large-scale accountability tests to accompany the Common Core State Standards. http://parconline.org.

- *Classroom Discussions in Math: A Teacher's Guide for Using Talk Moves to Support the Common Core and More, 3rd edition* (Chapin, O'Connor, and Anderson 2013) offers strategies for orchestrating productive student discourse around good problems.

- *5 Practices for Orchestrating Mathematics Discussions* (Smith and Stein 2011) presents a problem-based teaching model focused on productive student discourse.

- *Mathematics and Plausible Reasoning, Volume II: Patterns of Plausible Inference* (Polya 2009) discusses multiple dimensions of mathematical thinking, argumentation, and reasoning, generally at the high school level, from a widely respected expert on problem solving.

- *Principles and Standards for School Mathematics* (National Council of Teachers of Mathematics 2000) describes mathematical content and process standards, including the reasoning and communication standards discussed in this message.
- Common Core State Standards for Mathematics (NGA Office and CCSSO 2010) includes descriptions of the eight standards for mathematical practice.

Standard for Mathematical Practice 4

Model with mathematics. (Mathematical Modeling)

Mathematically proficient students can apply the mathematics they know to solve problems arising in everyday life, society, and the workplace. In early grades, this might be as simple as writing an addition equation to describe a situation. In middle grades, a student might apply proportional reasoning to plan a school event or analyze a problem in the community. By high school, a student might use geometry to solve a design problem or use a function to describe how one quantity of interest depends on another. Mathematically proficient students who can apply what they know are comfortable making assumptions and approximations to simplify a complicated situation, realizing that these may need revision later. They are able to identify important quantities in a practical situation and map their relationships using such tools as diagrams, two-way tables, graphs, flowcharts and formulas. They can analyze those relationships mathematically to draw conclusions. They routinely interpret their mathematical results in the context of the situation and reflect on whether the results make sense, possibly improving the model if it has not served its purpose.

—Common Core State Standards for Mathematics (NGA Center and CCSSO 2010, 7)

CONNECTIONS TO STANDARD FOR MATHEMATICAL PRACTICE 4

Mathematical modeling as described in Practice 4 calls for students to use mathematical thinking and reasoning to tackle a problem, often ill-defined, from the world outside of school and from beyond the field of pure mathematics. Thus it is connected to NCTM's reasoning and proof standard and to Practices 2 and 3 on reasoning, especially Practice 2. The emphasis in Practice 2 on quantitative reasoning relates to the kind of real-world scenarios addressed in this modeling practice, but in Practice 2, problems tend to be well defined, in comparison to the need to mathematize an ill-defined situation in Practice 4. Practice 4 begins with applying what students know about solving problems, but evolves into addressing a particular kind of problem—application problems. Thus it is closely related to Practice 1 on problem solving and NCTM's problem solving standard, as well as NCTM's connection standard related to connecting mathematics to the world outside of mathematics. Practice 4 brings together many mathematical habits of mind and calls for students to not only develop and use mathematical models, but also to report their results, often using technological tools. Thus we could further make a case for connecting this practice to Practices 5 (tools) and 6 (communicating precisely) and to NCTM's communication and representation standards, but these connections are not as central to Practice 4 as those mentioned above. (See Appendix F for a description of the five NCTM process standards [2000].)

35

Math in the *Real* Real World—Mathematical Modeling in School Mathematics

STANDARD FOR MATHEMATICAL PRACTICE 4:
MODEL WITH MATHEMATICS

What distinguishes modeling from other forms of applications of mathematics are (1) *explicit* attention at the beginning to the *process* of getting from the problem outside of mathematics to its mathematical formulation and (2) an explicit reconciliation between the mathematics and the real-world situation at the end. Throughout the modeling process, consideration is given to both the external world and the mathematics, and the results have to be both mathematically correct and reasonable in the real-world context.

—Applied mathematician and pioneer modeling
educator Dr. Henry Pollak ("A History of the Teaching
of Modeling," 2003, 649)

Summer Weather Shocker

by STEVE McCAULEY, WFAA, Posted on July 13, 2013 at 11:17 PM
www.wfaa.com/news/local/Summer-weather-shocker-cool-and-rainy-215408241.0html

This is such a bizarre forecast coming up for mid-July. After typical summertime highs hovering around 100 degrees in most of North Texas on Saturday, we'll be basking in temperatures in the low-to-mid 80s on Sunday. The mercury may even dip into the 70s during the afternoon in some spots. Why? A storm system will be moving due south from Oklahoma on Sunday morning. Showers and thunderstorms will continue into the evening hours and the raindrops will keep falling through Monday's morning rush hours. The rainfall is forecast to be especially heavy in the parched western zones of the WFAA viewing area.

(continued)

Looks like the computer models continue to have major problems in trying to determine just how much rain will occur with this backwards-moving low. One of the models now wants to bull's-eye the Dallas-Fort Worth area with six to ten inches of rain. Before you get too giddy, be aware that other models are going for less than one inch. This just goes to show you that when weather patterns move backwards, sometimes the computer models get cross-eyed and can't see straight.

Mathematical modeling, often with the help of computer simulations, is regularly used to analyze data and make predictions about all kinds of situations in our world. Models appear in stories about weather, finance, sports, health, and a variety of business and social enterprises. The weather story above came out during an unusual period of weather in Texas. The story provides a useful setting for beginning to understand where mathematical models come from and how they are used, and to explore how we can incorporate mathematical modeling into school mathematics programs in ways that help students use the mathematics they learn in meaningful ways outside of school.

A Closer Look at Mathematical Modeling

The word *model* is used in many ways in school mathematics and in the English language outside of school, sometimes causing confusion by what is meant by *mathematical modeling*. *Mathematical modeling* is different from *modeling* mathematics, in which a teacher or student might use blocks or cubes or rods or plastic bears to *model* (illustrate) a property, relationship, or operation in mathematics. It's also very different from the notion of a teacher or student *modeling* (showing or demonstrating) a process. Mathematical modeling lies at the intersection of mathematical problem solving and the real world.

In the quote at the beginning of this message, Henry Pollak (2003) reminds us that mathematical modeling is our way of using mathematics to make sense of the world and to inform a wide range of scientific, community, and business issues—from hurricane tracking to economic forecasts to planning an event to determining a family's budget. The *Summer Shocker* weather story offers an opportunity to see just what Pollak means. Based on data gathered over many years, and using scientific principles, scientists and mathematicians have created mathematical models that represent typical weather outcomes, such as for rain, temperature, and wind, based on particular combinations of atmospheric conditions, such as high or low pressure systems across a particular path or region, temperature inversions, tidal patterns, and so on. A model consists of one or more mathematical functions that represent the complex data as accurately as possible, while also taking into account physical laws and scientific principles. Finding the best function or combination of functions to represent such messy data is a complex task and the focus of the work of many scientists and applied mathematicians, made easier by the unprecedented power of today's technology. The fact that multiple models may be developed for the same situation, as reported in the weather story, demonstrates the complexity and challenges involved with high-level mathematical modeling, subject to the expertise, assumptions, and interpretation of the data by different individuals or teams. Once a model is developed, every time a particular meteorological situation arises, the model can help forecasters make predictions. When the situation actually occurs, we can see how accurate the predictions were, and more data can be added to the database in order to further refine or adjust the model.

Mathematical modeling can also be used for simpler kinds of problems, accessible to students. Determining a family's budget, planning a school event, or designing a play area on the school grounds might provide the setting for developing a mathematical model to address a problem in the real world of students.

MODELING ILL-DEFINED PROBLEMS AND SITUATIONS

A major distinction between a modeling situation and a more traditional word problem, even an applied word problem, is that modeling situations tend to be less well-defined than applied problems, especially problems that may appear at the end of a lesson or unit on a particular procedure that can be used to solve the problems. A modeling problem requires the important step of translating the situation into a more approachable mathematical form—turning an ill-defined problem into a mathematically well-defined one. One way of looking at this difference is that application, or applied, problems generally arise by considering an idea, concept or skill from within mathematics and finding ways to use that idea outside of mathematics, whereas modeling involves looking at a situation outside of mathematics and searching for ideas, concepts, relationships, or skills from within mathematics to help resolve questions about the situation (Dossey 2010). Two sample problems illustrate this difference. Consider the following applied (word) problem:

> *The local university has 4,500 students and 35% of the students recycle their plastic water bottles. If each student uses 9 water bottles per week, how many bottles do students recycle?*

Now consider a related modeling problem:

> *How much plastic is recycled at the university?*

The second problem is not yet a problem ready to be solved, even though there is a question asked. Rather, students would need to think about what would be involved in finding a solution, what assumptions would need to be made, and how a problem might be clearly posed: *Who besides students needs to be considered? Do they all consume the same number of bottles? How big are the bottles used in generating the data provided? How will we determine "how much plastic"? What other sources of plastic besides water bottles do we need to consider?* All of these questions and considerations would go into the first step of articulating a clear problem before finding a mathematical model to answer the question and solving the problem. And, as Pollak (2003) noted at the beginning of this message, in dealing with problems that are not well defined mathematically, mathematical modeling requires constantly checking back with the original situation to make sure the mathematical representation(s) we choose make sense.

MODELING FROM DATA

Models can be based on collected data—the more data available, the higher the likelihood that the model may closely represent the situation— or on physical laws, or both. For the *Summer Shocker* weather story, developers of the several models mentioned in the story likely relied on the massive databases of weather information maintained by the National Oceanic and Atmospheric Administration (NOAA; see the "More to Consider" section of this message) as well as on what they knew from scientific laws, such as those regarding heat, energy, and so on.

A mathematician once explained to me that, when looking at a small set of graphed data points, it may appear that the data are related in a roughly linear fashion. But if we were to *zoom out*—to pull back and look at a bigger section of the data from farther away—we might discover that the data appear to be better represented by a quadratic function. Pulling back even further and looking at more data, we might determine that an exponential function is a better fit. And it may well be that multiple functions are required to more closely represent the data. Determining the best model possible requires mathematicians to look at data from many perspectives and to tap into the formidable capabilities of their technological tools to find appropriate mathematical representations that most closely match the whole set of data.

Finding a model that accurately fits a set of data is not an exact process, and it's not unusual to find multiple models, especially for a weather situation, that may differ in their conclusions and predictions, sometimes dramatically, as in the *Summer Shocker* story. That weather story appeared on the fifth consecutive day that temperatures reached at least 100 degrees Fahrenheit. As predicted by the models, apparently in near agreement, the high temperature the next day was 81 degrees and the following day dropped to 74 degrees. The rain predictions were less certain, with the models offering widely different predictions, ranging from six or more inches to less than an inch. Actual rain totals over the three days of the storm were 1.61 inches, falling in between the most extreme predictions of the models (Steve McCauley, personal communication, 2014). The data from what actually occurred is now part of the weather database used to create the models, so the next models may or may not be in closer agreement. The nature and accuracy of mathematical models depends in part on the amount and accuracy of data used, but also on assumptions and considerations made in framing the problem initially. Thus it is possible to generate models that may vary slightly in their details, leading to sometimes different predictions. Because of advances in science and in our ability to gather accurate data, however, today's weather models are far more accurate and consistent than at any time in the past, and it's reasonable to expect that this trend will continue. In any case, we can expect to see such mathematical models hit the headlines again at some point, whether to alert us to an approaching storm, make a prediction about a trend in housing, analyze behavior of the stock market, or illuminate some other aspect of the real world.

A Classroom View of Mathematical Modeling

The kind of sophisticated mathematical models we've looked at so far involve mathematics far beyond what we teach in elementary and secondary schools. But understanding the nature of mathematical modeling at this high level can be helpful as we incorporate age-appropriate uses of modeling into our school mathematics programs.

In structuring classrooms to support the development of mathematical modeling, the most important consideration is making sure students have the opportunity to deal with increasingly diverse and unspecified problems. It's not enough for students to learn how to solve problems of a particular type—problems where they learn a procedure and then deal with problems that call for that particular procedure. They need opportunities to think about a situation and mess around with mathematical ideas that might prove useful in answering one or more questions about the situation, even if the questions aren't clear at the outset.

PROBLEM SOLVING AND MATHEMATICAL MODELING

The process for solving word problems in school mathematics, including application problems, is often based on four steps developed by

George Polya (2004). These steps have helped many students develop skill and confidence in dealing with word problems:

1. Understand the problem

2. Devise a plan

3. Carry out the plan

4. Look back

This process works well for the kinds of application problems students see most often in mathematics classrooms and textbooks. These four steps are consistent with the steps used in addressing a modeling problem, however they are just a subset of the complete process a modeling problem entails. There are two main differences, as described by Henry Pollak (2003) in the quote at the beginning of this message. The first is starting the mathematical modeling process with *interpreting* a real situation so that it is suitable for addressing with a mathematical model. The second difference comes at the end of the steps, after working with the problem. Students need to go back to the situation to *verify* or *validate* the results.

The second half of the description for Practice 4 provides a summary of the mathematical modeling process. The full text of the Common Core standards includes a discussion of the modeling process appropriate to high school mathematics (NGA Center and CCSSO 2010, 72–73), as well as including the below diagram representing the process. This visual model of the process can help us understand its iterative nature and the need to continually move between the mathematics and the situation.

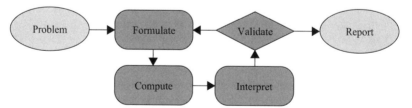

Source: Common Core State Standards for Mathematics (NGA Center and CCSSO 2010)

In writing about the description of mathematical modeling presented in the Common Core State Standards for Mathematics, respected mathematics education researcher Zalman Usiskin (2011) translated this modeling process into five steps leading up to the reporting of results, reflecting the process shown in the diagram:

1. simplify the real problem (Problem);

2. find a mathematical model for the simplified problem (Formulate);

3. solve the problem that is the mathematical model (Compute);

4. translate the solution back into the real world situation (Interpret); and

5. check whether the solution is feasible (Validate)—if not, go back to Step 1 or Step 2.

This process can be developed in age-appropriate ways from the early grades through high school. While it may be frustrating to some students and teachers that modeling problems tend to not be well defined, and that the process may require multiple attempts to arrive at a suitable model, students of any age can learn to develop mathematical models if we give them the right kinds of experiences. In particular, students need regular opportunities to deal with mathematizing situations—translating contextual situations into a mathematical representation of some kind—and continually verify whether the mathematics they've chosen makes sense for the problem they're addressing. They also need regular opportunities to present their solutions to other students, the teacher, or other adults (the "Report" part of the modeling cycle). The details of how to do this will vary, depending on the age and experience of the student.

CHOOSING MODELING TASKS

In some ways, appropriate tasks that call for mathematical modeling should meet similar criteria for any worthwhile mathematical task we choose in a problem-centered (upside-down) classroom. (See Message 12, "Upside-Down Teaching," and Message 32, "Problems Worth Solving," for discussions of such problems; Appendices A, B, and C contain a list of sources for tasks along with a tool for evaluating tasks.) But evaluating modeling tasks calls for a few additional criteria. In particular, whereas mathematical tasks in general might appear either in a real context or a purely mathematical context, modeling problems, by definition, must be related to the real world. Also, in order to allow for the full modeling process, there needs to be some lack of definition in the task as it is presented, calling for students to consider, discuss, clarify, and refine the task into a well-posed problem ready for mathematical work. Rita Borromeo Ferri (2013), a well-known modeling expert from Germany, has suggested that, in general, modeling tasks for school mathematics should be

- open,
- complex,
- authentic,
- realistic, and
- solvable through all phases of the modeling process (like the process described earlier in this message).

She further notes that, for secondary students, problems should

- originate from applied mathematics or from industry;
- be only a little simplified;
- often have no known solution;
- often present only a problematic situation, where students have to determine or develop a question which can be solved; and should
- lead to various possible problem definitions and solutions.

Ferri also suggests that students should not watch modeling being done, but should interact directly with a modeling problem themselves. She recommends that teachers not intervene in students' process prematurely, noting that students' experiences of helplessness and insecurity are, in fact, central to the process and a necessary phase.

Teachers in the United States may not want students to feel totally helpless and insecure, but it's worth considering that their constructive struggling may be an important aspect of implementing this practice. The kind of hard work and extended effort necessary to work through a modeling task like those described above also support a growth mindset of intelligence, where the struggle itself can help students become smarter (see Message 1, "Smarter Than We Think"). By working through a hard, even ill-defined problem, students also develop confidence and perseverance in their mathematical ability, supporting their growth in this practice and Practice 1 on problem solving.

ELEMENTARY STUDENTS AND MATHEMATICAL MODELING

Clearly, young children do not need to tackle problems involving weather or economic forecasts, and they may not implement the complete modeling process. But, they can definitely begin to learn to use at least some of the basic steps of the modeling process, and develop mathematical habits of mind that equip them to refine their use of the process over time. When approached in appropriate ways, word problems can be part of building a foundation for modeling, as students learn to consider a situation and write an addition or subtraction equation to represent it. It may well be that using blocks or cubes at this age to represent a situation might be part of the modeling process, but not as an end goal. Rather, young students might use such objects to act out a problem situation and help them arrive at a mathematical representation or mathematical model of the problem.

As students advance through the grades, they should have experience with more complex problems and also with posing problems. In learning to pose a problem, students are called on to think closely about a situation and to go back and forth between the situation and the mathematics—an important facet of the modeling process. It's also possible to give even young students an opportunity to deal with problems that may not be well formed and may not lend themselves to an instant decision about how to represent the situation mathematically. For example, a straightforward word problem designed for the upper elementary grades is discussed in Message 13, "Clueless":

> *A school has 295 students. School buses hold 25 students. How many school buses are needed to fit all the students?*

This problem could be transformed into a modeling problem by starting with the same information but asking the question (John Pelesko, personal communication, 2013):

> *How do we get all the students home?*

Now, students have to dig into the situation, ask questions, engage in discussion, and, eventually, try to generate a mathematically clear problem they can solve. In doing so, students may need to make assumptions, set certain conditions, clarify what they want to know, and try to narrow the scope of the problem: *Some students might get a ride home from a parent or friend. How many buses are available? What about students who stay after school for some reason?* This work and the related discussion help students mathatize the situation into a clear, well-formed problem and, eventually, come up with a solution.

MIDDLE SCHOOL STUDENTS AND MATHEMATICAL MODELING

As students move into middle school, they should be able to handle more complex problems, such as planning a school event or family outing or making a recommendation about a community issue, using proportional reasoning or other mathematics addressed at their grade level. In working with this kind of real situation, students can continue to refine their use of the five-step modeling cycle, first clarifying the problem and determining what mathematical question(s) to ask and then looking for a way to use mathematics to solve the problem, moving back and forth between the problem and the situation to verify that what they're doing makes sense, and reporting the results of their work after finding a solution.

HIGH SCHOOL STUDENTS AND MATHEMATICAL MODELING

By the time students are in high school, they are strengthening their set of algebraic, geometric, and statistical tools and are able to take on more sophisticated types of modeling problems. Perhaps they might propose a plan for a new recycling program or prepare a proposal to the school board for updating the technology in the school. The description of Practice 4 suggests that students might use geometry to model a design problem or use a function to represent the relationship between two monetary quantities. Throughout high school, modeling should be an element in every mathematics course, whether for STEM-intending students or students choosing a different direction for the future. The ability to deal with real problems, especially those that are not well formed, is an important life skill applicable to many careers and real situations, and it can help a student develop confidence and perseverance.

What Can We Do?

Mathematics education is clearly responsible for teaching how to use mathematics in everyday life and in intelligent citizenship, and let's not forget it.... So what really matters is learning and practicing the mathematical modeling process. The particular field of application—whether it is everyday life, being a good citizen, or understanding some piece of science—is less important than experiencing this thinking process.

—Dr. Henry Pollak ("What Is Mathematical Modeling?", 2014)

Most teachers and materials developers agree that we should focus our mathematics teaching on problem solving. But we can ramp up the level of mathematical thinking for our students by gradually including at least some problems that are less prescriptive and less specific to a particular procedure—problems from the world outside of math class that may not always be crystal clear. Developing the ability to make sense of such ill-formed problems, translate them into mathematically clear problems, find mathematical concepts or skills to address a problem, and possibly adjust the approach if it doesn't yield the desired results can be an exciting way for students to grow. In dealing with more and more problems that don't have immediately obvious solutions, students come to see themselves as smart math students who can use what they know in meaningful ways. When they see themselves that way, they're ready for whatever comes next.

Reflections and Discussion

FOR TEACHERS

- What issues or challenges does this message raise for you? In what ways do you agree with or disagree with the main points of the message?
- How is teaching for mathematical modeling different from teaching for problem solving?
- How often do you give students a chance to deal with problems that may not be "neat"—problems for which not only is the solution not obvious, but perhaps the problem lacks a clear mathematical question to answer?
- For the level you teach, how can you help students learn to model mathematically using the five steps described in this message?
- Look at the newspaper, evening news/weather shows, or the Internet to identify some models used in everyday life. How can you use these real examples of mathematical models to potentially motivate your students about using mathematics in the real world?
- What barriers do you see in implementing this practice? How can you address those barriers?

FOR FAMILIES

- What questions or issues does this message raise for you to discuss with your son or daughter, the teacher, or school leaders?
- Look at the newspaper, evening news/weather shows, or the Internet to identify some models used in everyday life. How can you use these real examples of mathematical models to potentially motivate your daughter or son about using mathematics in the real world?

FOR LEADERS AND POLICY MAKERS

- How does this message reinforce or challenge policies and decisions you have made or are considering?
- How much mathematical modeling do you think goes on in your school(s)? How can you increase attention to mathematical modeling?
- Look at the newspaper, evening news/weather shows, or the Internet to identify some models used in everyday life. How can you use these real examples of mathematical models to potentially motivate teachers and students about using mathematics in the real world?

RELATED MESSAGES

Smarter Than We Think

- Message 31, "Developing Mathematical Habits of Mind," and Message 40, "Mathematical Habits of Instruction," address overall mathematical thinking and reasoning.
- Message 32, "Problems Worth Solving," addresses the need for students to learn how to persevere in solving problems, including applied problems, similar to, but different from modeling problems.
- Message 12, "Upside-Down Teaching," advocates a problem-centered approach to teaching that fosters students' productive discourse, including explanations, justifications, and critiques of others' arguments.
- Message 9, "Learning to Work," emphasizes the importance of connecting mathematics to applications in the world outside of school.
- Message 33, "Making Sense of Mathematics," discusses the importance of developing quantitative reasoning ability, dealing with numerical situations in applied contexts, including modeling problems.

Faster Isn't Smarter

- Message 7, "Not Your Grandpa's Algebra," considers the importance of using applied problems and modeling in high school mathematics.
- Message 17, "Constructive Struggling," discusses the value of students struggling with challenging problems.

MORE TO CONSIDER

- *Connecting the NCTM Process Standards and the CCSSM Practices* (Koestler, Felton, Bieda, and Otten 2013) unpacks each practice and relates it to NCTM's five process standards, including sample classroom vignettes for elementary, middle, and high school.
- Common Core State Standards for Mathematics (NGA Center and CCSSO 2010) offers, in addition to the one-paragraph description of the modeling standard, a diagram of the modeling cycle (2010, 72) in the high school standards and discusses the integration of modeling across other standards.
- "Real World Math" (Coffee and Sherard 2012) is an online resource of published articles and supporting materials related to using middle grades mathematics in other content disciplines and real world situations. www.nctm.org/publications/worlds/default .aspx?id=34028.
- *Advanced Mathematical Decision Making (Student and Teacher Materials)* (Charles A. Dana Center at The University of Texas at Austin 2010b) offers materials for teaching a rigorous grade

12 mathematics course involving open-ended applications of mathematics related to statistics, finance, discrete mathematics and numerical reasoning. www.utdanacenter.org/amdm.

- "Mathematical Modeling in the School Curriculum" (Usiskin 2011) discusses the modeling process and how we can address mathematical modeling in school mathematics. http://ucsmp.uchicago.edu/resources/conferences/2011-09-11.

- "A History of the Teaching of Modeling" (Pollak, in Stanic and Kilpatrick 2003) presents an overview of mathematical modeling and looks at how it has been addressed in school mathematics over time.

- "What Is Mathematical Modeling?" (Pollak 2014) provides a concise overview of mathematical modeling and the role of modeling in kindergarten through grade 12 mathematics. http://dese.mo.gov/divimprove/curriculum/documents/cur-math-comcore-what-is-mathematical-modeling.pdf.

- "Habits of Mind: An Organizing Principle for a Mathematics Curriculum" (Cuoco, Goldenberg, and Mark 1996) argues for structuring mathematics programs around rich, sometimes ill-formed, problems and mathematical habits of mind.

- High School Mathematics and Its Applications (HIMAP) offers instructional modules for incorporating modeling lessons into high school mathematics. COMAP, www.comap.com/highschool/projects/himap.html.

- The Next Generation Science Standards (Achieve, Inc. 2014) includes an appendix connecting the science standards to the *Common Core State Standards for Mathematics*, with many of the connections demonstrating mathematical modeling. www.nextgenscience.org/next-generation-science-standards.

- "Model Data" (National Oceanic and Atmospheric Administration, www.ncdc.noaa.gov/data-access/model-data and NOAA Satellite and Information Service) and "NOAA National Operational Model Archive and Distribution System (NOMADS)" (National Environmental Satellite, Data, and Information Service [NESDIS], http://nomads.ncdc.noaa.gov/) are sites operated by NOAA that offer examples of mathematical modeling of weather with links to various related pages.

- *Principles and Standards for School Mathematics* (National Council of Teachers of Mathematics 2000) describes mathematical content and process standards, including those listed in this message.

Special appreciation to John Pelesko and Zalman Usiskin for their insights and suggestions on an earlier version of this message.

Standard for Mathematical Practice 5

Use appropriate tools strategically. (Strategic Use of Tools)

Mathematically proficient students consider the available tools when solving a mathematical problem. These tools might include pencil and paper, concrete models, a ruler, a protractor, a calculator, a spreadsheet, a computer algebra system, a statistical package, or dynamic geometry software. Proficient students are sufficiently familiar with tools appropriate for their grade or course to make sound decisions about when each of these tools might be helpful, recognizing both the insight to be gained and their limitations. For example, mathematically proficient high school students analyze graphs of functions and solutions generated using a graphing calculator. They detect possible errors by strategically using estimation and other mathematical knowledge. When making mathematical models, they know that technology can enable them to visualize the results of varying assumptions, explore consequences, and compare predictions with data. Mathematically proficient students at various grade levels are able to identify relevant external mathematical resources, such as digital content located on a website, and use them to pose or solve problems. They are able to use technological tools to explore and deepen their understanding of concepts.

—Common Core State Standards for Mathematics (NGA Center and CCSSO, 2010, 7)

CONNECTIONS TO STANDARD FOR MATHEMATICAL PRACTICE 5

Practice 5 reminds us that selecting and using tools is part of what it means to do mathematics. The practice notes that any of a number of tools allow us to solve problems and to represent situations and formulate and refine related mathematical models, connecting this practice to Practice 1 on problem solving and Practice 4 on mathematical modeling, as well as NCTM's problem solving and representation standards. The emphasis of this practice is not just on knowing how to use the tools, which is important, but learning how to be smart about selecting appropriate tools at appropriate times for appropriate uses. (See Appendix F for a description of the five NCTM process standards [2000].)

MESSAGE

36

Building and Using a Mathematical Toolbox—High-Tech, Low-Tech, and Brain Power

STANDARD FOR MATHEMATICAL PRACTICE 5:
USE APPROPRIATE TOOLS STRATEGICALLY

We've got to find a way to make *this* fit into the hole for *this* using nothing but *that*.

—NASA technicians, *Apollo 13* (1995)

The astronauts in the film *Apollo 13* faced a seemingly daunting challenge when a problem in space caused massive damage to their capsule. All they had available were the standard tools— none of which were suitable for the unpredicted problem—and a variety of materials that happened to be on board for other purposes. The engineers and technicians on the ground in Houston gathered together for an emergency meeting with a pile of *stuff*—duplicates of everything the astronauts had with them, including not only their tools, but also odds and ends of materials, most of which were not designed for engineering or repair purposes. Their challenge, with time running out, was to determine the appropriate tool or tools to solve the astronauts' problem, or use the materials available to fashion a new tool. Finally, the on-ground team came up with a solution for a new kind of tool, and the astronauts were able to assemble the new tool with what they had available. They then returned safely to earth.

In mathematics teaching today, we have access to far more tools than the astronauts had at their disposal. Our day-to-day teaching may not be quite as dramatic as the Apollo 13 crisis, but our decisions about what tools we use in teaching and what tools our students need to know how to use are critical in helping students develop the strategic skill of knowing how and when to use the many tools available. This ability is an important mathematical habit of mind, allowing us to take on problems we couldn't otherwise solve and using the tools to reach higher levels of mathematical learning and thinking than we could otherwise reach.

A Closer Look at the Strategic Use of Tools

Listed in the description of Standard for Mathematical Practice 5 are a variety of technological tools, including calculators and a wide range of computer applications, from spreadsheets and computer algebra systems to statistical packages and dynamic geometry software. This list of technological tools quickly will become outdated as new types of devices and systems are invented and as new ways to use existing tools are discovered. Our understanding of this practice needs to be general enough to adapt to new tools as they appear over time.

Beyond technology, however, students also need experience using a variety of tools for developing mathematical concepts and solving problems, including tools for measurement (rulers, protractors, balance scales); tools for geometric constructions (compass, straightedge); and manipulative tools (counters, base ten blocks, algebra tiles). It can also be helpful for a student to use tiles or blocks as tools to visualize a problem situation and determine an appropriate problem-solving approach.

Perhaps the most low-tech tool of all is a pencil. There are times when a pencil is a more appropriate choice as a tool than a calculator or computer. Even more useful than a pencil may be the ability to estimate and do some calculations mentally. The ability to do mental mathematics, while not specifically mentioned in the description of this practice, is an important addition to this list, helpful for both estimation and for performing reasonable mental calculations. Mental mathematics may be the most important tool of all in our high-tech world, in that we always have it with us and it can help us determine whether we might need other tools, from pencil and paper to the most advanced mathematical software or devices available. Investing instructional time on mental math is always a good idea. Students benefit, and parents love it when their sons and daughters come home from school being able to do math in their heads.

KNOWING HOW AND KNOWING WHEN

As technology continues to evolve and provide us with new tools, new opportunities, and new challenges, the most lasting and transferable skill we can offer students is the ability to know when (and when not) to use an available tool. We want students to learn that there may be times when doing without a high-tech tool may be faster and more appropriate for the task at hand. For example, it makes no sense whatsoever to reach for a calculator to multiply a number by 10 or to estimate how much water a container will hold. The first takes far longer than remembering and recalling the mental calculation. The second is silly, since it isn't possible to use a calculator for that task, except perhaps as a dip stick. And there are certain important mathematical skills we might want students to know how to do at least some of the time mentally, with a pencil and paper, or with low-tech measurement or construction tools—sketching the graph of a line on a well-labeled pair of axes, constructing a right angle, knowing multiples of ten, measuring the length of a line segment or the size of an angle, being able to solve a one- or two-step equation. If teachers don't address the topic of strategically selecting what tool to use and when to use it, students will be left on their own outside of school to make such decisions, often reaching for an inefficient or inappropriate tool.

Meanwhile, educators need to stay as current as possible with knowing what tools can help students advance their mathematical thinking and learning and how to use those tools appropriately at their grade level. Some tools allow us to solve problems more efficiently, performing calculations or constructions that might otherwise be time-consuming. But some tools are so innovative that they open new pathways for solving problems or allow access to mathematical topics not accessible without them. Matrices, for example, become a powerful tool for solving algebraic equations when students know how to use them, and being able to use matrices allows students to take on higher-level problems such as dealing with the inventory of jeans in a retail store—various sizes and lengths—or organizing a supply chain for a mail-order business—keeping track of certain materials for a custom-made computer case as orders are filled. And using a computer algebra system can open the door to understanding a wide range of functions in advanced algebra, much as graphing calculators allow students to visualize and deal with more diverse forms of algebraic functions than they could address without the technology.

We should never assume, however, that we know the best way for students to use a tool like a graphing calculator or a particular piece of software. NCTM past president, high school mathematics teacher, and teacher educator Gail Burrill once said in 1997 to a group of teachers that when she gave her high school students graphing calculators, they started coming up with ways of approaching problems that she had never imagined. We can help students learn ways of using tools, but then we must accept that they will likely devise other ways to solve problems and may even invent new types of problems to solve.

A NOTE ON EQUITY

One of the biggest issues in helping all students learn how and when to use technology is that many students, because of their socio-economic status, may not have the same access to tools outside of school as other more affluent students. And some schools will be less well equipped with technology than others. It's critical that schools be a place where everyone can access powerful mathematical tools; it can't be left entirely to students to provide their technology. Thus, it's important for communities and schools to discuss how to make access to technological tools a reality, especially in areas where students may not be likely to have these tools at home.

A Classroom View of the Strategic Use of Tools

The first step in helping students learn to be strategic in their use of tools is to give them experience using the tools—manipulative materials, measurement/construction tools, numerical and graphing calculators, dynamic geometry software, spreadsheets, computer algebra systems, or tools yet to be invented—so they know what the tool can help them do. Beyond familiarity and experience, however, students need opportunities to use tools to solve problems. Part of the process of solving problems involves selecting which tools, if any, might be useful and appropriate. This can be especially important for something like manipulative materials that students use to learn concepts. Students may have used place-value blocks or algebra tiles, for example, in learning about numerical or symbolic expressions. But they may not realize these tools can be helpful in solving problems, either mathematical or contextual. Teachers can remind students that the tools are available as they undertake a problem. More importantly, teachers can facilitate discussion about how different students used the manipulatives and in what ways they found them helpful (or not helpful).

With respect to technology, we should be careful not to think that students should have to prove they don't need it (demonstrating they can perform every type of computation or symbol manipulation the calculator or device can do) before being allowed to use it. On the contrary, sometimes using a technological approach can help students see enough instances of a mathematical concept or idea that they begin to make generalizations about the concept, and sometimes they may even invent mathematically sophisticated uses of the technology. Even as we help students learn to use technology, we need to also help them learn to use their brains. We need to teach them mental math skills, from knowing combinations that add up to ten or one hundred to knowing how to calculate 10 percent or 15 percent of any amount to being able to estimate the volume of a box. And we are not yet ready to retire all of our pencils—although electronic tablets may be making a move in that direction—and certainly not ready to retire the essential tools of measurement and geometry.

A student's effective and strategic use of a particular tool will evolve over time, as students become more comfortable with the tool and also more comfortable with the mathematics the tool supports. In the case of some tools, like manipulative materials or calculators, at some point students may realize that using the tool is less efficient than performing an operation mentally or grabbing a pencil. The ability to strategically select and use tools means a student is able to decide when and how to use it and, most of all, when not to.

What Can We Do?

The astronauts on Apollo 13 ended up using a combination of high-tech and low-tech tools to solve their life-or-death problem. But mostly, they used the combined brain power of a team of people working together to determine how to use the tools they had in the most efficient and helpful way possible. With all of the tools available today for mathematics teachers and students, the most important job teachers have is to give students experience with the tools and engage students in purposeful conversations about the tools' uses and limitations. Talk about when it makes sense to use a tool and when it might be more cumbersome or time consuming to use it than to do something the *old-fashioned way*. Talk about what happens if a new tool were to come along that might make a person consider replacing another way of solving a problem. Under what conditions might it make sense to jump on the opportunity to use a new tool? Under what conditions might a person want to step back and consider other issues? If teachers don't take the time to invest in helping students become wise consumers of a growing onslaught of tools, the next time a student sees a shiny new thing outside of school, like a tip calculator or new app for a mobile phone, the student may be seduced by its glitz and forget that he or she knows how to multiply 7×10 or compute 10 percent or 15 percent without it.

Reflections and Discussion

FOR TEACHERS

- What issues or challenges does this message raise for you? In what ways do you agree with or disagree with the main points of the message?

- What tools do your students routinely use in your classroom? Have they had conversations about when using certain tools makes sense and when it might make more sense to do something mentally or with a pencil or low-tech measurement tool?

- What kind of access do you provide students to technology appropriate for your grade level?

- Do you have prerequisite pencil-and-paper skills expected before students are allowed to use technology tools? If so, are you open to discussing with colleagues whether withholding access to the tools helps or hinders students?

- What barriers do you see in implementing this practice? How can you address those barriers?

FOR FAMILIES

- What questions or issues does this message raise for you to discuss with your son or daughter, the teacher, or school leaders?

- What tools do you use in your every day life to do mathematics (spreadsheets, calculators, computer programs, electronic devices, and so on)? Can you have conversations with your daughter or son to discuss how you use the tools and when you choose not to?

- What is your son's or daughter's teacher's philosophy of using technological tools in the classroom? Are there school policies about their use? How can you find out more about how these tools are used in class?

- How open are you to the possibility that it might help your daughter or son learn mathematics selectively using a calculator, graphing calculator, spreadsheet, or other type of computer software or electronic device?

FOR LEADERS AND POLICY MAKERS

- How does this message reinforce or challenge policies and decisions you have made or are considering?

- What is your policy on using mathematical technology in the classroom, including calculators, graphing calculators, spreadsheets (on calculators, computers, phones, and so on), dynamic geometric software, computer algebra systems, statistical packages, and new tools that may not yet be available (or invented)?

- How can you support teachers in helping students make wise choices about using technological tools as part of their mathematics learning?

- How do you ensure access to appropriate tools for all students, regardless of their economic situation at home?

RELATED MESSAGES

Smarter Than We Think

- Message 31, "Developing Mathematical Habits of Mind," and Message 40, "Mathematical Habits of Instruction," address overall mathematical thinking and reasoning.

- Message 24, "Beyond PowerPoint," looks at the role of technology in the classroom.

- Message 28, "Bringing Testing into the 21st Century," reminds us of the need to incorporate appropriate tools in tests that target mathematical thinking.

Faster Isn't Smarter

- Message 24, "Do It in Your Head," emphasizes the importance of teaching students mental math.

- Message 30, "Crystal's Calculator," relates the story of a student who made the choice to learn pencil and paper computation skills after having access to a calculator in algebra.

- Message 5, "Technology Is a Tool," looks at the impact of technology on our choices of what and how we teach.
- Message 20, "Putting Calculators in Their Place," considers the role of calculators in the mathematics classroom, especially at the elementary level.

MORE TO CONSIDER

- *Connecting the NCTM Process Standards and the CCSSM Practices* (Koestler, Felton, Bieda, and Otten 2013) unpacks each practice and relates it to NCTM's five process standards, including sample classroom vignettes for elementary, middle, and high school.
- "The Ethics of Using Computer Algebra Systems (CAS) in High School Mathematics" (Usiskin 2012) discusses the strategic and ethical uses of this advanced technology in school mathematics. http://ucsmp.uchicago.edu/resources/conferences/2012-03-01.
- *Focus in High School Mathematics: Technology to Support Reasoning and Sense Making* (Dick and Hollebrands 2011), part of NCTM's series Reasoning and Sense-Making in High School, describes the use of various technologies in support of teaching high school mathematics with a focus on reasoning and sense-making.
- "Using Calculators for Teaching and Learning Mathematics" (Ronau, Rakes, Bush, Driskell, Niess, and Pugalee 2011) summarizes research and recommendations on using calculators in school mathematics.
- "African Americans and Technology Use: A Demographic Portrait" (Smith 2014) offers a demographic look at technology use. http://pewinternet.org/Reports/2014/African-American-Tech-Use.aspx.
- "Who's Not Online and Why" (Zickuhr 2013) is a study of who uses the Internet. http://pewinternet.org/Reports/2013/Non-internet-users.aspx.
- "How Teachers Are Using Technology at Home and in Their Classrooms" (Purcell, Heaps, Buchanan, and Friedrich 2013) reports how and where teachers use technology. http://pewinternet.org/Reports/2013/Teachers-and-technology.aspx.
- *Principles and Standards for School Mathematics* (National Council of Teachers of Mathematics 2000) describes mathematical content

and process standards, including those in this message; also includes a principle on technology.

• Common Core State Standards for Mathematics (NGA Center and CCSSO 2010) includes descriptions of the eight standards for mathematical practice.

Standard for Mathematical Practice 6

Attend to precision. (Precision)

Mathematically proficient students try to communicate precisely to others. They try to use clear definitions in discussion with others and in their own reasoning. They state the meaning of the symbols they choose, including using the equal sign consistently and appropriately. They are careful about specifying units of measure, and labeling axes to clarify the correspondence with quantities in a problem. They calculate accurately and efficiently, express numerical answers with a degree of precision appropriate for the problem context. In the elementary grades, students give carefully formulated explanations to each other. By the time they reach high school they have learned to examine claims and make explicit use of definitions.

—Common Core State Standards for Mathematics (NGA Center and CCSSO, 2010, 8)

CONNECTIONS TO STANDARD FOR MATHEMATICAL PRACTICE 6

Standard for Mathematical Practice 6 emphasizes communicating one's thinking with precision, attending to careful use of mathematical language and definitions in reasoning and discussions (NGA Center and CCSSO 2010, 7). This practice relates primarily to NCTM's communication standard (2000). Because of the need for accurate language in reasoning, this practice also relates closely to Practice 3 involving making reasoned arguments, explicitly calling for students as early as the elementary grades to "give carefully formulated explanations to each other." Practices 3 and 6 together offer an important focus for instruction around rich student discourse and the development and communication of reasoning. Thus, Practice 6 also relates to Practice 2 on abstract and quantitative reasoning, as well as to NCTM's reasoning and proof standard. Because using mathematical definitions and terminology precisely is so central to understanding and using mathematics overall, this notion of precision is an important element of mathematical structure, connecting this practice to Practice 7. Practice 6 also involves paying attention to precision when using measurement tools, constructing coordinate graphs, and performing computations when an exact answer is necessary using pencil and paper, a calculator, or mentally, thus connecting this practice closely to the mathematical content standards involving measurement, graphing, and computation. (See Appendix F for a description of the five NCTM process standards [2000].)

37

Communicating Mathematically— Using Precision in Language and More

STANDARD FOR MATHEMATICAL PRACTICE 6: ATTEND TO PRECISION

Traveling through hyperspace isn't like dusting crops, boy! Without precise calculations, we could fly right through a star or bounce too close to a supernova and that'd end your trip real quick, wouldn't it?

—Han Solo, to Luke Skywalker in *Star Wars* (Lucas 1977)

Han Solo's quote reminds us of the importance of precision in computation, not only in space travel but also in the many ways we use mathematics. It also reminds us of the importance of communicating mathematics precisely in order for those calculations to serve their particular purpose; to have calculated correctly but not have transmitted the results accurately could have been disastrous for Han Solo's voyage in *Star Wars*. Standard for Mathematical Practice 6 emphasizes the importance of considering precision in many aspects of mathematics, including computation, but especially calls attention to the importance of precision in how we communicate mathematical ideas, reasoning, and thinking.

There's a thin line between being precise and focusing so intently on getting the right answer that a person becomes immobilized or discouraged about dealing with any new mathematical idea or problem. Many adults can remember getting math problems almost right except for one careless error resulting in a *wrong* answer; too many such experiences can affect a person's willingness to try another problem and can even lead to hating math. Today we see school mathematics as consisting not only of computation, but also of many types of mathematical concepts and many types of problems, some of which may involve estimation and some of which may have multiple answers. Most of all, today we know that communicating is a critical part of mathematics—being able to explain, justify, even argue using symbols, words, and language correctly to convey mathematical ideas and reasoning, whether to guide a space ship, plan a school event, or design a city's budget.

Even as we embrace this broad vision of mathematics, the need for precision remains an important part of our work with students, not only in obvious situations like performing a calculation or using a measurement tool, but also in learning how to communicate precisely with mathematics and about mathematics. We don't want students to be afraid to make a mistake; on the contrary, making mistakes can be a powerful lever for student discussion and improved learning. We do, however, want students to continually work toward using their emerging mathematical language accurately, realizing that specific words mean something particular in mathematics and that the words can be used in certain ways to communicate mathematical ideas in explanations and arguments.

A Closer Look at Precision

Practice 6 focuses on precision in communication, but also addresses precision more broadly. The notion of precision is reflected somewhat differently in the various parts of mathematics where it appears.

PRECISION WITH COMMUNICATION

Students can begin developing mathematical language from their first experience with mathematics. Teachers can help students build precise mathematical language by using correct terms when introducing mathematical ideas. For example, students can learn the word *equation* rather than *number sentence*, not only building their precise mathematical vocabulary, but also avoiding confusion; young children learning grammar at the same time may wonder where the upper-case letter and period go in the number "sentence."

By being careful with our own language and communication, we can also avoid teaching 'temporary mathematics' that may need to be later undone. It may lead to future problems both with mathematical understanding and with the use of precise language to tell students, for example, that in solving an equation, "You can move a number to the other side and change its sign." As another example, teaching children that rectangles and squares are two distinct shapes can interfere with their use of the definitions of these terms and potentially interfere with their ability to communicate precisely about such figures, not to mention that someone will have to *unteach* what students have previously learned when we later let students know that a square is a particular kind of rectangle.

NCTM's communication standard suggests that by middle school, students should understand mathematical definitions and be able to use them and communicate with them, and that this use should become pervasive in high school. But the standard also reminds us that the path to precise language should not be rushed. It begins with students expressing their ideas and their thinking in their own words, progressing to using age-appropriate mathematically accurate language and, eventually, more formal and precise mathematical definitions and terms. As we give students opportunities to talk in math class, we validate their role in learning and we help them clarify their thinking. Engaging them in this way also allows students to be ready to learn the nature of mathematical language we want them eventually to use.

PRECISION WITH SYMBOLS

Communicating precisely with symbols is also an important aspect of Practice 6. Students should learn early on the proper use of the *equal* (=) sign to assert that two quantities have the same value. Many young children come to think that the equal sign means to give the answer to what lies to the left of the sign. By using this symbol in more than just that one way—perhaps showing equations where what is missing appears on the left side of the sign or as part of an expression—we help students hone their ability to communicate using symbols. Attending to the meaning of the symbols we use is important throughout a student's education. It may be helpful for students to think of the *greater than* (>) or *less than* (<) signs as being the jaw of a hungry alligator (I still do sometimes), but eventually they need to formalize their use of these symbols and incorporate them into a growing collection of symbols that precisely communicate mathematical ideas and thinking. And students need to consider different ways symbols may be used in entering information and interpreting results with a graphing calculator, spreadsheet, and so on.

PRECISION WITH GRAPHS

Practice 6 also calls for students to be precise in constructing graphs, paying attention to the units used, labeling axes correctly, and so on. Even when using graphing technology, it's important for students to pay attention to units on axes (as reflected in their choice of window) and to generate graphs with whatever level of precision is called for by the mathematical task or problem at hand.

PRECISION WITH MEASUREMENT

Practice 6 also reminds us of the need for precision when students use measurement tools. Precision starts with using the tools correctly, such as lining up a ruler with the *0* as the beginning of what is being measured, not the end of the ruler or the *1*, or being able to use a balance scale accurately or read a measuring cup. This practice also reminds us of the importance of being careful in specifying specific units of measurement when reporting the solution to a problem.

Precision in measurement is important in many industries, and it has appeared in and out of the mathematics curriculum over the years, usually at about the middle school level. In science and business, some tasks require measurements to a high degree of precision, defined as part of the task. But there is only a certain level of precision that any tool can generate, so matching the tool to the task is important. I recently noticed that a new bathroom scale gave decimal weights to the tenth of a pound. Yet, on further investigation, it turned out that measurements only appear in even-numbered tenths (95.2, 95.4, 95.6, and so on). *What is the precision of the scale? Is it to the nearest tenth of a pound? The nearest fifth of a pound? And can a bathroom scale accurately distinguish between 95.2 pounds and 95.4 pounds?* These are important kinds of questions students can discuss as they develop their understanding of precision in measurement.

PRECISION WITH COMPUTATION

The description of Practice 6 is intentionally brief, but it clearly implies that students need to be aware of and pay attention to precision in their work throughout mathematics. The most obvious area for such precision goes back to the beginning of this message—getting the right answer to a computation problem. Sometimes an exact answer is not called for in a problem, such as looking for an estimate of the cost of a project in advance. Sometimes a range of answers may answer a problem. But sometimes an exact answer is needed, either to a straight computation problem or an extended task that involves computation. In such cases, we want to encourage students to be precise—to be careful and not make careless mistakes. But here, again, we don't want to make students so worried about making a mistake that it hinders their confidence and willingness to persevere with a hard problem.

Finally, precision is required when using a calculator or other technology to perform a calculation. Students need to learn that it's possible to make a careless mistake on a calculator, just as it is with a pencil and paper. By balancing the value of learning from mistakes and the importance of accuracy, we can help students grow in their ability to reasonably attend to precision.

A Classroom View of Precision

As with many abilities, the best way to learn precision—in communication, graphing, measurement, or computation—is to use it. Students need many opportunities to communicate their mathematical ideas and thinking, ideally in a classroom routinely focused on student discourse around worthwhile mathematical tasks. Such discourse closely supports the development of Practice 3 on constructing and evaluating arguments. Students don't need to be corrected on every word they say or write, but they need to grow year by year in their level of sophistication and expertise with respect to their use of mathematical definitions and terms, both orally and in writing.

Likewise, students need opportunities to sketch simple graphs on a number line and coordinate plane and opportunities to use graphing technology to build more complex coordinate graphs, with discussions about their choice of units, labeling of axes, parameters for the graphing window, and so on. They need opportunities to measure real objects and discuss precision with respect to their work with measurement.

And, yes, students need opportunities to perform computations, including those with pencil and paper. In today's high-tech world, students need to spend far less time on such computations than in the past, perhaps working with reasonable-sized divisors and manageable fractions. For more complex computations, they need to pay attention to precision when using a calculator or other technology. They can use estimation skills and their understanding of numbers and operations to make sure the answer generated by the technology makes sense. They can also use their skills to translate the results provided by their tools into an appropriate answer to the problem being solved. Finally, they can benefit from attending to precision as they develop at least some mental math skills.

HELPING ENGLISH LANGUAGE LEARNERS COMMUNICATE

Developing and using precise mathematical language can be a special challenge for students whose first language is not English. These students, especially, need opportunities to express themselves in explaining their thinking and asking questions of others, such as those offered when we organize teaching around good problems. We must be alert, however, to words that may be used in confusing ways in English communication of mathematical ideas. As students first learn English, they depend on what they hear. Words like *write* and *right* can be confusing, and the many ways we use the words can add to the confusion (*Write* your name in the upper *right* corner of your paper. Draw a *right* angle on your paper. Yes, *right* there. That's *right*!) (Leiva 2007). By being aware of potential confusion, teachers can head off the confusion by engaging students in discussions specifically related to the words being used.

As we let students talk and explain, we can gain insights into which words they may misunderstand or misinterpret.

Teachers can also help students, both English language learners and students who are proficient in English, develop precise mathematical language by discussing words they hear outside of math class that may have a different meaning in mathematics—similar, area, function, right, factor—or that may have somewhat loose meanings when used outside of mathematics—"Everyone get in a circle." "I'd like a brownie square."

COMMUNICATING ON *THE TEST*

Fortunately, the two major testing systems being developed to assess the Common Core State Standards for Mathematics have indicated Practices 3 and 6 as primary targets. Students will be expected to explain their thinking and construct logical arguments (Practice 3), communicating with mathematical precision appropriate to their grade level (Practice 6). This emphasis on mathematical practices in testing can positively influence classroom practice, supporting teachers in helping students develop their ability to communicate mathematically by actually doing it.

What Can We Do?

Striving for precision should not involve taking on a right/wrong attitude or becoming overly punitive with respect to mistakes. We now know that as students make mistakes and strive to correct them, their brains will continue to grow. Nevertheless, precision is an important consideration for students' work in measurement, geometry, number, and other mathematical content areas, as part of their development of mathematical habits of mind.

Likewise, with respect to communication, it's important to find a balance between letting students develop their language naturally and helping them become more precise in their language. Neither we nor they can lose sight of the vision of mathematics as a discipline grounded in order, structure, and agreements about how we communicate ideas precisely. A major part of our role as educators is to make sure students understand these agreements and know how to use mathematical language appropriately to both develop and express their thinking.

There are aspects of this practice that will be reflected any time a student communicates his or her thinking about properties, patterns, or problems. If we can internalize the intention of this practice to encourage students to pay attention to the language and numbers they use, we can help students improve their ability to convey their ideas effectively and with increasing precision and support their learning of mathematics content at the same time.

Reflections and Discussion

FOR TEACHERS

- What issues or challenges does this message raise for you? In what ways do you agree with or disagree with the main points of the message?
- How do you help students develop and refine their language? Do you give them frequent opportunities to explain and justify?
- How can you help students calculate more accurately without discouraging them about possibly making mistakes?
- How can you support students to learn from their mistakes even as they improve their level of precision?
- Are there words you use at your level that might be considered *temporary mathematics*—needing to be retaught or untaught later? If so, what keeps you from using more precise words instead, and are you open to considering the possibility that students might be able to learn the more precise words?
- What barriers do you see in implementing this practice? How can you address those barriers?

FOR FAMILIES

- What questions or issues does this message raise for you to discuss with your daughter or son, the teacher, or school leaders?
- How can you give your son or daughter the opportunity to tell you in good mathematical language what he or she is learning in school or to explain a problem he or she just solved?
- How can you use mistakes as part of the learning process to help your daughter or son be more accurate in graphing, measuring, or computing?

FOR LEADERS AND POLICY MAKERS

- How does this message reinforce or challenge policies and decisions you have made or are considering?
- How well do your teachers organize their classrooms to stimulate discussion and explanation and to help their students learn to communicate precisely? How can you help them find time to do this?
- How can you help your teachers learn appropriate mathematical terms and definitions if they are not mathematics experts?
- How can you connect your teachers with teachers at other grade levels to discuss how language evolves over the grades and to share ideas for helping students develop their ability to communicate mathematically?
- How well do your teachers balance a focus on accuracy and precision with strategies for students to learn from mistakes? How can you make finding this balance a priority?

RELATED MESSAGES

Smarter Than We Think

- Message 31, "Developing Mathematical Habits of Mind," and Message 40, "Mathematical Habits of Instruction," address overall mathematical thinking and reasoning.
- Message 2, "¿Habla matemáticas?," offers suggestions for supporting the development of mathematical language with students whose first language is not English.
- Message 28, "Bringing Testing into the 21st Century," advocates including on high-stakes tests questions that call for students to communicate their thinking.
- Message 12, "Upside-Down Teaching," offers a vision of teaching around problems with extensive opportunities for student discourse and language development.
- Message 8, "Oops!," reminds us of the value of making mistakes on the road to right answers.

Faster Isn't Smarter

- Message 22, "We Don't Care About the Answer," reminds us that arriving at correct answers is important, even as we focus on the process used to arrive at the answer.
- Message 17, "Constructive Struggling," advocates a teaching approach that calls for students to communicate their thinking about challenging problems.

MORE TO CONSIDER

- *Connecting the NCTM Process Standards and the CCSSM Practices* (Koestler, Felton, Bieda, and Otten 2013) unpacks each practice and relates it to NCTM's five process standards, including sample classroom vignettes for elementary, middle, and high school.
- *Promoting Purposeful Discourse: Teacher Research in Secondary Math Classrooms* (Eisenmann and Cirillo 2009) describes cases of secondary teachers promoting students' mathematical discussions and conversations.
- *The Young Child and Mathematics, 2nd Edition* (Copley 2010) discusses the development of mathematical knowledge and thinking with young children, including the development of precise mathematical language.
- *The Problem with Math Is English: A Language-Focused Approach to Helping All Students Develop a Deeper Understanding of Mathematics* (Molina 2012) reminds us of the importance of

mathematical communication and makes recommendations about developing communication skills with students whose first language is not English.

- "Using Two Languages When Learning Mathematics: How Can Research Help Us Understand Mathematics Learners Who Use Two Languages?" (Moschkovich 2009) summarizes research on students learning mathematics in a language other than their native language.

- *Classroom Discussions in Math: A Teacher's Guide for Using Talk Moves to Support the Common Core and More, 3rd edition* (Chapin, O'Connor, and Anderson 2013) describes how to orchestrate discussions that incorporate precision in communication.

- *Principles and Standards for School Mathematics* (National Council of Teachers of Mathematics 2000) describes mathematical content and process standards, including those listed above.

- Common Core State Standards for Mathematics (NGA Center and CCSSO 2010) includes descriptions of the eight standards for mathematical practice.

Consider also the resources listed in Message 2, "¿Habla matemáticas?," related to supporting mathematical language development for English language learners, including those listed below.

- *The Problem with Math Is English: A Language-Focused Approach to Helping All Students Develop a Deeper Understanding of Mathematics* (Molina 2012) focuses on language development as a key for helping students learn mathematics.

- *Supporting English Language Learners in Math Class, Grades K–2* (Bresser, Melanese, and Sphar 2008a) provides lessons and other resources for supporting English language learners in mathematics.

- *Supporting English Language Learners in Math Class, Grades 3–5* (Bresser, Melanese, and Sphar 2008b) provides lessons and other resources for supporting English language learners in mathematics.

- *Supporting English Language Learners in Math Class, Grades 6–8* (Melanese, Chung, and Forbes 2011) provides lessons and other resources for supporting English language learners in mathematics.

- *Supporting English Language Learners in Math Class: A Multimedia Professional Learning Resource, Grades K–5* (Bresser, Felux, and Melanese 2009) includes two books, DVD with video, professional development and coaching resources, and a *Facilitator's Guide* for supporting the mathematics learning of students whose first language is not English.

Standard for Mathematical Practice 7

Look for and make use of structure. (Mathematical Structure)

Mathematically proficient students look closely to discern a pattern or structure. Young students, for example, might notice that three and seven more is the same amount as seven and three more, or they may sort a collection of shapes according to how many sides the shapes have. Later, students will see 7×8 equals the well-remembered $7 \times 5 + 7 \times 3$, in preparation for learning about the distributive property. In the expression $x^2 + 9x + 14$, older students can see the 14 as 2×7 and the 9 as $2 + 7$. They recognize the significance of an existing line in a geometric figure and can use the strategy of drawing an auxiliary line for solving problems. They also can step back for an overview and shift perspective. They can see complicated things, such as some algebraic expressions, as single objects or as being composed of several objects. For example, they can see $5 - 3(x - y)^2$ as 5 minus a positive number times a square and use that to realize that its value cannot be more than 5 for any real numbers x and y.

—Common Core State Standards for Mathematics (NGA Center and CCSSO 2010, 8)

CONNECTIONS TO STANDARD FOR MATHEMATICAL PRACTICE 7

Given that it relates to the structure of mathematics overall, Standard for Mathematical Practice 7 influences all of the other seven practices (NGA Center and CCSSO 2010, 8). In particular, Practice 8 on regularity in repeated reasoning addresses the value of noticing patterns and making generalizations based on those patterns, an important element of mathematical structure common to Practice 7. Also, Practice 6 on communicating precisely emphasizes the importance of using precise mathematical definitions, an important element of mathematical structure.

While mathematical structure is not the explicit focus of any one of the NCTM process standards, it is especially evident in NCTM's connections standard, as mathematical structure connects big ideas across grade levels and content strands. NCTM's reasoning and proof standard is based on the principles of logic and reasoning that form the structural operating system for the discipline of mathematics, thus also connecting Practice 7 to this standard. (See Appendix F for a description of the five NCTM process standards [2000].)

38

Building Things— Helping Students Understand and Use Mathematical Structure

STANDARD FOR MATHEMATICAL PRACTICE 7:
LOOK FOR AND MAKE USE OF STRUCTURE

The building that inspired memories of my engineer father—and this message.

My father built things. I thought about him one recent morning as I drove by a building nearing completion not far from where I live. I saw two workers in hard hats looking up at the sprawling five-story structure composed of Texas limestone, bricks, and large expanses of glass. Maybe they were discussing some aspect of the work that still needed attention. I imagined them driving by that building in ten or twenty years and knowing that they had created something significant, even beautiful, that likely would stand for decades. They might remember having placed *that* rock where it needed to go or having built *that* balcony where someone might be standing just then.

My father must have experienced feelings like that. He was a civil engineer, working in different jobs in the private and public sector over the years, often overseeing construction. During his long career, he led the design and/or building of fire stations, gas stations, a water treatment facility, a manufacturing plant, an elementary school, housing subdivisions, bridges and numerous roads, as well as many other projects. I was always proud to see what he had built. In his later years, I enjoyed driving down a road with my parents and having my father (or more often my mother) point out some tangible evidence of his diverse and distinguished career.

In teaching mathematics, we also have the opportunity to build things. Or, more appropriately, we have the opportunity to help students build things. What they build can have lasting effects for their future. In particular, we can help them build a strong working knowledge of mathematics and ways of thinking mathematically grounded in mathematical structure. From that foundation, students can learn the mathematics they need and the mathematical habits of mind that will let them use what they learn.

A Closer Look at Mathematical Structure

We can think of mathematical structure in some ways like the structural elements of a building if we could x-ray it (to see inside) and also look at it from different perspectives (from a satellite photograph or from standing inside). The building can be seen as a whole, but also as its parts (rooms, corridors, balconies, storage facilities), and its various utility systems (plumbing, electricity, air conditioning, heat, and so on). The building can't exist without careful attention to how its parts are built and what they're built of, all following certain *guiding principles*—scientifically known properties about engineering and agreements about rules that need to be followed (like building codes or neighborhood covenants and restrictions) that specify to the architect how it will be built. A detailed materials list contains *precise specifications* for what materials will be used for each element of the building, from bricks to sheetrock to flooring to windows. In the blueprints for the building, the engineers consider all of these principles and offer views of the building from all different angles and perspectives, leaving as little room for error as possible.

BUILDING A STRUCTURAL UNDERSTANDING OF MATHEMATICS

In some ways, mathematical structure may be more complex than the construction of a building; many mathematicians spend their careers immersed in the study of its structure. Making sense of structure and how it relates to our day-to-day teaching can be challenging, to say the least. However, paying attention to mathematical structure can reap significant benefits for students, both in terms of how they understand certain mathematical concepts, and also in terms of how they can build new learning on what they already know. It's a bit of a stretch to use the metaphor of the structure of a building to talk about mathematical structure. But the overall idea of thinking of mathematics as a big building consisting of structural elements and built on *guiding principles* and *precise specifications* can help us begin to dig into the notion of mathematical structure—something that is present in everything we do in mathematics. At the risk of oversimplifying, the *properties* central to our work with numbers—commutative, associative, distributive—are part of the set of guiding principles for the discipline of mathematics. The *words and conventions* for how we write numbers, expressions, and equations and how we use mathematical vocabulary are part of the *specifications* for mathematics, and precision in how we define and use them is important (as noted in Practice 6).

Other structural characteristics can help us understand how mathematics works; we can uncover such characteristics by looking at mathematics through its *patterns* and *big ideas*. As we help students learn to pay attention to both the patterns and the big ideas of mathematics, and help them make and articulate generalizations based on what they see, they can begin to develop an important mathematical habit of mind—seeing what mathematics looks like beyond simply the problem at hand. Let's take a look at how patterns and big ideas can be helpful in understanding and using mathematical structure.

Patterns as a Structural Tool

Mathematics is a science of *patterns*, and we can use patterns as a springboard to explore many aspects of mathematics. When we help students learn to look for and notice patterns, and articulate what they see, it can give them insights into important elements of mathematical structure. They can build on patterns they notice (*I keep getting the same answer when I add the same numbers in a different order*) to make generalizations and to understand unifying properties of numbers (*it doesn't matter in which order I add two numbers*). The power behind this practice is that an understanding and appreciation of mathematical structure can give students the ability to use an important property like the distributive property to help them deal with otherwise complicated problems or numerical or symbolic expressions, such as algebraic expressions they may encounter in advanced mathematics courses.

Big Ideas as a Structural Tool

Mathematics consists of many concepts and procedures. Students can begin to make sense of all these concepts and procedures by thinking of them in bigger chunks, by focusing on *big ideas*.

Some *big ideas* are connecting themes that recur over and over again in mathematics—ideas like equivalence and proportionality. These ideas can span multiple grade levels, evolving into more sophisticated forms from year to year. Equivalence, for example, is an important big idea that is central to the study of mathematics every year. The idea of equivalent quantities is a critical structural element of mathematics, so investing in students' understanding of what the equal (=) sign means and how they can use it to express relationships will pay off time and time again.

Some *big ideas* might connect many related ideas, concepts, and skills, such as understanding the big idea of multiplication (what multiplication means, what it might look like in a problem, its relationship to other operations, how it connects with area, how to use an algorithm to multiply, and so on). When students can tie many concepts, skills, and problems together by understanding a big idea like multiplication that connects them, they are far more likely to be able to recognize a multiplication situation when they encounter one in a problem or a new concept to be learned.

THE BIG PICTURE OF MATHEMATICS AS A DISCIPLINE

Ideally, students will come to see mathematics as a discipline that consists of different types of numbers, letters, and geometric objects, guided by unifying principles and properties grounded in reasoning. We can help students notice these elements and come to see mathematics as a whole, along with its structural elements, including precise definitions and rules of logic and reasoning that govern the way we communicate with and about mathematics. An understanding of mathematical structure can serve as a foundation for students to develop computational proficiency, conceptual understanding of numerical and symbolic ideas, spatial sense to deal with geometric objects, and their ability to apply what they have learned in solving increasingly complex problems, both routine and non-routine. Their ability to learn these fundamental skills, grasp these basic understandings, make sense of these objects, and tackle these problems is based on understanding the characteristics of numbers and objects and on knowing how mathematical properties and symbols can represent relationships among those numbers and objects—all major aspects of mathematical structure.

Using Mathematical Structure: Moving from Numbers to Symbols

For example, as elementary students notice patterns related to even and odd numbers (*they alternate, they end with certain digits, there's always 1 left over when I match up pairs in an odd number, certain things happen when we do operations on two evens or two odds or one of each*), these observations can help them make generalizations about important mathematical features of these numbers (*seeing that every even number is a multiple of 2*). These generalizations then plant seeds of algebraic reasoning, a branch of mathematics based on symbolically generalizing work with numbers. As students expand their use of numbers to include fractions and decimals, they can attend to the relationship between the rational number system and the whole number system. *What rules still hold? In what ways do the numbers behave the same or differently from those I already know?* We can also help them notice the recurrence of the same fundamental properties of this expanded number system as they routinely, but consciously and explicitly, use the distributive, associative, and commutative properties with a growing set of numbers.

Making the transition from numbers to algebra provides one of the most fertile areas in which to notice and use mathematical structure. The properties students have become accustomed to using in their work with real numbers become familiar and useful tools as they begin to deal with symbolic expressions that follow the same properties, but with letters and numbers instead of just numbers. Thus, they can use this important element of mathematical structure—its properties—to help them build new learning on what they already know.

*Using Mathematical Structure: Making Sense of Problems
and Concepts*

One of the most useful mathematical habits of mind we can help students develop (and develop ourselves) is the ability to *zoom out* and *zoom in*. This ability is especially useful in helping students pay attention to the structure in new concepts they're learning or new problems they're trying to solve. In algebra, for example, as students begin to notice the structure within expressions, sometimes they can zoom out to see past what is inside parentheses and notice that a long expression or equation is basically about the difference of two squares or a linear relationship. And sometimes they can zoom in to see what's inside parentheses and realize that they know how to deal with that part of the expression or equation. In both cases, students can build on what they notice about the big picture or the smaller component to determine possible parameters for the values they may be looking for (noticing, for example, that the value of something squared in an expression is going to be positive, regardless of whether the "something" is positive or negative).

A Classroom View of Mathematical Structure

The best way for students to develop mathematical habits of mind in general, including the ability to notice and use mathematical structure, is to offer plenty of opportunities to work on problems that go beyond straightforward application of a recently learned procedure, concept, or rule. Thus, selecting worthwhile mathematical tasks that call for students to think deeply about a mathematical idea or problem situation is the starting point of implementing this practice. Much as with the other practices, it's not necessary (or appropriate) to try to find a task specifically about mathematical structure. Rather, we want to infuse an awareness of structure into students' work on whatever task they're tackling, regardless of the mathematics content involved.

STARTING WITH PATTERNS

Patterns are a central element of mathematics and an important tool in learning mathematics. We want students to notice and use what they notice about patterns to help them uncover mathematical structure and learn new mathematics. In the classroom, students need frequent opportunities to look for and notice patterns, discuss with each other what they notice, and make conjectures—informed guesses—about the mathematical principles the patterns might represent. Planning for such explorations and purposeful discussions should be central in organizing instruction for the development of an understanding of structure.

TEACHING FOR BIG IDEAS

An important part of a teacher's job is to help students increasingly build their knowledge and understanding of big ideas within and across years, making sure they realize what's important about the big idea each time it appears and helping students learn appropriately accurate language about, and facility with, the big idea that will carry over the next time they see it. Recognizing a unifying big idea appearing in a new context or problem—such as equivalence, proportionality, unit rates, the nature of variation between two related quantities—can help students use the structural characteristics of the big idea to help them make sense of the new context or problem in relationship to other contexts or problems where the big idea appears. As students notice recurring elements of mathematical structure (such as the nature of equivalence or the special way quantities relate to one another in a proportional relationship) and connect what they notice to a new situation, their observations can help them tackle problems involving mathematical procedures they haven't yet learned.

TEACHING FOR ZOOMING OUT AND ZOOMING IN

So much of mathematical structure becomes clear only when students can step back from the immediate procedure or problem at hand to see a pattern or big idea that might help them make sense of what they're working on. So learning to *zoom out* (to notice the whole problem or discuss the overarching concept behind a procedure), and then *zoom in* (to deal with a part of the problem or particular procedure they're learning) is a central mathematical habit of mind represented in this practice (and in other practices). Students need many opportunities to zoom out and zoom in, and teachers can structure classrooms to keep students thinking about this technique for identifying the underlying structure of a concept, skill, or problem and for using what they notice about the structure to make sense of the concept, master the skill, or solve the problem.

DISCUSSION AND PURPOSEFUL QUESTIONING

It's not enough to create a problem-centered classroom, even if the problems are wonderful. The real learning, especially related to understanding the underlying structure of mathematics, happens during productive discourse among students and between students and the teacher. A classroom discussion that helps students internalize mathematical structure depends largely on the teacher's facilitation of the discussion and purposeful questioning of students: asking them what they notice, asking them to look for and articulate patterns, asking where they've seen a certain pattern before, asking them what's the same and what's different about problems or situations, and so on.

In a classroom focused on this kind of purposeful interaction around a problem, the teacher's role is critical and both the professional strategies required and the mental abilities we want students to learn can be subtle. Without necessarily telling students what is important in advance, the teacher's choice of questions can stimulate students to think about the examples of structure that may be revealed in their work (*What do you notice is the same about these two problems? Where have you seen this idea before? Why do you think it turned out the same either way the two of you did it?*). And helping students connect their productive discussion to more formal mathematical properties, principles, algorithms, etc. is essential in order to equip them to use what they've learned in the future. The teacher can help students understand and use correct mathematical terminology and notation as they learn, for example, that the distributive property can be expressed as: $a(b + c) = ab + ac$. Connecting this kind of formalization to what students have explored is important (and hard to do without simply telling them the rules). Such connections can help them build a growing understanding of the structure of mathematics as a discipline, something that should become deeper over the years. Learning to ask questions that cause students to think and pay attention to important structural elements, and helping students realize and formalize what they have learned, are new skills for most teachers and possibly the most important skills for a teacher of mathematics to learn.

What Can We Do?

From working with young children to help them notice patterns and articulate what those patterns represent, to helping elementary students make sense of basic operations, to supporting older students as they extend their knowledge of numbers to rational, irrational, and eventually complex numbers, we must be aware of the mathematical structure in what we're teaching. Not only must we be aware, but we must be purposeful in helping students learn appropriate language to talk about the elements of structure they uncover or recognize.

The legacy some people leave in this world is concrete and visible—like buildings, roads, or bridges. The legacy teachers leave is their students. An important way we can equip our students to be able to use their years of mathematics learning after they leave us is to help them understand mathematics as a discipline built on rules, properties, definitions, assumptions, and unifying ideas they can call on whenever they encounter a mathematical situation, whether they know in advance how to deal with that situation or not.

Reflections and Discussion

FOR TEACHERS

- What issues or challenges does this message raise for you? In what ways do you agree with or disagree with the main points of the message?
- How do you think the idea of a building relates to mathematical structure?
- What do you consider *big ideas* for your level—ideas that cross grade levels (like equivalence, proportionality, or the use of the distributive property) or that illuminate a major mathematical concept (like multiplication)? How can you work with your colleagues above and below your grade level to enhance your work on these big ideas?
- What elements of mathematical structure do you see in the level you teach? How well do you understand those elements and where can you go to learn more?

- How well do you think students understand those elements? What can you do to enhance their understanding?
- What questions can you ask students to support them in understanding the structure of mathematics?
- What barriers do you see in implementing this practice? How can you address those barriers?

FOR FAMILIES

- What questions or issues does this message raise for you to discuss with your daughter or son, the teacher, or school leaders?
- How do you think the idea of a building relates to mathematical structure?
- What questions can you ask your son or daughter to support him or her in understanding the structure of mathematics?

FOR LEADERS AND POLICY MAKERS

- How does this message reinforce or challenge policies and decisions you have made or are considering?
- What do you think a *big idea* is in mathematics? How can you support teachers working across grade levels to understand big ideas and connecting principles, relating their understanding to improving student learning?
- How can you help teachers learn more about mathematical structure?

RELATED MESSAGES

Smarter Than We Think

- Message 31, "Developing Mathematical Habits of Mind," and Message 40, "Mathematical Habits of Instruction," address overall mathematical thinking and reasoning.
- Message 39, "Patterns with a Purpose," discusses Standard for Mathematical Practice 8 on regularity in reasoning, closely related to Practice 7.
- Message 14, "Effectiveness and Efficiency," discusses teaching mathematics well the first time a topic is taught in depth, including a discussion of the role of mathematical structure in teaching effectively and efficiently.

Faster Isn't Smarter

- Message 14, "Balance Is Basic," reminds us of the need to address concepts, skills, and problem solving in a comprehensive mathematics program that includes attention to mathematical structure.

MORE TO CONSIDER

- *Connecting the NCTM Process Standards and the CCSSM Practices* (Koestler, Felton, Bieda, and Otten 2013) unpacks each practice and relates it to NCTM's five process standards, including sample classroom vignettes for elementary, middle, and high school.

- *Number and Operations, Part 3: Reasoning Algebraically About Operations Casebook* (Schifter, Bastable, Russell, and Monk 2008) offers ideas for helping students make generalizations from their work with numbers, with a focus on mathematical properties, as they prepare for studying algebra. (See also other books in the Developing Mathematical Ideas series.)

- *Patterns, Functions, and Change Casebook* (Schifter, Bastable, and Russell 2008) focuses on noticing repeating patterns and number sequences and looks at making the transition to work with functions. (See also other books in the Developing Mathematical Ideas series.)

- *Connecting Arithmetic to Algebra* (Russell, Schifter, and Bastable 2011) uses the transition to algebra as a platform for addressing properties, generalizations, and reasoning, considering their role in mathematical structure.

- *Number Talks: Helping Children Build Mental Math and Computation Strategies, Grades K–5, Updated with Common Core Connections* (Parrish 2010, 2014) offers strategies grounded in mathematical structure for developing computational skills.

- *Thinking Mathematically: Integrating Arithmetic and Algebra in Elementary School* (Carpenter, Franke, and Levi 2003) focuses on the structure of number systems and algebra to offer strategies to help students make the transition from arithmetic to algebra.

- "Mathematics, Mathematicians, and Mathematics Education" (Bass 2005) is an assessment by mathematician Hyman Bass of the role mathematicians can play in improving school mathematics, including working with educators to incorporate the structure of mathematics into school programs.

- *Principles and Standards for School Mathematics* (National Council of Teachers of Mathematics 2000) describes mathematical content and process standards, including those listed in this message.
- Common Core State Standards for Mathematics (NGA Center and CCSSO 2010) includes descriptions of the eight standards for mathematical practice.

Standard for Mathematical Practice 8

Look for and express regularity in repeated reasoning. (Repeated Reasoning)

Mathematically proficient students notice if calculations are repeated, and look both for general methods and for shortcuts. Upper elementary students might notice when dividing 25 by 11 that they are repeating the same calculations over and over again, and conclude they have a repeating decimal. By paying attention to the calculation of slope as they repeatedly check whether points are on the line through (1, 2) with slope 3, middle school students might abstract the equation $(y - 2)/(x - 1) = 3$. Noticing the regularity in the way terms cancel when expanding $(x - 1)(x + 1)$, $(x - 1)(x^2 + x + 1)$, and $(x - 1)$ $(x^3 + x^2 + x + 1)$ might lead them to the general formula for the sum of a geometric series. As they work to solve a problem, mathematically proficient students maintain oversight of the process, while attending to the details. They continually evaluate the reasonableness of their intermediate results.

—Common Core State Standards for Mathematics (NGA Center and CCSSO 2010, 8)

CONNECTIONS TO STANDARD FOR MATHEMATICAL PRACTICE 8

Practice 8 focuses on a particular type of mathematical reasoning that can be used to build computational proficiency with understanding—reasoning based on repetition or regularity in a mathematical situation. This connection to reasoning relates Practice 8 to Practice 2 on reasoning abstractly and quantitatively and to NCTM's reasoning and proof standard. But seeing regularity depends on noticing patterns and making generalizations based on those patterns. The dependence on patterns ties Practice 8 closely to Practice 7 on mathematical structure. The emphasis on building understanding of the repeated process(es) lying behind an operation blends procedural fluency with conceptual fluency in connecting Practice 8 to these two mathematical proficiency strands in *Adding It Up* (National Research Council 2001). The last part of the Common Core's description of this Practice focuses on solving problems, keeping an eye on the process as well as the details and continually checking the reasonableness of the intermediate results. This emphasis brings the set of mathematical practices full circle to where they started, connecting Practice 8 to Practice 1 on problem solving and to NCTM's standard on problem solving. (See Appendix F for a description of the five NCTM process standards [2000].)

39

Patterns with a Purpose—Seeing and Reasoning from Regularity

STANDARD FOR MATHEMATICAL PRACTICE 8:
LOOK FOR AND EXPRESS REGULARITY IN
REPEATED REASONING

Mathematics is an exploratory science that seeks to understand every kind of pattern—patterns that occur in nature, patterns invented by the human mind, and even patterns created by other patterns.... To grow mathematically, children must be exposed to a rich variety of patterns appropriate to their own lives through which they can see variety, regularity, and interconnections.

—Lynn Steen (*On the Shoulders of Giants*, 1990)

When I taught mathematics in French in Burkina Faso (West Africa), one of the first limitations I discovered in my French vocabulary was that I couldn't find a word for *pattern* in the way I was accustomed to using the word in teaching mathematics. Every word the dictionary listed meant a slightly different kind of pattern, and my French-speaking African teaching colleagues looked at me strangely when I tried to talk about what I was trying to say. Since that experience, I've had an opportunity to speak to more French-speaking mathematics educators. Finally, a friend from France suggested that the word he uses for talking about patterns in teaching mathematics is *régularité*. This is one of those French words that has an immediately obvious English translation: *regularity*. The dictionary lists as synonyms for *regularity: recurring, periodic, rhythmic*. I thought about it and realized that, while not exactly encompassing everything I might consider a mathematical pattern, there certainly are a lot of patterns that arise from noticing regularity, specifically regularity generated by some kind of repetition. Practice 8 deals with the regularity we observe as we consider and reason about some kind of repeated pattern.

A Closer Look at Repeated Reasoning

One of the recommendations in NCTM's 1989 *Curriculum and Evaluation Standards for School Mathematics* was to incorporate algebraic thinking into the mathematics curriculum from prekindergarten through grade 12. A key element of this focus was the use of patterns so that students can begin to develop generalizations from those patterns, and many programs developed in the wake of these standards incorporated the heavy use of patterns. Some mathematicians and others have criticized programs arising from NCTM's standards after observing that, for much of the time in elementary mathematics classrooms, children seemed to be simply exploring patterns and being asked to repeat or extend them. These critics are correct that students are just playing with beads (or macaroni or plastic bears or ...) *if* children only have the opportunity to see, explore, and enjoy many kinds of patterns, but are not given the opportunity to connect what they notice about the patterns with the important mathematical structures those patterns represent. Such connections are critical; this kind of activity can represent the beginning of a process leading to articulating a generalization, developing reasoning skills, and planting seeds of algebraic thinking. It necessitates explicit attention to mathematical connections, representations, and justifications. Playing with patterns has to have a purpose, and that purpose is grounded in noticing and building on the regularity of repeated reasoning. At some point, the patterns have to be connected to the mathematical ideas they represent, or they have no business taking up students' time.

NOTICING REPEATED REASONING

The heart of Practice 8 lies in noticing a repeated pattern when performing certain numerical and symbolic calculations, reasoning from that pattern, and extending it to other kinds of numbers. This kind of *repeated reasoning* can be helpful in learning simple facts as well as performing calculations with large numbers. In terms of basic facts, we can set up students to notice a pattern of 'one more' just by being purposeful in how we present facts to students for exploration. As they are beginning to learn about addition, for example, if we offer them a sequence of additions to model, we might present to them $7 + 5$, then $7 + 6$, then $7 + 7$, and so on. The regularity in this pattern is obvious to adults, but for children it can be a wonderful discovery: that adding a number one bigger than the number added the previous time leads to an answer that's one more than the previous answer.

REPEATED REASONING AND COMPUTATION

One of the most useful examples of the use of repeated reasoning is the development of computational procedures—algorithms—with whole numbers. The Common Core standards reflect a trend of the last two decades or more as states have chosen to reduce the number of objectives, targets, or standards addressed in their state standards at each grade level, at least for the elementary and middle grades. The motivation for, and rationale behind, this shift is that we need less repetition in content and skills from grade to grade in order to focus more attention on the content we teach each year. And one of the best ways to reduce the repetition in the curriculum is to compress the teaching of basic computational procedures—algorithms—into fewer years with less redundancy. Compressing computation begins with helping students use the regularity—the repetition of the pattern—in the algorithm to generalize the procedures they're learning so that they don't need to learn the same procedures over and over again.

Compressing the teaching of computation is only possible if we tap into students' ability to generalize from the patterns that form the basis of our computational algorithms. In other words, we want to teach students a procedure that works for all kinds of numbers, not just for three-digit numbers this year, four-digit numbers the next year, and so on. For example, in performing multi-digit addition or subtraction, there is no need to reteach the algorithm year after year, each year with more digits than the year before, if students recognize that the same regrouping pattern used in the hundreds place also works in the thousands place. And the ten thousands place. And the hundred thousands place. And the billions place. Students can see regularity as they perform these calculations, and we can help them articulate, generalize about, and express this pattern by making sure they have opportunities to talk about what they see and show what they notice. Young children, especially, love the idea that they can apply something they just learned to deal with really large numbers. It makes them feel like math geniuses!

The same concept can be applied to looking at other types of computation—with fractions, decimals, binomials, and so on. The power of a computational algorithm is that it represents a generalized procedure built on regularity, the focus of this Practice. Likewise, students can use repeated reasoning when noticing regular patterns as they identify computational shortcuts, such as appending zeroes to a number when multiplying by ten, one hundred, one thousand, and so on. Essentially, students can extend patterns they learn with simple cases to many more cases if we help them focus on noticing what repeats or recurs within the procedure as numbers or expressions get larger or more complicated.

REPEATED REASONING BEYOND COMPUTATION

Practice 8 also reminds us that repeated reasoning appears in many parts of mathematics outside of computation. Students working with equivalent ratios might discover a pattern in a table, leading to observations about unit rates and, maybe eventually, to seeing the underlying notions of slope. As they measure angles in polygons and represent their findings, they might be able to see or understand a pattern about the sum of the angles. By looking for regularity and patterns, whether students discover the patterns themselves or see other students' patterns, they begin to make sense of the underlying mathematical principles and properties.

A Classroom View of Repeated Reasoning

Organizing the classroom around rich problems provides a good platform from which to teach mathematics with an emphasis on mathematical habits of mind and the standards for mathematical practice. Such a classroom is described as part of Message 12 , "Upside-Down Teaching." In this model, students work on a problem they may or may not know how to solve in advance, often working in groups or pairs. Then, the teacher orchestrates a classroom discussion about how students approached the problem, what gave them difficulty, and what they noticed or observed in the process, with students making their case and discussing other students' observations. The teacher asks questions and facilitates the discussion to help students come away knowing the intended mathematics. This classroom model clearly addresses the aspect of this practice dealing with solving problems, especially as the teacher helps students learn to zoom out and pay attention to the big picture—what the problem is all about—while also helping them learn to zoom in and focus on the details of the problem.

The practice of looking for and expressing regularity as a tool for developing facility in performing computations or simplifying expressions may seem at first glance not to fit in the classroom described above. The tasks used in a problem-centered classroom, however, need not be limited to contextual problems; purely mathematical tasks can serve as the focus of such a lesson, targeting some concept or skill in mathematics such as developing the procedure or algorithm for performing multidigit multiplication. This kind of algorithm can be developed by engaging students in well-structured activities designed for that specific purpose (Van de Walle and Lovin 2005a). Students can still be very engaged in the task, often using manipulative materials to visualize the regrouping process with the necessary operation. The key to learning what we want them to learn, as always in such a classroom, is to engage students in productive discourse about what they're doing. (*Why did you trade those for that? How did you decide to put a 1 there? Does your picture show the same thing as Virginia's?*)

Moving from numerical thinking to algebraic thinking is an important transition that can be facilitated if we have given students experience with repeated reasoning in the context of computation. This transition can only happen when teachers thoughtfully orchestrate classroom discourse around students' explorations of the operations, asking questions about what students notice, having them draw pictures, make diagrams, construct tables, and write equations to represent the patterns they notice and asking students to make and justify conjectures about what might happen if the pattern were extended.

It's interesting that the last part of Practice 8 comes back to where the set of practices begins—solving problems. The benefits of looking for regular patterns extend to how students approach problems. Encourage students to *zoom out* and *zoom in* (see Message 38, "Building Things") to pay attention to the process(es) they use within their big view of the problem, while also attending to the details. As students progress through the grades, they refine their problem-solving abilities and learn to pay attention to intermediate goals and results in solving increasingly complex problems. As they do so, they can gain helpful insights into a problem by looking at different representations of parts of the problem. They continually check to see if the representations and results they are generating make sense within the structure of the problem, a critical skill that reflects one of the most important mathematical habits of mind we want for students—to make sense of mathematics as students are learning it and to be able to use mathematics to solve problems.

What Can We Do?

Looking for patterns trains the mind to search out and discover the similarities that bind seemingly unrelated information together in a whole. A child who expects things to "make sense" *looks* for the *sense* in things and from this sense develops understanding. A child who does not see patterns often does not *expect* things to make sense and sees all events as discrete, separate, and unrelated.

—Mary Baratta-Lorton (*Mathematics Their Way,* 1995)

The description of Practice 8 helps us zoom out ourselves—noticing the regularity we may have come to take for granted in an algorithm or rule can help us focus on big ideas in our teaching. We can practice exactly what we hope students will do, not only zooming out to sometimes pay attention to the big idea, but also sometimes zooming in to focus on the details that repeat within a particular process or principle.

Effective teachers make sure that students have opportunities for productive discourse, asking questions that lead them to notice regularity and articulate the repeating patterns they see and constantly asking students to explain and justify their conjectures about what they notice. Effective teachers also make sure students always know what mathematics they have learned in a lesson, sometimes bringing formalization to students' more informal discoveries, discussions, conjectures, and generalizations. The goal of Practice 8 is a worthy goal for the mathematical thinking and learning of all students—to develop mathematical habits of mind that help students both learn important mathematical content and skills, and also help them make sense of what they learn and apply it to solve problems.

Reflections and Discussion

FOR TEACHERS

- What issues or challenges does this message raise for you? In what ways do you agree with or disagree with the main points of the message?

- Consider Message 38, "Building Things," in conjunction with this message. How does the use of patterns in looking for repeated reasoning fit with the use of patterns to generate properties and principles within the broader arena of mathematical structure?

- What barriers do you see in implementing this practice? How can you address those barriers?

FOR FAMILIES

- What questions or issues does this message raise for you to discuss with your daughter or son, the teacher, or school leaders?

- If you help your son or daughter with schoolwork in mathematics, how can you help him or her notice the commonalities in the processes he or she uses—that, for example, the regrouping process in adding and subtracting large numbers is the same for each digit, no matter how large the numbers?

- As your young daughter or son helps at home, what patterns can you find together to discuss?

FOR LEADERS AND POLICY MAKERS

- How does this message reinforce or challenge policies and decisions you have made or are considering?
- How well do your teachers use patterns to support students in learning important mathematics?
- When teachers have a real need to learn or relearn content in a way that supports the development of the mathematical practices, what can you do to support and help them?

RELATED MESSAGES

Smarter Than We Think

- Message 31, "Developing Mathematical Habits of Mind," and Message 40, "Mathematical Habits of Instruction," address overall mathematical thinking and reasoning.
- Message 38, "Building Things," discusses Practice 7 on mathematical structure but closely relates to Practice 8.
- Message 14, "Effectiveness and Efficiency," reiterates the need to teach effectively the first time, something more readily accomplished when students are encouraged to look for regularity in patterns.

Faster Isn't Smarter

- Message 23, "The Power of Patterns," considers the role of patterns in school mathematics.
- Message 33, "Engaged in What?," reminds us that simply being involved in activities (or simply exploring patterns) does not necessarily lead to powerful mathematics learning.

MORE TO CONSIDER

- *Connecting the NCTM Process Standards and the CCSSM Practices* (Koestler, Felton, Bieda, and Otten 2013) unpacks each practice and relates it to NCTM's five process standards, including sample classroom vignettes for elementary, middle, and high school.
- *Patterns, Functions, and Change Casebook* (Schifter, Bastable, and Russell 2008) focuses on noticing repeating patterns and number sequences and looks at transitioning to work with functions. (See also other books in the Developing Mathematical Ideas series.)
- *Connecting Arithmetic to Algebra* (Russell, Schifter, and Bastable 2011) uses the transition to algebra as a platform for addressing

properties, generalizations, and reasoning, considering their role in mathematical structure.

- *Lessons for Algebraic Thinking: Grades K–2* (von Rotz and Burns 2002) offers lessons that demonstrate how to help students move from numerical to algebraic thinking building on work with patterns.

- *Lessons for Algebraic Thinking: Grades 3–5* (Wickett, Kharas, and Burns 2002) offers lessons that demonstrate how to help students move from numerical to algebraic thinking building on work with patterns.

- *Lessons for Algebraic Thinking, Grades 6–8* (Lawrence and Hennessy 2002) offers lessons that demonstrate how to help students move from numerical to algebraic thinking building on work with patterns.

- *Thinking Mathematically: Integrating Arithmetic and Algebra in Elementary School* (Carpenter, Franke, and Levi 2003) looks at research and classroom practice that helps students move from numerical thinking to algebraic thinking.

- *Adding It Up: Helping Children Learn Mathematics* (National Research Council 2001) describes five strands of mathematical proficiency and summarizes related research.

- *Principles and Standards for School Mathematics* (National Council of Teachers of Mathematics 2000) describes mathematical content and process standards, including the reasoning and proof standard discussed in this message.

- Common Core State Standards for Mathematics (NGA Center and CCSSO 2010) includes descriptions of the eight standards for mathematical practice.

40

Mathematical Habits of Instruction— Helping Students Become Smarter Than They Think

REFLECTIONS, THEMES, AND IMPLICATIONS

> The habits of mind should not ordinarily be the explicit objects of our teaching; rather, each student should internalize them as they do mathematics. Part of the way this outcome can be achieved is by teachers modeling the very habits we want students to develop.
>
> —Kenneth Levasseur and Al Cuoco ("Mathematical Habits of Mind" in Schoen and Charles 2003)

Whether called *mathematical habits of mind, practices, processes, proficiency,* or some other term, our mathematical goals for instruction today universally call for helping students learn how to think, reason, and communicate mathematically as they make sense of and solve problems. Mathematical habits of mind cannot be taught as content to be learned. Rather, teaching to develop mathematical habits of mind calls for shifts in the way we teach mathematical knowledge and skills, infusing opportunities to develop the habits of mind as a routine part of our instruction. Some of those shifts may be subtle and some may be substantial, depending on where we start.

From considering the mathematical practices elaborated in the Common Core State Standards for Mathematics, we can enrich our vision of what it takes to teach so that students develop the ways of thinking and reasoning described in the Common Core, in NCTM's standards (NCTM 2000), and advocated in writings about mathematical habits of mind over the past twenty years. We can develop a real picture of what I call *mathematical habits of instruction.*

The Premise

My basic premise underlying mathematical habits of instruction is:

> What students need for their future is as much about
> **how they think** as about **what they know**,
> and helping students succeed in
> learning what they need is as much about
> **how we teach** as about **what we teach**.

There is more agreement today than ever before about what mathematics we should teach. Increasingly that agreement includes teaching students to think, reason, question, make sense of, understand, and use what they know to solve all kinds of problems, both vague and clearly posed. If we want to pay attention to how students think, then we must also pay attention to how teachers teach. The starting point for teaching in ways that support students' thinking is a classroom environment based on worthwhile mathematical problems and committed to rich student discussion.

Recurring Themes

Mathematical habits of mind cannot be reduced to a list of discrete skills. Rather, the processes and proficiencies used to describe these habits overlap and intertwine to produce a rich picture of mathematical thinking and related skills. In considering such habits of mind through the lens of the Standards for Mathematical Practice in the Common Core State Standards, this same kind of interdependent, overlapping picture of mathematical thinking emerges. Some recurring themes arise from these practices that can shape our vision of mathematical habits of instruction. The resulting themes may not come from one particular practice or another, but their importance to our vision of instruction transcends the category(-ies) within which they fall. Let us revisit a few of those themes. (See also Messages 31 through 39, which look at mathematical habits of mind overall and at each of the eight Standards for Mathematical Practice.) To support students in developing mathematical habits of mind, we should consider:

- a teaching model focused on problems and discourse;
- regular monitoring of student learning with formative assessment;
- learning to zoom out, zoom in, and go back and forth;
- the use of technological tools in support of mathematical thinking and learning; and
- attention to mathematical structure, including building on patterns and developing precision.

A FOCUS ON PROBLEMS AND DISCOURSE

The most universal theme in the discussion of essentially every mathematical practice is the need to create a classroom environment centered on worthwhile problems, student discussion, and interaction. The upside-down teaching model presented in Message 12, "Upside-Down Teaching," describes just this kind of environment—one that is centered on worthwhile tasks and focused on student explanations, justifications, conjectures, questioning, arguments, thinking, and reasoning. The model clearly reflects a priority on the Common Core's Standard for Mathematical Practices 1 (problem solving), 3 (constructing and evaluating arguments), and 6 (communicating with precision). Focusing on these three elements of mathematical thinking is an appropriate platform upon which to build mathematical understanding across content and across practices and processes.

Problems selected can include some that are in purely mathematical contexts, some that are applied, and some that call for mathematical modeling. The classroom can be focused on challenging students to think beyond what they know, thus allowing them to not only learn new mathematics, but also to grow their intelligence, as described in Message 1, "Smarter Than We Think," on a growth model of intelligence. A problem-centered approach to teaching and a commitment to having students discuss, explain, and communicate their thinking is the most important mathematical habit of instruction.

A COMMITMENT TO FORMATIVE ASSESSMENT

Teaching to develop mathematical habits of mind and grow students' intelligence requires teachers to pay attention to, and keep track of, how well students are learning and developing their thinking skills. As described in Message 19, "How to Know What They Know," this means using both formal and informal types of assessment, especially formative assessment, on a regular basis. The most useful kind of formative assessment is that which allows students to explain their thinking, either orally or in writing. Not only does the practice of regular formative assessment help teachers know whether a student is *getting it*, it also helps the student clarify his or her thinking, adjust it if necessary, or solidify it if it holds up to the scrutiny of being shared and discussed with peers and the teacher. And the act of explaining helps a student develop an important mathematical habit of mind in itself.

ZOOMING OUT, ZOOMING IN, AND GOING BACK AND FORTH

Zooming out and zooming in, as well as going back and forth between different aspects of a concept or problem, are recurring themes we see across the Common Core's Standards for Mathematical Practice and in nearly every description of mathematical habits of mind (see Message 31,

"Developing Mathematical Habits of Mind"). We want students to be able to zoom out by stepping back and looking at the big picture of a problem or a mathematical concept, while also being able to decide when to zoom in and focus on the particular details they might attend to at the moment. We want them to continually go back and forth between representations and between the mathematical process and the context of the situation, if in an applied setting. We want them constantly stepping aside and reflecting on whether their work and their answers (or intermediate answers along the way) make sense with respect to the mathematical task or contextual situation.

Likewise, teachers need to be able to zoom out and think about the big picture toward which they're teaching, with respect to a problem or concept, while zooming in as needed to help students focus on the particular parts of the concept or problem. We want teachers to go back and forth between the various ideas or parts of a problem, helping students make connections within and across problems and mathematical ideas. As teachers reflect on their work and students' progress in learning what we want them to learn, they can continually fine-tune their practice and help students also attend to reflecting on their learning.

USING TECHNOLOGY WISELY

Specifically mentioned in Practice 5, and related to other mathematical habits of mind involving representation and communication, the use of appropriate tools, especially technology, is an important aspect of developing mathematical thinking. (See also Message 24, "Beyond PowerPoint.") In particular, students need to learn how to use appropriate tools that support their mathematics learning, including technological tools that allow for computation, graphing, symbol manipulation, statistical calculations, and geometric constructions. Of even more importance than knowing *how* to use these tools is being able to determine *when* it's appropriate to use them and *which* tool to select for a particular purpose. Helping students learn to be wise decision makers about technology and other tools can strengthen their confidence as mathematical problem solvers and independent thinkers and make them less likely to grab an electronic device for an inappropriate purpose.

PAYING ATTENTION TO MATHEMATICAL STRUCTURE

Mathematical structure is pervasive in the Common Core's Standards for Mathematical Practice and has been in discussions of mathematical habits of mind for many years. Structure is not only explicitly called out in Practice 7, but is reflected in Practice 8 on repeated reasoning, Practice 2 and 3 on reasoning, and Practice 6 on precision, as well as influencing the other practices on problem solving, modeling, and the use and selection of tools. Likewise, mathematical structure is highlighted in the vision of mathematics proficiency described in *Adding It Up* (National Research Council 2001 and throughout the standards of the National Council of Teachers of Mathematics. Cuoco, Goldenberg, and Mark (2010), Stanley and Walukiewicz (2004), and others infuse mathematical structure throughout their discussions of habits of mind.

For teaching, this means that we need to constantly pay attention to the properties of mathematics and to the structures within problems and tasks. We can help students notice, generalize, and articulate the use of properties as they develop their number and operation sense. As we help them learn to look for and generalize patterns across topics and problems, and zoom out and zoom in to focus on the big picture and the parts of that picture, our students can begin to internalize a sense of the structure of mathematics. We can help students find connections among problems and build new learning on the structural elements they already know. We can support their growing understanding of mathematical structure by using precision in our own language (at an age-appropriate level for our students) and by being consistent in our use of correct terms and mathematical language. Using mathematical language precisely depends on a strong understanding of mathematics. Many teachers may benefit from addressing mathematical language in professional development programs, not as a separate subject of study, but as a thread that runs throughout related learning experiences.

What Can We Do?

First and foremost, we can keep the mathematical habits of mind at the forefront of our attention, demonstrating them ourselves as we invest in an upside-down teaching model. And we can commit to regularly paying attention to (formatively assessing) how well students are learning what we want them to learn and how well they are developing the thinking skills we want them to develop. Within this framework, we can incorporate the use of technology and decision-making skills related to that use, even as students develop other critical elements of mathematical thinking. We can be sure never to lose sight of the structure and nature of mathematics as a discipline with rules and order, even as we create an environment of rich exploration in which we value mistakes as much as successes. We can see the power of developing mathematical habits of mind and we have a vision of what mathematical habits of instruction look like; now our job is to make that vision a reality in our classrooms.

There's a lot for teachers to know about mathematics and mathematical thinking. And there's a lot for teachers to learn about teaching to develop mathematical thinkers. Committing to ongoing professional learning is critical to being able to successfully implement the kind of teaching described in this book. Success does not mean doing everything the ideal way all at once; success is a journey toward ever refining one's teaching practice to continually move closer to the goal. There simply is no other profession where this kind of commitment to lifelong learning is so important. Our students' lives depend on it.

Reflections and Discussion

FOR TEACHERS

- What issues or challenges does this message raise for you? In what ways do you agree with or disagree with the main points of the message?

- If you don't already teach primarily using a problem-centered approach, what challenges do you see in trying to move closer to "Upside-Down Teaching," as described in Message 12? How might you (and your colleagues) address those challenges?

- How well do you zoom in and out, as described throughout Messages 32–39 on the Standards for Mathematical Practice, when planning your instruction and when working with students, looking at the big picture, and going back to smaller details?

- How comfortable are you with having your students use (and using yourself) the kind of powerful technology described in Message 24, "Beyond PowerPoint," and Message 36, "Building and Using a Mathematical Toolbox"—technology that can calculate, graph, perform algebraic manipulation, manage statistical data, or do geometric constructions? How can you engage in discussion with colleagues about the potential benefits and likely changes to teaching if you were to use such tools more effectively?

- What kinds of formative assessment do you use to keep track of student learning on a day-to-day basis? What ways do your colleagues use? How can you share best practices to improve student learning?

- How do you attend to your own professional learning?

FOR FAMILIES

- What questions or issues does this message raise for you to discuss with your daughter or son, the teacher, or school leaders?

- How open are you to the possibility that your son or daughter might be given a problem for which he or she does not know all the steps to do in advance? What possible benefits can you see from having your son or daughter tackle problems that push his or her thinking?

- How open are you to the possibility that ways of thinking might be as important a goal in math class as a list of mathematical concepts and skills?

FOR LEADERS AND POLICY MAKERS

- How does this message reinforce or challenge policies and decisions you have made or are considering?
- How can you support a problem-centered classroom focused on student discourse? How consistent is such a model with your teacher evaluation system? What changes or modifications to your evaluation system or other policies might encourage teachers to try such an approach?
- How can you support the wise use of technology in mathematics classrooms, including technology that can calculate, graph, perform symbol manipulation, manage statistical data, or do geometric constructions?
- How can you help teachers find effective ways of continuing to grow that are accessible to them, including opportunities to refine their teaching practice and assessment techniques?

RELATED MESSAGES

Smarter Than We Think

- Message 31, "Developing Mathematical Habits of Mind," describes the mathematical habits of mind we want students to learn.
- Messages 32 through 39 address the eight Standards for Mathematical Practice in the Common Core State Standards and discuss each standard with respect to overall mathematical habits of mind.
- Message 12, "Upside-Down Teaching," offers a problem-centered approach to mathematics teaching that supports the development of mathematical understanding and mathematical habits of mind.
- Message 1, "Smarter Than We Think," describes a growth model of intelligence that reminds us that all students can learn to think mathematically.
- Message 19, "How to Know What They Know," discusses the importance of formative assessment in support of student mathematics learning.
- Message 24, "Beyond PowerPoint," presents considerations in using technology appropriately as a key instructional tool in support of student thinking.

Faster Isn't Smarter

- Message 17, "Constructive Struggling," suggests that offering students challenging problems with orchestrated opportunities for discourse can help students develop mathematical understanding and mathematical habits of mind.
- Message 1, "Math for a Flattening World," reminds us of the need for thinking, reasoning, and problem solving skills in our changing world.

MORE TO CONSIDER

There is much to learn and to discover in your professional journey. The "More to Consider" section in every message in this book offers additional resources, with complete information listed in Resources and References at the end of the book. Consider starting with Message 31 on developing mathematical habits of mind and going back through the book for other messages that interest you. Also see the various lists of resources offered in the Appendices. Set a goal to enrich your learning by checking out a new resource every so often. The first two resources that follow provide some additional thinking about mathematics teaching overall and may be especially interesting in thinking about mathematical habits of instruction. The third resource is an article on teaching for mathematical habits of mind, the source for the opening quote. The last resource is a rubric for considering student growth on the practices.

- "What Math Knowledge Does Teaching Require?" (Thames and Ball 2010) considers the deep mathematical knowledge, unique to teaching, that can help teachers help students develop mathematical understanding and mathematical habits of mind.
- "Takeaways from Math Methods: How Will You Teach Effectively?" (Bay-Williams 2014) is a short blog entry highlighting three key steps toward teaching for mathematical understanding and mathematical habits of mind, designed for preservice teachers, but relevant for any teacher of mathematics.
- "Mathematical Habits of Mind" (Levasseur and Cuoco 2003) looks at teaching to develop mathematical habits of mind.
- "Standards of Student Practice in Mathematics Proficiency Matrix" (Hull, Balka, and Miles 2011) offers a tool to help teachers monitor student progress with the Common Core Standards for Mathematical Practice. www.mathleadership.com/sitebuildercontent/sitebuilderfiles/standardsoftudentpracticeinmathematicsproficiencymatrix.pdf.

Afterword

I remember vividly an episode from my second year of teaching, more than forty years ago now. An older teacher, Gladys, was talking with me after school one day late in the fall. With tears in her eyes, she told me that she had come to the decision that this would be her last year of teaching. She said she couldn't take it any more—that students were rowdy and not engaged, that parents were demanding, and that administrators seemed to expect the impossible. She felt that things had gotten much more difficult since she had started teaching in the 1940s and it was even more challenging than ever to make a difference.

Today, I talk with many teachers, and I visit many schools. I see exciting things happening in classrooms across the nation, and I see teachers working together in real, effective professional learning communities focused on continually refining their teaching by thinking deeply about their students' learning. I am exhilarated by what both students and teachers accomplish under often challenging conditions. But I also continue to encounter discouraged teachers like Gladys. In their frustration, some teachers leave teaching; others retreat to a superficial teaching model, essentially giving up on the good stuff.

The challenges teachers face are not unique to mathematics. However mathematics, in particular, seems to lie at the center of many issues. Political and other barriers deter teachers from teaching mathematics in ways their students need, in turn frustrating teachers (like Gladys) and disengaging students.

As I reflect on my career and as I ponder the future, I offer five recommendations. If this nation were to dedicate money, time, and energy in these five areas, maybe we would finally realize our untapped potential. Maybe we would bring student learning to levels that will make us proud and, in so doing, give hard-working teachers the pride, satisfaction, and recognition they so deeply crave and deserve. My five recommendations involve:

- Teachers
- Expectations
- Courses
- Tests
- Long-Term Commitment

Teachers

It sometimes seems that policy makers believe that the way to improve mathematics teaching is to replace teachers with technology. We see increased interest in online or programmed courses, many involving a student sitting alone at a computer or on a hand-held device. There's much a person can learn individually with well-designed technology. But there's no replacement for a knowledgeable human teacher. A teacher is critical in helping students learn how to think and record their thinking, justify their reasoning, and discuss their approaches to solving a problem or their observations about a mathematical situation; this kind of professional teaching can't be done by a computer and it can't be done from a script. I'm convinced that the most significant learning comes when a teacher engages students in working through challenging problems, facilitating discussion, and connecting students' work to the mathematical outcomes we want them to learn. In order for this to happen, every student needs to be taught mathematics by someone who knows the discipline deeply and who understands how to elicit student thinking. This has always been a goal in secondary classrooms, but in order for this to happen in elementary classrooms, we need to agree that students should be taught mathematics by mathematics specialists from the earliest grades in elementary school, whether this means ensuring that all elementary teachers become experts in mathematics or whether it means staffing schools with mathematics specialists responsible for all mathematics teaching. And, regardless of the grade level, teachers need strong support from administrators, communities, and policy makers, including opportunities to learn, grow, and collaborate in a nurturing environment.

Expectations

We know that tracking students into separate groups by perceived ability level does not work as a tool for raising achievement at any grade level, in spite of the surface appeal of doing so (Burris, Corbett, Welner, and Bezoza 2009; Oakes 1987). It may seem like teaching might be a bit easier with students who are in about the same place *if* our only goal were to tell students rules and hope they would remember them. However, there is no evidence that tracking works even for this purpose. If we adopt a teaching model based on engaging students with problems and facilitating their discussion of solutions with each other, all students can benefit from the multiple perspectives, talents, and insights that are possible in heterogeneously grouped classrooms. At the high school level, we need to maintain our commitment to high academic standards for all students, so that their high school diploma is a sign of a significant education that prepares them for the future. Current graduation requirements in most states support this notion by calling for three or more years of academic (nonremedial) mathematics for all students, regardless of their future paths (which few of them really know, in spite of what they or their parents might think). Some states are getting nervous, however, that this may be too much to expect. A few states are considering pulling back into tracked diplomas, where some students would receive a lower level of mathematics preparation, leaving them ill prepared for life after high school as they move into either post-secondary education or directly into the world of work. I suggest that we need to maintain a commitment to high-quality mathematics for all students through a level about equal to Algebra 2, even if that particular course may not be the right course, as addressed in my next recommendation.

Courses

Over the past few decades, we've seen a lot of progress in focusing the U.S. mathematics curriculum around fewer topics to allow greater depth of instruction and learning. But that progress has primarily been limited to the kindergarten through grade 8 curriculum. We seem to have largely ignored the fact that our high school mathematics program remains too crowded. We cling to old notions of what courses students should take (Algebra 1, Geometry, Algebra 2) and what content those courses should address. It's time to recognize that technology exists to help with many symbol-manipulative tasks and that not every student needs to know how to simplify radical expressions or solve rational equations with a pencil and paper. Maybe we can seize the opportunity technology allows to rethink the mathematics all students need and make room for increasingly important topics. We've now seen several good examples of integrated high school curricula from innovative developers and from states that offer relevant, engaging experiences for all students. Now it's time to get together and forge a new organizational structure for high school mathematics courses. The first step toward this goal is to agree on key topics that might appear in first, second, and third years of a common, integrated high school mathematics course sequence for all students—a sequence that not only addresses important ideas from algebra and geometry, but that also incorporates mathematical modeling and authentic applications and pays significant attention to statistics and finance. Then let us explore a few high-quality options for the fourth year (or fifth year, depending on when students start) of high school mathematics study in order to prepare students for different paths after high school. One of those paths should aim toward calculus, and other paths might focus on quantitative reasoning or further study of statistics, finance, or applied mathematics. But all paths should involve rigorous experiences in both mathematics and broader academic areas, calling for students to gather evidence, work collaboratively and independently, discuss and defend their point of view, and present their findings orally and in writing.

Tests

Accountability is a good thing. As educators, we should insist on being held accountable for producing results—showing the public that our students are learning challenging mathematics. However, many of the assessments we have seen during most of the last three decades have tended to be shallow, assessing only the knowledge and skills that could be addressed in single-answer, multiple-choice questions. A few states tried deeper kinds of tests during this time, requiring open-ended problem solving, but essentially all such efforts were eventually abandoned because of the time and money required for development, administration, and scoring. Now it's time to seriously take on the challenge of developing and using tests that target what we say we value—thinking, reasoning, and solving problems—and it's time to reconsider what we do with the information we gather on such tests. It's time for sanity in using tests and interpreting test scores so that tests don't get in the way of the exciting, engaging kind of mathematics teaching that leads to the learning we say we value.

Long-Term Commitment

Teachers care more than anything else whether their students learn. And most teachers work hard for the good of their students. They also generally work hard to implement whatever new program is imposed on them. Unfortunately, their conscientious hard work is too often discounted, dismissed, discarded, and sometimes even demeaned over nothing more than political whim. We've seen it happen in state after state and district after district. One state's example is particularly heartbreaking and sadly illuminating. The state developed and implemented over a multiyear timeline a comprehensive set of kindergarten through grade 12 mathematics standards, including integrated high school mathematics standards. It seemed they did everything right. They put in place a multiyear implementation plan, and they supported teachers in many ways, including professional development. Nevertheless, just as high school students were starting to reap the benefits of several years of hard work by teachers, a few key state-level leaders left their roles. New leaders, possibly encouraged by the voices of a small number of reluctant and vocal teachers and parents who objected to change, initiated a dramatic redirection purposefully leading to the eventual demise of this excellent program. Students became the casualties of yet another come-and-gone reform effort. Teachers, both those who initially supported the program and those who hadn't, became more convinced than ever that their work didn't matter and wondered why they should ever agree to participate in the next new program.

We have an opportunity right now to seize on the interest of states to share common standards. A universal theme we see in high-performing countries is their use of a set of national standards as a focal point. In the United States, we may not want a single set of national standards required in all fifty states. But when we agree voluntarily on commonalities in goals, as with the Common Core State Standards for Mathematics, we can make huge strides in how we reach those common goals. As we recognize issues or areas where those goals might need to be modified, we can make adjustments. But the absolute worst thing policy makers can do is to ask educators to invest their very precious time and energy to implement new common standards, only to change direction a couple of years later. Let us commvit to work together to revisit, reconsider, and revise standards periodically. Let's also commit to not abandoning this or any program prematurely, only to bring in something different. We owe it to teachers and students to persevere and adapt, even when it gets challenging, working toward solutions that give every student the opportunity to learn high-level mathematics.

What Next?

We know how to teach mathematics well, and, fortunately, many wonderful teachers do just that every day. If you know effective teachers still giving their heart and soul to their profession and their students, thank them. If you're involved in decision-making or policy making, consider listening to the voices of teachers and students. Let's not argue about what to do and, instead, work together, listening to each other and building on what we all bring to the conversation, as we put students' best interests first. In summary, let's consider taking at least these actions:

- *Teachers*: Ensure that elementary students are taught mathematics by a teacher knowledgeable about mathematics and about how to teach it in ways that help students learn to think, reason, and solve problems; this means using math specialists from the earliest grades.

- *Expectations*: Maintain high expectations for all students at all grade levels, and commit to maintaining rigorous high school mathematics requirements so that all students are prepared for life after graduation.

- *Courses*: Adapt to the changing needs of our students by restructuring high school courses, beginning with agreeing on what major topics should be included in each of three years of a new integrated high school mathematics course sequence not organized around separate courses for algebra and geometry.

- *Tests*: Create and continually improve new tests that address mathematical thinking and reasoning, and be careful not to use test scores for purposes they were never intended for.

- *Long-Term Commitment*: Commit to maintaining progress on implementing current initiatives (including in particular, the Common Core State Standards), making adjustments, and committing to regularly revisiting and revising standards and tests.

Everywhere I look, I see potential—for students, teachers, schools, and communities—potential we can realize if we work together. Our students' future is in our hands—and, as a colleague noted to me, our future is in theirs.

MESSAGES RELATED TO THE FIVE RECOMMENDATIONS

The Recommendation	*Smarter Than We Think*	*Faster Isn't Smarter*
Teachers	12: Upside-Down Teaching 14: Effectiveness and Efficiency 24: Beyond PowerPoint 29: Finding Great Teachers	33: Engaged in What? 17: Constructive Struggling 37: Boring! 38: Ten Kinds of Wonderful
Expectations	1: Smarter Than We Think 8: Oops! 27: Fixing High School 31: Developing Mathematical Habits of Mind	1: Math for a Flattening World 2: Untapped Potential 8: More Math, More Dropouts?
Courses	16: Let It Go . . . 27: Fixing High School	7: Not Your Grandpa's Algebra 8: More Math, More Dropouts?
Tests	19: How to Know What They Know 28: Bringing Testing into the 21st Century	11: Weighing Hens 19: Embracing Accountability 35: Putting Testing in Perspective
Long-Term Commitment	23: Common Sense and the Common Core 30: Walking the Walk 40: Mathematical Habits of Instruction	4: Good Old Days 12: Beyond Band-Aids and Bandwagons 26: Beyond Pockets of Wonderfulness 13: Seek First to Understand

Acknowledgments

To adequately acknowledge the many people who have influenced my thinking and writing over the years, including helping me shape this book, would require many pages. I will try to be as concise as I know how.

I must offer a renewed thank-you to the National Council of Teachers of Mathematics (NCTM). Not only did the Council and all of its members honor me with the best job I've ever had, the NCTM staff provided tremendous personal and professional support during my presidency and the writing of my earlier "Messages."

I'm tremendously fortunate and honored that Jo Boaler wrote the Foreword for this book. Among her many talents, Jo brings incredible expertise on some of the most important themes in the book, especially regarding a growth mindset and related implications of that mindset on how we structure mathematics classrooms.

I have been privileged to work again with the wonderful Math Solutions staff. Once more, my incredible editor Jamie Cross committed everything she had to move this book forward on an accelerated timeline. Jamie continually found ways to enhance the end result, while encouraging me at every step along the way. She helps me evolve as an author, working by my side to bring a slightly different perspective that always improves my work. Denise Botelho and her production team did an outstanding job of juggling timelines and managing all the component parts of the book, while accommodating (or sometimes influencing) my personal whims and quirks to produce what I think is a beautiful book. Joan Carlson has been invaluable as a friend and colleague in a role somewhere between sounding board and editorial advisor. Her insights and suggestions ring true as a voice from the classroom and beyond. And Carolyn Felux continues to be my brilliant, insightful, supportive friend, offering critical input at key points in the process.

I want to thank a small group of friends and colleagues who responded to queries and offered their insights into various pieces and parts of this book. Lisa Brown, Amanda Jansen, Susan Hull, Eric Robinson, Jo Boaler, and Cindy Chapman reviewed various parts of the book and offered their insights. Nita Copley, Janie Schielack, Matt Larson, Dick Stanley, and Kathy Heid all responded on almost no notice to help me clarify my thinking about one issue or another. All five of these friends and colleagues helped move me past being stuck at particularly crucial times. Zalman Usiskin reviewed early drafts of two messages and offered rich insights, extended discussion, and guidance to strengthen them. And applied mathematician John Pelesko, my new friend, helped me make sense of mathematical modeling, from his summer presentation designed for teachers to his timely responses to my frequent emails and questions as I was writing. I appreciate John's reviews of early drafts of messages on modeling and mathematical structure and his suggestions for improving them.

I'd like to offer special thanks to the fifteen professional friends who provided input into creating Appendix A on sources for tasks. Listing them alphabetically, thanks to Skip Fennell, Jim Fey, David Foster, Johnny Lott, Jon Manon, John Pelesko, Jamila Riser, Ingrid Ristroph, Dick Schaeffer, Alan Schoenfeld, Mike Shaughnessy, Beth Skipper, Dick Stanley, LuAnn Weynand, and Ellen Whitesides.

Numerous other professional friends and colleagues have helped me grow as a teacher and as a leader over many years. Some of these individuals are named or mentioned anonymously in various messages and others have contributed to what I know and think without attribution. If you recognize something in one of the messages that sounds like your thoughts or advice, it probably is. If you wonder about something you read in a message that may describe a less than positive attitude, event or situation, I'm sure it's from or about someone else.

And of course, I must thank the thousands of students, teachers, and other educators and leaders with whom I have interacted during my career, in and outside of the classroom, across the nation and around the globe. Over and over, I see teachers who struggle with how to do their job the best way they can. I salute those who are willing to venture outside of their comfort zone as they try new approaches and constantly work to improve their craft so that all of their students might reach their full potential. The policy makers and community members I have encountered care deeply that all of their students learn. Many of them take hard stands in visionary directions and make incredible commitments in the face of significant challenges, just because it's the right thing for students, even while others push for policies that may have unintended consequences interfering with that goal. Then there are the students. Whether mine or yours, they continue to inspire me with how much they can accomplish—many in spite of significant barriers. They sometimes soar and sometimes struggle, always reminding us of their potential to reach great heights if we just do our part on their behalf.

I have been blessed on many levels by the people and circumstances in my life. Most especially, I thank my family for their continued support as I have pursued the rich and often divergent paths that life continues to present. I thank my daughters, my sisters, and my parents for their patience and understanding over the years, and I am grateful every day for their love and support. I thank my grandchildren for reminding me over and over again why we do the work we do.

Finally, I thank Robert. I could not have written this book without his continual encouragement and unwavering support.

Appendices

Sources for Finding Worthwhile Problems and Tasks

The following sources offer problems or tasks that can generate student exploration and provide opportunities for discussion and learning. Some tasks are correlated to the Common Core State Standards (CCSS) and/or standards from the National Council of Teachers of Mathematics (NCTM). This Appendix consists of three sections: collections of tasks and lessons, comprehensive curriculum projects centered on extended problems or tasks, and student problem-solving tools.

COLLECTIONS OF TASKS AND LESSONS

- *Illustrative Mathematics Project* (Institute for Mathematics and Education, University of Arizona) is a growing collection of resources related to the CCSS, including tasks, videos, lessons and curriculum modules. www.illustrativemathematics.org.

- *Mathalicious* is a collection of lesson plans based on tasks that can lead to productive discourse, correlated to the CCSS and including both teacher and student resources. www.mathalicious.com.

- *Emergent Math* is a blog by Geoff Kralls that includes a rich set of curriculum maps and related tasks for problem-based learning related to the CCSS for grades 6 through high school. http://emergentmath.com.

- The Dan Meyer Blog includes tasks and related recommendations for teaching students around rich problems. http://blog.mrmeyer.com. See also Dan Meyer's Google site with links to a large set of problems/tasks, both his own and those of others. http://docs .google.com/spreadsheet/pub?key=0AjIqyKM9d7ZYdEhtR3BJM mdBWnM2YWxWYVM1UWowTEE&output=html.

- *Math Landing: Resources for Elementary Math Specialists and Teachers* (Maryland Public Television [and partner organizations]) is a collection of lessons and other resources for classrooms and professional development, targeted at elementary teachers and correlated to the CCSS and NCTM standards. www.mathlanding.org.

- *Mathematics Assessment Project* (University of California at Berkeley and the Shell Center) is a collection of tasks, lessons and other mathematics resources for in-depth assessment and classroom instruction for grades 6 through 12. http://map.mathshell.org/materials/index.php.

- *Fostering Algebraic Thinking: A Guide for Teachers, Grades 6–10* (Driscoll 1999) includes problems and strategies for developing algebraic thinking.

- *Achieve the Core* (Student Achievement Partners) includes resources for teaching the CCSS, including sample instructional and assessment tasks. www.achievethecore.org.

- *Yummy Math* is a collection of tasks in real-world contexts, correlated to the Common Core Standards for Mathematical Practice (NGA Center and CCSSO 2010, 6) and the NCTM process standards. www.yummymath.com.

- *Inside Mathematics* (Noyce Foundation and the Silicon Valley Mathematics Initiative) offers various resources correlated to the CCSS for teaching mathematics, including sample tasks and related videos and discussions. www.insidemathematics.org.

- *NRICH: Enriching Mathematics* is a site from the United Kingdom with sample tasks matched to curriculum objectives. http://nrich.maths.org/frontpage.

- *STatistics Education Web* (STEW) and American Statistical Association (ASA) offer resources for teaching statistics across the grades, including tasks and lessons correlated to the CCSS and NCTM standards, as well as ASA's *Guidelines for Assessment and Instruction in Statistics Education* (GAISE). www.amstat.org/education/stew.

- Consortium for the Advancement of Undergraduate Statistics Education (CAUSE) offers resources for undergraduate statistics that include tasks appropriate for secondary school. www.causeweb.org.

- *Reasoning and Sense-Making Task Library* (National Council of Teachers of Mathematics) is a collection of high school tasks supporting NCTM's *Focus on Reasoning and Sense-Making* initiative. www.nctm.org/rsmtasks.

- *Real-World Math: Articles, Lesson Plans, and Activities for the Middle Grades* (National Council of Teachers of Mathematics) is a collection of articles and tasks with connections to other disciplines and real-world applications. www.nctm.org/publications/worlds.

- *NCTM Illuminations* is NCTM's task website, which offers tasks (many with applets and technological support) that can be used in the classroom. http://illuminations.nctm.org.
- *Annenberg Learner* offers lessons structured around in-depth problems. www.learner.org.
- University of Delaware Math Circle offers problems from past sessions for use in the classroom. www.udmathcircle.org.
- National Association of Math Circles offer *Problems from Math Circles* activities that can be used in the classroom. www.mathcircles.org.
- *Great Tasks Project* (National Council of Supervisors of Mathematics) offers ample tasks to support the CCSSM. www.mathedleadership.org/ccss/greattasks.html.
- *Teaching Children Mathematics* (National Council of Teachers of Mathematics) has a "Problem Solvers" section that offers a problem for classroom use each month. See also *Teaching Children Mathematics*'s "Math Tasks to Talk About" blog, a forum for sharing and discussing rich problems. www.nctm.org/publications/blog/blog.aspx?blogid=599514.
- *Problems to Ponder* (National Council of Teachers of Mathematics and Mike Shaughnessy) offers a series of challenging problems published in the NCTM President's Corner from 2010 through 2012. www.nctm.org/about/content.aspx?id=26070.
- *Thinking Mathematically* (Mason, Burton, and Stacey 2010) includes great problems for in-depth problem-solving experiences.
- *Algebra for Athletes* (Bauer 2007) presents fairly short problems, which could be extended into longer-term tasks, in a wide range of athletic contexts, including ice-skating and field hockey.

COMPREHENSIVE CURRICULUM PROJECTS

Elementary School

- *Everyday Mathematics*, http://ucsmp.uchicago.edu
- Investigations in Number, Data and Space, http://investigations.terc.edu *and* www.scottforesman.com/tours/investigations
- *Math Trailblazers*, http://mymathtrailblazers.com *and* www.kendallhunt.com/mtb3

Middle School

- *Connected Mathematics Project* (CMP), http://connectedmath.msu.edu *and* www.pearsonschool.com/index.cfm?locator=PS1yJe
- *Mathematics in Context* (MiC), http://mathincontext.eb.com
- *MathScape*: Seeing and Thinking Mathematically, www2.edc.org/mathscape/phil/default.asp *and* www.glencoe.com/sec/math/mathscape/index.php
- *Middle Grades Math Thematics*, www.classzone.com/books/math_thematics1 *and* www.classzone.com/books/math_thematics2 *and* www.classzone.com/books/math_thematics3
- *Middle School Mathematics Through Applications Project* (MMAP), http://scil.stanford.edu/research/projects/mmap.html

High School

- *Contemporary Mathematics in Context* (CORE-Plus), http://www.wmich.edu/cpmp
- Interactive Mathematics Program (IMP), http://mathimp.org
- *Mathematics: Modeling our World* (ARISE), www.comap.com/highschool/projects/arise.html
- *SIMMS Integrated Mathematics*, www.montana.edu wwwsimms *and* www.kendallhunt.com/simms
- *UCSMP mathematics*, http://ucsmp.uchicago.edu

STUDENT PROBLEM-SOLVING TOOLS

- Thinking About Thinking: Problem-Solving Tool is a tool to help students organize their thinking and persevere in solving a problem. http://learningandtheadolescentmind.org/resources_02.html (right-hand column).
- Thinking About Thinking: Self-Reflection Tool is a tool for students to use to reflect on their thinking after working on a problem. http://learningandtheadolescentmind.org/resources_02.html (right-hand column).

Selecting and Evaluating Worthwhile Tasks

The following sources offer criteria or rubrics for evaluating and selecting worthwhile tasks that provide opportunities for mathematical exploration, discourse, and significant learning.

- "Why Is Teaching with Problem Solving Important to Student Learning?" (Lester and Cai 2010) includes criteria for worthwhile problems for developing problem solving and thinking skills. www.nctm.org/news/content.aspx?id=25713.

- "Selecting and Creating Mathematical Tasks: From Research to Practice" (Smith and Stein 1998); the rubric presented in this article is also reprinted in *5 Practices for Orchestrating Productive Mathematics Discussions* (Smith and Stein 2011).

- The Illustrative Mathematics Project collects and evaluates tasks that exemplify the *Common Core State Standards for Mathematics*. https://docs.google.com/file /d/0B7UDDaSOTTwkcWRJZjRGNWFWTWs/edit?pli=1.

- "A Designer Speaks: Challenges in U.S. Mathematics Education Through a Curriculum Developer Lens" (Lappan and Phillips 2009) includes criteria for evaluating worthwhile mathematics tasks. www.educationaldesigner.org/ed/volume1/issue3/article11.

- "Transition to Common Core" is a wikispace that includes a rubric from Bay-Williams and Van de Walle as well as other criteria and sources for tasks and tools. https://transitiontocommoncore .wikispaces.hcpss.org/Worthwhile+Math+Tasks.

- Resources to Supplement Rubric: Implementing Standards for Mathematical Practice, IAS Park City Mathematics Institute offers criteria for selecting tasks to support the Common Core Standards for Mathematical Practice. (See Appendix C for the related rubric.) http://mathforum.org/pcmi/hstp/resources/rubric.

C Rubric for Implementing the Common Core Standards for Mathematical Practice

Background

The rubric on the following pages was created as a tool for teachers to monitor and reflect on their teaching practice with respect to the mathematical practices described in the Common Core State Standards. The rubric was developed by participants at the 2011 Summer School Teachers Program, Institute for Advanced Study Park City Mathematics Institute and is reprinted with permission.

Using the Rubric

Review each row corresponding to a mathematical practice. Use the boxes to mark the appropriate description for your task or teacher action. The task descriptors can be used primarily *as* you develop your lesson to make sure your classroom tasks help cultivate the mathematical practices. The teacher descriptors, however, can be used *during* or *after* the lesson to evaluate how the task was carried out. The column titled "proficient" describes the expected norm for task and teacher action while the column titled "exemplary" includes all features of the proficient column and more. A teacher who is exemplary is meeting criteria in *both* the proficient and exemplary columns.

PRACTICE	NEEDS IMPROVEMENT	EMERGING (teacher does thinking)	PROFICIENT (teacher mostly models)	EXEMPLARY (students take ownership)
Make sense of problems and persevere in solving them.	*Task:* ☐ Is strictly procedural. ☐ Does not require students to check solutions for errors. *Teacher:* ☐ Does not allow for wait time; asks leading questions to rush through task. ☐ Does not encourage students to individually process the tasks. ☐ Is focused solely on answers rather than processes and reasoning.	*Task:* ☐ Is overly scaffolded or procedurally "obvious." ☐ Requires students to check answers by plugging in numbers. *Teacher:* ☐ Allots too much or too little time to complete task. ☐ Encourages students to individually complete tasks, but does not ask them to evaluate the processes used. ☐ Explains the reasons behind procedural steps. ☐ Does not check errors publicly.	*Task:* ☐ Is cognitively demanding. ☐ Has more than one entry point. ☐ Requires a balance of procedural fluency and conceptual understanding. ☐ Requires students to check solutions for errors using one other solution path. *Teacher:* ☐ Allows ample time for all students to struggle with task. ☐ Expects students to evaluate processes implicitly. ☐ Models making sense of the task (given situation) and the proposed solution.	*Task:* ☐ Allows for multiple entry points and solution paths. ☐ Requires students to defend and justify their solution by comparing multiple solution paths. *Teacher:* ☐ Differentiates to keep advanced students challenged during work time. ☐ Integrates time for explicit meta-cognition. ☐ Expects students to make sense of the task and the proposed solution.

(continued)

PRACTICE	NEEDS IMPROVEMENT	EMERGING (teacher does thinking)	PROFICIENT (teacher mostly models)	EXEMPLARY (students take ownership)
Reason abstractly and quantitatively.	*Task:* □ Lacks context. □ Does not make use of multiple representations or solution paths. *Teacher:* □ Does not expect students to interpret representations. □ Expects students to memorize procedures with no connection to meaning.	*Task:* □ Is embedded in a contrived context. *Teacher:* □ Expects students to model and interpret tasks using a single representation. □ Explains connections between procedures and meaning.	*Task:* □ Has realistic context. □ Requires students to frame solutions in a context. □ Has solutions that can be expressed with multiple representations. *Teacher:* □ Expects students to interpret and model using multiple representations. □ Provides structure for students to connect algebraic procedures to contextual meaning. □ Links mathematical solution with a question's answer.	*Task:* □ Has relevant realistic context. *Teacher:* □ Expects students to interpret, model, and connect multiple representations. □ Prompts students to articulate connections between algebraic procedures and contextual meaning.
Construct viable arguments and critique the reasoning of others.	*Task:* □ Is either ambiguously stated or too easy. *Teacher:* □ Does not ask students to present arguments or solutions.	*Task:* □ Is not at the appropriate level. *Teacher:* □ Does not help students differentiate between assumptions and logical conjectures.	*Task:* □ Avoids single steps or routine algorithms. *Teacher:* □ Identifies students' assumptions. □ Models evaluation of student arguments.	*Teacher:* □ Helps students differentiate between assumptions and logical conjectures. □ Prompts students to evaluate peer arguments.

PRACTICE	NEEDS IMPROVEMENT	EMERGING (teacher does thinking)	PROFICIENT (teacher mostly models)	EXEMPLARY (students take ownership)
	❑ Expects students to follow a given solution path without opportunities to make conjectures.	❑ Asks students to present arguments but not to evaluate them. ❑ Allows students to make conjectures without justification.	❑ Asks students to explain their conjectures.	❑ Expects students to formally justify the validity of their conjectures.
Model with mathematics.	*Task:* ❑ Requires students to identify variables and to perform necessary computations. *Teacher:* ❑ Identifies appropriate variables and procedures for students. ❑ Does not discuss appropriateness of model.	*Task:* ❑ Requires students to identify variables and to compute and interpret results. *Teacher:* ❑ Verifies that students have identified appropriate variables and procedures. ❑ Explains the appropriateness of model.	*Task:* ❑ Requires students to identify variables, compute and interpret results, and report findings using a mixture of representations. ❑ Illustrates the relevance of the mathematics involved. ❑ Requires students to identify extraneous or missing information. *Teacher:* ❑ Asks questions to help students identify appropriate variables and procedures. ❑ Facilitates discussions in evaluating the appropriateness of model.	*Task:* ❑ Requires students to identify variables, compute and interpret results, report findings, and justify the reasonableness of their results and procedures within context of the task. *Teacher:* ❑ Expects students to justify their choice of variables and procedures. ❑ Gives students opportunity to evaluate the appropriateness of model.

(continued)

PRACTICE	NEEDS IMPROVEMENT	EMERGING (teacher does thinking)	PROFICIENT (teacher mostly models)	EXEMPLARY (students take ownership)
Use appropriate tools strategically.	*Task:* ❑ Does not incorporate additional learning tools. *Teacher:* ❑ Does not incorporate additional learning tools.	*Task:* ❑ Lends itself to one learning tool. ❑ Does not involve mental computations or estimation. *Teacher:* ❑ Demonstrates use of appropriate learning tool.	*Task:* ❑ Lends itself to multiple learning tools. ❑ Gives students opportunity to develop fluency in mental computations. *Teacher:* ❑ Chooses appropriate learning tools for student use. ❑ Models error checking by estimation.	*Task:* ❑ Requires multiple learning tools (i.e., graph paper, calculator, manipulatives). ❑ Requires students to demonstrate fluency in mental computations. *Teacher:* ❑ Allows students to choose appropriate learning tools. ❑ Creatively finds appropriate alternatives where tools are not available.
Attend to precision.	*Task:* ❑ Gives imprecise instructions. *Teacher:* ❑ Does not intervene when students are being imprecise. ❑ Does not point out instances when students fail to address the question completely or directly.	*Task:* ❑ Has overly detailed or wordy instructions. *Teacher:* ❑ Inconsistently intervenes when students are imprecise. ❑ Identifies incomplete responses but does not require student to formulate further response.	*Task:* ❑ Has precise instructions. *Teacher:* ❑ Consistently demands precision in communication and in mathematical solutions. ❑ Identifies incomplete responses and asks student to revise their response.	*Task:* ❑ Includes assessment criteria for communication of ideas. *Teacher:* ❑ Demands and models precision in communication and in mathematical solutions. ❑ Encourages students to identify when others are not addressing the question completely.

PRACTICE	NEEDS IMPROVEMENT	EMERGING (teacher does thinking)	PROFICIENT (teacher mostly models)	EXEMPLARY (students take ownership)
Look for and make use of structure.	*Task:* ❑ Requires students to automatically apply an algorithm to a task without evaluating its appropriateness. *Teacher:* ❑ Does not recognize students for developing efficient approaches to the task. ❑ Requires students to apply the same algorithm to a task although there may be other approaches.	*Task:* ❑ Requires students to analyze a task before automatically applying an algorithm. *Teacher:* ❑ Identifies individual students' efficient approaches, but does not expand understanding to the rest of the class. ❑ Demonstrates the same algorithm to all related tasks although there may be other more effective approaches.	*Task:* ❑ Requires students to analyze a task and identify more than one approach to the problem. *Teacher:* ❑ Facilitates all students in developing reasonable and efficient ways to accurately perform basic operations. ❑ Continuously questions students about the reasonableness of their intermediate results.	*Task:* ❑ Requires students to identify the most efficient solution to the task. *Teacher:* ❑ Prompts students to identify mathematical structure of the task in order to identify the most effective solution path. ❑ Encourages students to justify their choice of algorithm or solution path.

(continued)

PRACTICE	NEEDS IMPROVEMENT	EMERGING (teacher does thinking)	PROFICIENT (teacher mostly models)	EXEMPLARY (students take ownership)
Look for and express regularity in repeated reasoning.	*Task:* ❑ Is disconnected from prior and future concepts. ❑ Has no logical progression that leads to pattern recognition. *Teacher:* ❑ Does not show evidence of understanding the hierarchy within concepts. ❑ Presents or examines task in isolation.	*Task:* ❑ Is overly repetitive or has gaps that do not allow for development of a pattern. *Teacher:* ❑ Hides or does not draw connections to prior or future concepts.	*Task:* ❑ Reviews prior knowledge and requires cumulative understanding. ❑ Lends itself to developing a pattern or structure. *Teacher:* ❑ Connects concept to prior and future concepts to help students develop an understanding of procedural shortcuts. ❑ Demonstrates connections between tasks.	*Task:* ❑ Addresses and connects to prior knowledge in a non-routine way. ❑ Requires recognition of pattern or structure to be completed. *Teacher:* ❑ Encourages students to connect task to prior concepts and tasks. ❑ Prompts students to generate exploratory questions based on current task. ❑ Encourages students to monitor each other's intermediate results.

Source: http://mathforum.org/pcmi/hstp/resources/rubric/. Reprinted with permission of PCMI and IAS.

APPENDIX

D Essential Library

There are many other wonderful books that would be great to have in your professional library; these are a few selected resources that particularly relate to the messages in this book.

TEACHING STUDENTS FOR UNDERSTANDING AND PROFICIENCY

- *About Teaching Mathematics: A K–8 Resource, 3rd edition* (Burns 2007) offers insights and lessons for creating engaging classrooms and fostering mathematical thinking.
- *Thinking Mathematically: Integrating Arithmetic and Algebra in Elementary School* (Carpenter, Franke, and Levi 2003) offers insights into using elements of algebra to help make sense of arithmetic and vice versa.
- *Classroom Discussions in Math: A Teacher's Guide for Using Talk Moves to Support the Common Core and More, 3rd edition* (Chapin, O'Connor, and Anderson 2013) presents practical strategies for orchestrating student discussions around worthwhile problems to help students learn to think and communicate mathematically.
- *Fostering Algebraic Thinking: A Guide for Teachers, Grades 6–10* (Driscoll 1999) provides strategies for helping students make sense of algebraic concepts and ideas.
- *Teaching Mathematics Through Problem Solving: Prekindergarten– Grade 6* (Lester and Charles 2003) is one of a two-book series designed for teachers to help students make sense of problems and learn mathematical concepts.
- *Principles to Actions: Ensuring Mathematical Success for All* (National Council of Teachers of Mathematics, 2014) describes NCTM's principles (NCTM 2000) and offers practical strategies for overcoming barriers to their implementation.

- *Teaching Mathematics Through Problem Solving: Grades 6–12* (Schoen and Charles 2003) is one of a two-book series designed for teachers to help students make sense of problems and learn mathematical concepts.
- *5 Practices for Orchestrating Mathematics Discussions* (Smith and Stein 2011) discusses powerful practices to produce effective mathematics classrooms with high levels of student engagement.

MATHEMATICS FOR TEACHERS

- *Math Matters: Understanding the Math You Teach, Grades K–8, 2nd edition* (Chapin and Johnson 2006) is an excellent resource for mathematics content related to elementary and middle school.
- Essential Understandings is a series of sixteen books for teachers on critical content areas of prekindergarten through grade 12 mathematics; related teaching resources are available. Published by the National Council of Teachers of Mathematics.
- *Elementary and Middle School Mathematics: Teaching Developmentally, 7th edition* (Van de Walle, Karp, and Bay-Williams 2009) is the comprehensive go-to reference on mathematics content and strategies for elementary and middle school.

ASSESSMENT

- *INFORMative Assessment: Formative Assessment to Improve Math Achievement, Grades K–6* (Joyner and Muri 2011) offers practical, doable strategies for paying attention to how well students are learning the mathematics we want them to learn.

- *INFORMative Assessment: Formative Assessment to Improve Mathematics Achievement, Middle and High School* (Joyner and Bright forthcoming) offers practical, doable strategies for paying attention to how well stuents are learning the mathematics we want them to learn.

- *Mathematics Assessment: Myths, Models Good Questions, and Practical Suggestions* (Stenmark 1991) is a concise timeless resource offering simple, practical tips for assessing students' mathematical thinking, including sample rubrics and other resources.

HISTORICAL PERSPECTIVES FROM THE PROFESSION

- *Curriculum and Evaluation Standards for School Mathematics* (National Council of Teachers of Mathematics 1989) provides the first professional recommendations for what should be included in school mathematics programs.

- *Everybody Counts: A Report to the Nation on the Future of Mathematics Education* (Mathematical Sciences Education Board and National Research Council 1989) makes a compelling case for providing all students with appropriate, challenging, relevant mathematics, remarkable for its lasting insights that continue to resonate today.
- *Professional Standards for Teaching Mathematics* (National Council of Teachers of Mathematics 1991) is still the best description of what a professional teacher of mathematics does in the classroom and how leaders can support professional teachers.
- *Principles and Standards for School Mathematics* (National Council of Teachers of Mathematics 2000) updates NCTM's mathematics standards, including both principles for school mathematics and process and content standards for what should be taught.
- *Adding It Up: Helping Children Learn Mathematics* (National Research Council 2001) considers and defines five dimensions of mathematical proficiency and discusses implications of research studies to inform and guide teaching practice toward student proficiency.
- Common Core State Standards for Mathematics (NGA Center and CCSSO 2010) presents the mathematics standards currently in use in the majority of states.

APPENDIX

Research Summaries and Statements on Key Issues in School Mathematics

NATIONAL COUNCIL OF TEACHERS OF MATHEMATICS RESEARCH BRIEFS

The following research briefs present concise summaries of research on questions often asked by practitioners and policy makers. NCTM also offers in-depth research analyses and short research clips related to some of the topics addressed in these briefs and continues to develop additional resources on these and other topics. Access online at: www.nctm.org/news/content.aspx?id=8468

- "What Can We Learn from Research?" (Reed 2008).
- "How Can Teachers and Schools Use Data Effectively?" (Schleppenbach 2010).
- "Research on Students' Thinking and Reasoning About Averages and Measures of Center" (National Council of Teachers of Mathematics 2007).
- "What Does Research Say the Benefits of Discussion in Mathematics Class Are?" (Cirillo 2013b).
- "What Are Some Strategies for Facilitating Productive Classroom Discussions?" (Cirillo 2013a).
- "Involving Latino and Latina Parents in Their Children's Mathematics Education" (Civil and Menéndez 2010).
- "A Brief History of Homework in the United States" (Cooper 2008a).
- "Effective Homework Assignments" (Cooper 2008b).
- "Homework: What the Research Says" (Cooper 2008c).
- "Mathematics Professional Development" (Doerr, Goldsmith, and Lewis 2010).
- "Teaching Ratio and Proportion in the Middle Grades" (Ellis 2013).

- "Effective Strategies for Teaching Students with Difficulties in Mathematics" (Gersten and Clarke 2007a).
- "What Are the Characteristics of Students with Learning Difficulties in Mathematics?" (Gersten and Clarke 2007b).
- "Effective Teaching for the Development of Skill and Conceptual Understanding of Number: What Is Most Effective?" (Hiebert and Grouws 2007).
- "What Do We Know About the Teaching and Learning of Algebra in the Elementary Grades?" (Kieran 2007b).
- "What Do Students Struggle with When First Introduced to Algebra Symbols?" (Kieran 2007a).
- "Why Is Teaching with Problem Solving Important to Student Learning?" (Lester and Cai 2010).
- "Mathematics Specialists and Mathematics Coaches: What Does the Research Say?" (McGatha 2009).
- "Using Two Languages When Learning Mathematics: How Can Research Help Us Understand Mathematics Learners Who Use Two Languages?" (Moschkovich 2009).
- "Using Calculators for Teaching and Learning Mathematics" (Ronau, Rakes, Bush, Driskell, Niess, and Pugalee 2011).
- "What Do We Know About Students' Thinking and Reasoning About Variability in Data?" (Shaughnessy 2008).
- "Selecting the Right Curriculum" (Stein 2007).
- "Five "Key Strategies" for Effective Formative Assessment" (Wiliam 2007a).
- "What Does Research Say the Benefits of Formative Assessment Are?" (Wiliam 2007b).

NATIONAL COUNCIL OF SUPERVISORS OF MATHEMATICS IMPROVING STUDENT ACHIEVEMENT SERIES: RESEARCH-INFORMED ANSWERS FOR MATHEMATICS EDUCATION LEADERS

The following statements represent professional positions from NCSM, a group of mathematics education leaders. The statements include research background on the positions stated. Access online at: www.mathedleadership.org/resources/position.html

- "Improving Student Achievement by Leading *Effective and Collaborative* Teams of Mathematics Teachers" (2007a).
- "Improving Student Achievement by Leading Sustained Professional Learning for Mathematics Content and Pedagogical Knowledge Development" (2007b).

- "Improving Student Achievement by Leading the Pursuit of a Vision for Equity" (2008a).
- "Improving Student Achievement in Mathematics for Students with Special Needs" (2008b).
- "Improving Student Achievement in Mathematics by Addressing the Needs of English Language Learners" (2009a).
- "Improving Student Achievement in Mathematics by Leading Highly Effective Assessment Practices" (2009b).
- "Improving Student Achievement in Mathematics by Promoting Positive Self-Belief" (2010).
- "Improving Student Achievement in Mathematics by Systematically Integrating Effective Technology" (2011).
- "Improving Student Achievement in Mathematics by Expanding Learning Opportunities for the Young" (2012a).
- "Improving Student Achievement in Mathematics by Expanding Opportunities for Our Most Promising Students of Mathematics" (2012b).
- "Improving Student Achievement by Infusing Highly Effective Instructional Strategies into RTI Tier I Instruction" (2013a).
- "Improving Student Achievement in Mathematics by Using Manipulatives with Classroom Instruction" (2013b).

National Council of Teachers of Mathematics Process Standards

The following process standards relate closely to the Standards for Mathematical Practice in the Common Core State Standards, as noted in Messages 31–40. The K–12 statements of the standards are shown below, as presented in *Principles and Standards for School Mathematics* (NCTM 2000). More detailed description and grade-band standards can be found in the full document.

PROBLEM SOLVING

Instructional programs from prekindergarten through grade 12 should enable all students to:

- Build new mathematical knowledge through problem solving
- Solve problems that arise in mathematics and in other contexts
- Apply and adapt a variety of appropriate strategies to solve problems
- Monitor and reflect on the process of mathematical problem solving

REASONING AND PROOF

Instructional programs from prekindergarten through grade 12 should enable all students to:

- Recognize reasoning and proof as fundamental aspects of mathematics
- Make and investigate mathematical conjectures
- Develop and evaluate mathematical arguments and proofs
- Select and use various types of reasoning and methods of proof

COMMUNICATION

Instructional programs from prekindergarten through grade 12 should enable all students to:

- Organize and consolidate their mathematical thinking through communication
- Communicate their mathematical thinking coherently and clearly to peers, teachers, and others
- Analyze and evaluate the mathematical thinking and strategies of others
- Use the language of mathematics to express mathematical ideas precisely

CONNECTIONS

Instructional programs from prekindergarten through grade 12 should enable all students to:

- Recognize and use connections among mathematical ideas
- Understand how mathematical ideas interconnect and build on one another to produce a coherent whole
- Recognize and apply mathematics in contexts outside of mathematics

REPRESENTATION

Instructional programs from prekindergarten through grade 12 should enable all students to:

- Create and use representations to organize, record, and communicate mathematical ideas
- Select, apply, and translate among mathematical representations to solve problems
- Use representations to model and interpret physical, social, and mathematical phenomena

Resources and References

Achieve, Inc. *Closing the Expectations Gap 2013*. www.achieve.org/closing-expectations-gap-report. *27*

Achieve, Inc. "CCSS-CTE Classroom Tasks." www.achieve.org/ccsscte-classroom-tasks. *8*

Achieve, Inc. 2014. "The Next Generation Science Standards." www.nextgenscience.org/next-generation-science-standards. *35*

Anderson, Nancy Canavan, and Lanie Schuster. 2005. *Good Questions for Math Teaching: Why Ask Them and What to Ask, Grades 5–8*. Sausalito, CA: Math Solutions. *19*

Association of Mathematics Teacher Educators, Association of State Supervisors of Mathematics, National Council of Supervisors of Mathematics, and National Council of Teachers of Mathematics. 2010. "The Role of Elementary Mathematics Specialists in the Teaching and Learning of Mathematics." Joint Position Statement. Authors. *13, 14*

August, Diane, Maria Carlo, Cheryl Dressler, and Catherine Snow. 2005. "The Critical Role of Vocabulary Development for English Language Learners." *Learning Disabilities Research and Practice* 20 (1): 50–57. *2*

Bain, Ken. 2004. *What the Best College Teachers Do*. Cambridge, MA: Harvard University Press. *29*

Baker, Stephen. 2006. "Math Will Rock Your World." *BusinessWeek* (January 22). www.businessweek.com/printer/articles/201872-math-will-rock-your-world?type=old_article. *27*

Baratta-Lorton, Mary. 1995. *Mathematics Their Way: An Activity-Centered Mathematics Program for Early Childhood Education, 20th Anniversary Edition*. New York: Addison-Wesley. *39*

Bass, Hyman. 2005. "Mathematics, Mathematicians, and Mathematics Education." *Bulletin of the American Mathematical Society* 42 (4): 417–30. *14, 31, 38*

Bauer, Cameron. 2007. *Algebra for Athletes*. New York: Nova Science Publishers.

Bay-Williams, Jennifer M. 2014. "Takeaways from Math Methods: How Will You Teach Effectively?" San Rafael, CA: Edutopia. *12, 40*

Beaupré, Dan, Simone Bloom Nathan, and Anne Kaplan. "Testing Our Schools: A Guide for Parents." PBS Frontline. www.pbs.org/wgbh/pages/frontline/shows/schools/etc/guide.html. *28*

Benson, Peter L. 2008. *Sparks: How Parents Can Ignite the Hidden Strengths of Teenagers*. New York: Jossey-Bass. *1*

Binet, A. 1909. *Modern Ideas About Children* (transl. in 1975 by Suzanne Heisler). Menlo Park, CA: PUB. *1*

Bird, Brad, and Jan Pinkava. 2007. *Ratatouille*. Emeryville, CA: Pixar Animation Studios. DVD. *29*

Blanton, Maria, Linda Levi, Terry Crites, Barbara Dougherty, and Rose Mary Zbiek. 2011. *Developing Essential Understanding of Algebraic Thinking for Teaching Mathematics in Grades 3–5*. Reston, VA: National Council of Teachers of Mathematics. *17*

Boaler, Jo. 1993. "The Role of Contexts in the Mathematics Classroom: Do They Make Mathematics More 'Real'?" *For the Learning of Mathematics* 13 (2): 12–17. *12*

———. 2008. *What's Math Got to Do with It? Helping Children Learn to Love Their Least Favorite Subject—and Why It's Important for America*. New York: Viking. *18*

———. 2013a. "Ability and Mathematics: The Mindset Revolution That Is Reshaping Education." *FORUM* 55 (1): 143–52. *1*

———. 2013b. "The Stereotypes That Distort How Americans Teach and Learn Mathematics." *The Atlantic* (November 12). www.theatlantic.com/education /archive/2013/11/the-stereotypes-that-distort-how-americans-teach-and-learn-math/281303.

———. 2013. "Educ115N: How to Learn Math" Stanford University, https://class .stanford.edu/courses/Education/EDUC115N/How_to_Learn_Math/about (also see www.youcubed.org). *1, 8, 10, 13*

Bresser, Rusty. 2003. "Helping English-Language Learners Develop Computational Fluency." *Teaching Children Mathematics* 9 (6): 294–99. *2*

Bresser, Rusty, Carolyn Felux, and Kathy Melanese. 2009. *Supporting English Language Learners in Math Class: A Multimedia Professional Learning Resource, Grades K–5*. Sausalito, CA: Math Solutions. *2, 37*

Bresser, Rusty, Kathy Melanese, and Christine Sphar. 2008a. *Supporting English Language Learners in Math Class, Grades K–2*. Sausalito, CA: Math Solutions. *2, 37*

———. 2008b. *Supporting English Language Learners in Math Class, Grades 3–5*. Sausalito, CA: Math Solutions. *2, 37*

Briars, Diane, Harold Asturias, David Foster, and Mardi Gale. 2012. *Common Core Mathematics in a PLC at Work, Grades 6–8*. Reston, VA: National Council of Teachers of Mathematics. *23, 25*

Bridgeland, John M., John J. DiIulio, and Karen Burke Morison. 2006. *The Silent Epidemic: Perspectives of High School Dropouts*. Washington, DC: Civic Enterprises. *4, 6*

Brown, M. W. 1949. *The Important Book*. New York: HarperCollins. *5*

Brown, Roger. 2001. "Computer Algebra Systems and the Challenge of Assessment." *International Journal of Computer Algebra in Mathematics Education* 8 (4): 295–308. *28*

Brutlag, Dan. 2009. *Active Algebra, Grades 7–10: Strategies and Lessons for Successfully Teaching Linear Relationships*. Sausalito, CA: Math Solutions. *17*

Bryant, Cindy. 2013. "Growth Mindset and the Common Core Math Standards." *Edutopia Assessment Blog*, December 3. www.edutopia.org/blog /growth-mindset-common-core-math-cindy-bryant. *1, 31*

Burger, Edward B., and Michael Starbird. 2012. *The 5 Elements of Effective Thinking*. Princeton, NJ: Princeton University Press. *5, 8*

Burns, Marilyn. 1998. *Math: Facing an American Phobia*. Sausalito, CA: Math Solutions. *7, 10*

———. 2007. *About Teaching Mathematics: A K–8 Resource, 3rd edition*. Sausalito, CA: Math Solutions. *3*

———. "Math Reasoning Inventory." https://mathreasoninginventory.com. *13, 14*

Burrill, Gail. 1997. President's Address. National Council of Teachers of Mathematics annual meeting, April, Minneapolis, MN. *36*

Burris, Carol Corbett, Kevin G. Welner, and Jennifer Weiser Bezoza. 2009. "Universal Access to a Quality Education: Research and Recommendations for the Elimination of Curricular Stratification." Boulder, CO and Tempe, AZ: EPIC (Education and the Public Interest Center), University of Colorado at Boulder and EPRU (Education Policy Research Unit), Arizona State University. http://epicpolicy .org/publication/universal-access. *7*

Carpenter, Thomas P., John A. Dossey, and Julie L. Koehler, eds. 2004. *Classics in Mathematics Education Research*. Reston, VA: National Council of Teachers of Mathematics. *22*

Carpenter, Thomas P., Megan Loef Franke, and Linda Levi. 2003. *Thinking Mathematically: Integrating Arithmetic and Algebra in Elementary School*. Portsmouth, NH: Heinemann. *17, 31, 33, 38, 39*

Chapin, Suzanne H., and Art Johnson. 2006. *Math Matters: Understanding the Math You Teach, Grades K–8, 2nd edition.* Sausalito, CA: Math Solutions.

Chapin, Suzanne H., Catherine O'Connor, and Nancy Canavan Anderson. 2013. *Classroom Discussions in Math: A Teacher's Guide for Using Talk Moves to Support the Common Core and More, Grades K–6, 3rd edition.* Sausalito, CA: Math Solutions. *4, 8, 34, 37*

Chappell, Michaele F., and Tina Pateracki, eds. 2004. *Empowering the Beginning Teacher of Mathematics: Middle School.* Reston, VA: National Council of Teachers of Mathematics. *11*

Chappell, Michaele F., Jeffrey Choppin, and Jenny Salls, eds. 2004. *Empowering the Beginning Teacher of Mathematics: High School.* Reston, VA: National Council of Teachers of Mathematics. *11*

Chappell, Michaele F., Janie Schielack, and Sharon Zagorski, eds. 2004. *Empowering the Beginning Teacher of Mathematics in Elementary School.* Reston, VA: National Council of Teachers of Mathematics. *11*

Charles A. Dana Center at The University of Texas at Austin. 2012. "Learning and the Adolescent Mind." Austin: Charles A. Dana Center at the University of Texas at Austin. learningandtheadolescentmind.org. *1, 32*

———. 2010a. *Advanced Mathematical Decision Making (Advanced Quantitative Reasoning in Texas).* Austin: Charles A. Dana Center at The University of Texas at Austin. www.utdanacenter.org/amdm. *6, 27, 33*

———. 2010b. *Advanced Mathematical Decision Making (Student and Teacher Materials).* Austin, TX: Charles A. Dana Center at The University of Texas at Austin. www.utdanacenter.org/amdm. *6, 12, 35*

———. 2012a. "What Should I Look for in a Math Classroom?" Austin: Charles A. Dana Center at The University of Texas at Austin. www.utdanacenter.org /mathtoolkit/support/look.php. *7*

———. 2012b. "PARCC Prototyping Project." Austin: Charles A. Dana Center at The University of Texas at Austin. www.ccsstoolbox.org. *28, 31, 34*

Charles, Randall I., and Frank Lester, eds. 2010. *Teaching and Learning Mathematics: Translating Research for School Administrators.* Reston, VA: National Council of Teachers of Mathematics. *22*

Chingos, Matthew M., and Grover J. Whitehurst. 2011. *Class Size: What Research Says and What It Means for State Policy.* Washington, DC: The Brookings Institution. *20, 22*

Christen, Carol, and Richard N. Bolles. 2010. *What Color Is Your Parachute? For Teens: Discovering Yourself, Defining Your Future, 2nd edition.* Berkeley, CA: Ten Speed Press. *9*

Cirillo, Michelle. 2013a. "What Are Some Strategies for Facilitating Productive Classroom Discussions?" *Discussion Research Brief,* series ed. Sarah DeLeeuw. Reston, VA: National Council of Teachers of Mathematics. *12*

———. 2013b. "What Does Research Say the Benefits of Discussion in Mathematics Class Are?" *Discussion Research Brief,* series ed. Sarah DeLeeuw. Reston, VA: National Council of Teachers of Mathematics. *12*

Civil, Marta, and José María Menéndez. 2010. "Involving Latino and Latina Parents in Their Children's Mathematics Education." *Research Brief,* edited by Sarah DeLeeuw. Reston, VA: National Council of Teachers of Mathematics. *2, 7*

Clarke, David, Margarita Breed, and Sherry Fraser. 2004. "The Consequences of a Problem-Based Mathematics Curriculum." *The Mathematics Educator* 14 (2): 7–16. *12*

Clements, Douglas H., and Julie Sarama. 2009. *Learning and Teaching Early Math: The Learning Trajectories Approach.* New York: Routledge. *14, 16*

Clements, Douglas H., Julie Sarama, and Ann-Marie DiBiase, eds. 2007. *Engaging Young Children in Mathematics: Standards for Early Childhood Mathematics Education.* Mahwah, NJ: Lawrence Erlbaum. *16*

Coggins, Debra. 2007. *English Language Learners in the Mathematics Classroom.* Thousand Oaks, CA: Corwin Press.   2

Coffey, Margaret, and Wade Sherard. 2012. *Real World Math: Articles, Lesson Plans, and Activities for the Middle Grades.* Reston, VA: National Council of Teachers of Mathematics. www.nctm.org/publications/worlds/default.aspx?id=34028.   35

Collins, Allan, and Richard Halverson. 2009. *Rethinking Education in the Age of Technology: The Digital Revolution and Schooling in America.* New York: Teachers College Press.   24

Collins, Jim. 2001. *Good to Great: Why Some Companies Make the Leap ... and Others Don't.* New York: HarperCollins.   26, 30

———. 2005. *Good to Great and the Social Sectors: Why Business Thinking Is Not the Answer.* New York: HarperCollins.   26, 30

COMAP. www.comap.com.   27

COMAP. "High School Mathematics and Its Applications (HIMAP)." www.comap.com /highschool/projects/himap.html.   35

Cooper, Harris. 2008a. "A Brief History of Homework in the United States." *Research Brief*, edited by Judith Quander Reed. Reston, VA: National Council of Teachers of Mathematics.

———. 2008b. "Effective Homework Assignments." *Research Brief*, edited by Judith Quander Reed. Reston, VA: National Council of Teachers of Mathematics.

———. 2008c. "Homework: What the Research Says." *Research Brief*, edited by Judith Quander Reed. Reston, VA: National Council of Teachers of Mathematics.

Copley, Juanita V. 2010. *The Young Child and Mathematics, 2nd edition.* Washington, DC: National Association for the Education of Young Children and National Council of Teachers of Mathematics.   10, 14, 37

Costa, Arthur L., and Bena Kallick. 2008. *Learning and Leading with Habits of Mind: 16 Essential Characteristics for Success.* Alexandria, VA: Association for Supervision and Curriculum Development.   31

Crawford, Matthew B. 2009. *Shop Class as Soulcraft: An Inquiry into the Value of Work.* New York: Penguin Books.   8, 9, 27

———. 2010. "Contemporary Curriculum Issues: Organizing a Curriculum around Mathematical Habits of Mind." *Mathematics Teacher* 103 (9): 682–88.   31, 40

Cushman, Kathleen. 2005. *Fires in the Bathroom: Advice for Teachers from High School Students.* New York: New Press.   4, 6, 20

———. 2010. *Fires in the Mind: What Kids Can Tell Us About Motivation and Mastery.* San Francisco, CA: Jossey-Bass.   4, 6

Cushman, Kathleen, and Laura Rogers. 2008. *Fires in the Middle School Bathroom: Advice for Teachers from Middle Schoolers.* New York: The New Press.   4, 6, 20

Cyberchase. "Helping Kids Develop Positive Math Attitudes." www.pbs.org/parents /cyberchase/math-fun-more/using-math-at-home/positive-math-attitudes.   10

Daggett, Willard. 2013. "Preparing Students to Be College and Career Ready." Keynote address, 21st Annual Model Schools Conference, National Harbor, MD, June 30–July 3.   9

Damon, William. 2008. *The Path to Purpose: How Young People Find Their Calling in Life.* New York: Free Press.   1, 9

Daro, Phil. 2012. Presentation at Annual Association of State Supervisors of Mathematics, Philadelphia, PA, April 21.   18

Deyhle, D. 1989. "Pushouts and Pullouts: Navajo and Ute School Leavers." *Journal of Navajo Education* 6 (2): 44.   6

Dick, Thomas, and Karen Hollebrands, eds. 2011. *Focus in High School Mathematics: Technology to Support Reasoning and Sense Making.* Reston, VA: National Council of Teachers of Mathematics.   36

Doerr, Helen M., Lynn T. Goldsmith, and Catherine C. Lewis. 2010. "Mathematics Professional Development." *Research Brief*, edited by Sarah DeLeeuw. Reston, VA: National Council of Teachers of Mathematics.   21, 26, 29

Dong, Yu Ren. 2009. "Linking to Prior Learning." *Educational Leadership* 66 (7): 26–31. *2*

Dossey, John A. 2010. "Mathematical Modeling on the Catwalk: A Review of Modelling and Applications in Mathematics Education: The 14th ICMI Study." *Journal for Research in Mathematics Education* 41 (1): 88–95. *35*

Doyle, Arthur Conan. 1891. "The Adventure of the Speckled Band." *The Adventures of Sherlock Holmes.* Minneapolis, MN. *33*

Driscoll, Mark. 1999. *Fostering Algebraic Thinking: A Guide for Teachers, Grades 6–10.* Portsmouth, NH: Heinemann. *17*

Driscoll, Mark, Rachel Wing DiMatteo, Johannah Nikula, and Michael Egan. 2007. *Fostering Geometric Thinking: A Guide for Teachers, Grades 5–10.* Portsmouth, NH: Heinemann. *12*

DuFour, Rebecca, Robert Eaker, Gayle Harhanek, and Richard DuFour. 2004. *Whatever It Takes: How Professional Learning Communities Respond When Kids Don't Learn.* Bloomington, IN: Solution Tree. *25*

DuFour, Richard. 2004. "What Is a Professional Learning Community?" *Educational Leadership* 61 (8). *25*

DuFour, Richard, and Michael Fullan. 2013. *Cultures Built to Last: Systemic PLCs at Work.* Bloomington, IN: Solution Tree. *21, 25, 26*

DuFour, Richard, and Robert J. Marzano. 2011. *Leaders of Learning: How District, School, and Classroom Leaders Improve Student Achievement.* Bloomington, IN: Solution Tree. *25*

DuFour, Richard, Rebecca DuFour, Robert Eaker, and Thomas Many. 2010. *Learning by Doing: A Handbook for Professional Learning Communities at Work, 2nd edition.* Bloomington, IN: Solution Tree. *25*

Dweck, Carol S. 2000. *Self-Theories: Their Role in Motivation, Personality, and Development.* Philadelphia, PA: Psychology Press. *1*

———. 2006. *Mindset: The New Psychology of Success.* New York: Ballantine Books. *1, 7, 8, 15*

Eberly Center. "Assess Teaching and Learning." Carnegie Mellon University, www.cmu.edu/teaching/assessment/index.html. *19*

Education Development Center. "Implementing the Mathematical Practice Standards." Waltham, MA: Education Development Center. http://mathpractices.edc.org. *31*

———. "Mathematical Practice Institute." Waltham, MA: Education Development Center. https://mpi.edc.org. *31*

The Education Trust. 2013. *Shattering Expectations Series: Breaking the Glass Ceiling of Achievement for Low-Income Students and Students of Color.* Washington, DC: The Education Trust.

Elliot, Andrew J., and Carol S. Dweck. 2007. *Handbook of Competence and Motivation.* New York: The Guilford Press. *1*

Ellis, Amy. 2013. "Teaching Ratio and Proportion in the Middle Grades." *Research Brief,* edited by Sarah DeLeeuw. Reston, VA: National Council of Teachers of Mathematics.

FairTest. "The Case Against High Stakes Testing." Jamaica Plain, MA: The National Center for Fair and Open Testing (FairTest). www.fairtest.org/arn/caseagainst.html. *28*

Featherstone, Helen, Sandra Crespo, Lisa M. Jilk, Joy A. Oslund, Amy Noelle Parks, and Marcy B. Wood. 2011. *Smarter Together! Collaboration and Equity in the Elementary Math Classroom.* Reston, VA: National Council of Teachers of Mathematics. *10*

Ferri, Rita Borromeo. 2013. "Mathematical Modelling in School and in Teacher Education: Conceptions and Examples" PowerPoint Presentation. Kassel, Germany: University of Kassel. http://seminarios.conectaideas.com/ppt/Rita_Borromeo_Ferri_2.pdf. *35*

Fillingim, Jennifer G. and Angela T. Barlow. 2010. "From the Inside Out." *Teaching Children Mathematics* 17 (2): 80–88. *4, 10, 31*

Finkel, Donald L. 2000. *Teaching with Your Mouth Shut*. Portsmouth, NH: Boynton/ Cook. *12*

Friedman, Thomas L. 2007. *The World Is Flat 3.0: A Brief History of the Twenty-First Century*. New York: Picador. *12, 27*

———. 2011. *That Used to Be Us: How America Fell Behind in the World It Invented and How We Can Come Back*. New York: Farrar, Strauss, and Giroux. *12*

Fullan, Michael. 2003. *The Moral Imperative of School Leadership*. Thousand Oaks, CA: Corwin. *30*

———. 2007. *Leading in a Culture of Change*. San Francisco, CA: Jossey-Bass. *21, 26*

———. 2012a. *Change Forces: Probing the Depths of Educational Reform*. Levittown, PA: The Falmer Press. *21*

———. 2012b. *Stratosphere: Integrating Technology, Pedagogy, and Change Knowledge*. Don Mills, ON: Pearson Canada. *21, 24*

Functions, Statistics, and Trigonometry and Precalculus and Discrete Mathematics. UCSMP. http://ucsmp.uchicago.edu. *27*

Gardner, Howard. 2011. *Frames of Mind: The Theory of Multiple Intelligences*. New York: Basic Books. *29*

Gates, Bill. 2004. Speech at University of Illinois Urbana-Champaign, February 24. www.allthingswilliam.com/computers.html.

Gersten, Russell, and Benjamin S. Clarke. 2007a. "Effective Strategies for Teaching Students with Difficulties in Mathematics." *Research Brief*, edited by Judith Quander Reed. Reston, VA: National Council of Teachers of Mathematics.

———. 2007b. "What Are the Characteristics of Students with Learning Difficulties in Mathematics?" *Research Brief*, edited by Judith Quander Reed. Reston, VA: National Council of Teachers of Mathematics.

Ginsburg, Alan, Steven Leinwand, and Katie Decker. 2009. *Informing Grades 1–6 Mathematics Standards Development: What Can Be Learned from High-Performing Hong Kong, Korea, and Singapore?* Washington, DC: American Institutes for Research. *18*

Ginsburg, Alan, Geneise Cooke, Steven Leinwand, Jay Noell, and Elizabeth Pollock. 2005. *Reassessing U.S. International Mathematics Performance: New Findings from the 2003 TIMSS and PISA*. Washington, DC: American Institutes of Research. *18*

Gladwell, Malcolm. 2008. *Outliers: The Story of Success*. New York: Little, Brown and Company. *1, 29*

———. 2013. *David and Goliath: Underdogs, Misfits, and the Art of Battling Giants*. New York: Little, Brown and Company. *1, 29*

Gojak, Linda M. 2011. *What's Your Math Problem? Getting to the Heart of Teaching Problem Solving*. Huntington Beach, CA: Shell Education. *12, 32*

———. 2013a. "Are We Obsessed with Assessment?" *NCTM Summing Up* (November 4). *19*

———. 2013b. "The Power of a Good Mistake." *NCTM Summing Up* (January 8). *8*

Goldberg, Marsha Serling, and Sonia Feldman. 2003. *Teachers with Class: True Stories of Great Teachers*. Kansas City: Andrews McMeel Publishing. *11*

Gould, Stephen J. 2012. *The Mismeasure of Man, revised and expanded edition*. New York: W. W. Norton. *1, 22, 28*

Hall, Gene E., and Shirley M. Hord. 2005. *Implementing Change: Patterns, Principles and Potholes, 2nd edition*. Boston, MA: Allyn and Bacon. *21, 26*

Halmos, Paul R. 1985. *I Want to Be a Mathematician*. Washington, DC: Mathematical Association of America. *8, 12, 31*

Harel, Gershon. 2008. "What Is Mathematics? A Pedagogical Answer to a Philosophical Question." In *Proof and Other Dilemmas: Mathematics and Philosophy*, edited by Bonnie Gold and Roger Simons, 265–90. Washington, DC: Mathematical Association of America. *31*

Hargreaves, Andy, and Michael Fullan. 2013. *Professional Capital: Transforming Teaching in Every School*. New York: Teachers College Press. *29, 30*

Hart, Lynn C., Alice S. Alston, and Aki Murata, eds. 2011. *Lesson Study Research and Practice in Mathematics Education: Learning Together.* New York: Springer. *30*

Hartog, Martin D., and Patricia A. Brosnan. 1994 (September). "Doing Mathematics with Your Child." *ERIC/CSMEE Digest.* Columbus, OH: ERIC Clearinghouse for Science Mathematics and Environmental Education. http://files.eric.ed.gov/fulltext /ED372967. *7*

Herbel-Eisenmann, Beth, and Michelle Cirillo. 2009. *Promoting Purposeful Discourse: Teacher Research in Secondary Math Classrooms.* Reston, VA: National Council of Teachers of Mathematics. *37*

Herman, Joan L., and Jamal Abedi. 2004. "Issues in Assessing English Language Learners' Opportunity to Learn Mathematics." Los Angeles: Center for the Study of Evaluation (CSE) National Center for Research on Evaluation, Standards, and Student Testing (CRESST), Graduate School of Education and Information Studies University of California, Los Angeles. *2*

Hiebert, James, and Douglas A. Grouws. 2007. "Effective Teaching for the Development of Skill and Conceptual Understanding of Number: What Is Most Effective?" *Research Brief*, edited by Judith Quander Reed. Reston, VA: National Council of Teachers of Mathematics. *13, 14, 16*

Hirsch, Christian R., Arthur F. Coxford, James T. Fey, and Harold L. Schoen. 1995. "Teaching Sensible Mathematics in Sense-Making Ways with the CPMP." *Mathematics Teacher* 88 (8): 694–700. *12*

Hord, Shirley M., William L. Rutherford, Leslie Huling-Austin, and Gene E. Hall. 1987. *Taking Charge of Change.* Alexandria, VA: ASCD. *21, 26*

Hughes, Colin. "Project Euler." http://projecteuler.net. *24*

Hull, Susan Hudson. 2000. "Teachers' Mathematical Understanding of Proportionality: Links to Curriculum, Professional Development, and Support." Unpublished doctoral dissertation. Austin: The University of Texas at Austin. *21*

Hull, Ted H., Don S. Balka, and Ruth Harbin Miles. 2011. "Standards of Student Practice in Mathematics Proficiency Matrix. http://mathleadership.com/sitebuildercontent /sitebuilderfiles/standardsoftudentpracticeinmathematicsproficiencymatrix.pdf.

———. 2011. *Visible Thinking in the K–8 Mathematics Classroom.* Thousand Oaks, CA: Corwin. *40*

Hull, Ted H., Ruth Harbin Miles, and Don S. Balka. 2012. *The Common Core Mathematics Standards: Transforming Practice Through Team Leadership.* Thousand Oaks, CA: Corwin. *21, 23, 25, 26*

Islas, Dana. 2011. *How to Assess While You Teach Math: Formative Assessment Practices and Lessons, Grades K-2: A Multimedia Professional Learning Resource.* Sausalito, CA: Math Solutions. *19*

Islas, Dana, and Beth Terry. Forthcoming. *How to Assess While You Teach Math: Formative Assessment Practices and Lessons, Grades 3–5: A Multimedia Professional Learning Resource.* Sausalito, CA: Math Solutions. *19*

Joyner, Jeanne, and George Bright. Forthcoming. *INFORMative Assessment: Formative Assessment to Improve Mathematics Achievement, Middle and High School.* Sausalito, CA: Math Solutions. *19*

Joyner, Jeane M., and Mari Muri. 2011. *INFORMative Assessment: Formative Assessment to Improve Math Achievement, Grades K–6.* Sausalito, CA: Math Solutions. *14, 19*

Kanold, Timothy, and Matthew Larson. 2012. *Common Core Mathematics in a PLC at Work, Leader's Guide.* Reston, VA: National Council of Teachers of Mathematics. *23, 25, 29*

Keller, Helen, and William Gibson. 1962. *The Miracle Worker.* DVD. Directed by Arthur Penn. Los Angeles, CA: MGM. *5*

Kennedy, Mary. 2006. *Inside Teaching: How Classroom Life Undermines Reform.* Cambridge, MA: Harvard University Press. *20*

Kieran, Carolyn. 2007a. "What Do Students Struggle with When First Introduced to Algebra Symbols?" *Research Brief*, edited by Judith Quander Reed. Reston, VA: National Council of Teachers of Mathematics. *17*

———. 2007b. "What Do We Know About the Teaching and Learning of Algebra in the Elementary Grades?" *Research Brief*, edited by Judith Quander Reed. Reston, VA: National Council of Teachers of Mathematics. *17*

Kilgannon, Corey. October 24, 2007. "A High School Under the Hood." *The New York Times*. www.nytimes.com/2007/10/24/automobiles/autospecial/24school.html?pagewanted=all&_r=1&. *9*

Kilpatrick, Jeremy, W. Gary Martin, and Deborah Schifter, eds. 2003. *A Research Companion to Principles and Standards for School Mathematics*. Reston, VA: National Council of Teachers of Mathematics. *22*

King, Emma J. "Cosmologist: Science Career Video: The Mathematician Who Can't Add Up." http://vega.org.uk/video/programme/89. *3*

Lovell, Jim, Jeffrey Kluger, William Broyles Jr., and Al Reinert. 1995. *Apollo 13*. Directed by Ron Howard. Burbank, CA: NBC Universal. DVD. *36*

Koestler, Courtney, Mathew D. Felton, Kristen N. Bieda, and Samuel Otten. 2013. *Connecting the NCTM Process Standards and the CCSSM Practices*. Reston, VA: National Council of Teachers of Mathematics. *31, 32, 33, 34, 35, 36, 37, 38, 39*

Kohn, Alfie. 2000. *The Case Against Standardized Testing: Raising the Scores, Ruining the Schools*. Portsmouth, NH: Heinemann. *28*

Lambdin, Diana V., and Frank Lester, eds. 2010. *Teaching and Learning Mathematics: Translating Research for Elementary School Teachers*. Reston, VA: National Council of Teachers of Mathematics. *22*

Langrall, Cynthia W., ed. 2006. *Teachers Engaged in Research: Inquiry in Mathematics Classrooms, Grades 3-5*. Series ed. Denise S. Mewborn. Reston, VA: National Council of Teachers of Mathematics. *22*

Lappan, Glenda, and Elizabeth Phillips. 2009. "A Designer Speaks: Challenges in U.S. Mathematics Education Through a Curriculum Developer Lens." *Educational Designer* 1 (3). www.educationaldesigner.org/ed/volume1/issue3/article11. *12*

Larson, Matthew, Francis (Skip) Fennell, Thomasenia Adams, Juli Dixon, Beth Kobett, Jonathan Wray, and Timothy Kanold. 2012a. *Common Core Mathematics in a PLC at Work, Grades K–2*. Reston, VA: National Council of Teachers of Mathematics. *23, 25*

———. 2012b. *Common Core Mathematics in a PLC at Work, Grades 3–5*. Reston, VA: National Council of Teachers of Mathematics. *23, 25*

Lawrence, Ann, and Charlie Hennessy. 2002. *Lessons for Algebraic Thinking, Grades 6–8*. Sausalito, CA: Math Solutions. *17, 39*

Lawrence Hall of Science/EQUALS. Family Math. Berkeley: University of California, Lawrence Hall of Science. http://lawrencehallofscience.org/equals. *7, 32*

Lehrer, Jonah. 2012. *Imagine: How Creativity Works*. New York: Houghton Mifflin Harcourt. *5*

Leinwand, Steven. 2012. *Sensible Mathematics: A Guide for School Leaders in the Era of Common Core State Standards, 2nd edition*. Portsmouth, NH: Heinemann. *21, 26*

Lester, Frank K., Jr., ed. 2007. *Second Handbook of Research on Mathematics Teaching and Learning*. Volumes 1 and 2. Reston, VA: National Council of Teachers of Mathematics. *22*

Lester, Frank, and Jinfa Cai, 2010. "Why Is Teaching with Problem Solving Important to Student Learning?" *Problem Solving Research Brief*, series ed. Judith Quander Reed. Reston, VA: National Council of Teachers of Mathematics. *12, 32*

Lester, Frank, and Randall I. Charles, eds. 2003. *Teaching Mathematics Through Problem Solving: Prekindergarten–Grade 6*. Reston, VA: National Council of Teachers of Mathematics. *32*

Levasseur, Kenneth, and Al Cuoco. 2003. "Mathematical Habits of Mind." In *Teaching Mathematics Through Problem Solving: Grades 6–12*, edited by Harold L. Schoen and Randall I. Charles. Reston, VA: National Council of Teachers of Mathematics. *40*

Leiva, Mariam. 2007. "Connections for Equity: Math, Language, Culture, and Context." www.nctm.org/conferences/flipbooks/sandiego/sddailynewsfriday/files /new%20issue%202%20lr.pdf.

Lilburn, Pat, and Alex Ciurak. 2010. *Investigations, Tasks, and Rubrics to Teach and Assess Math, Grades 1–6.* Sausalito, CA: Math Solutions. *19*

Lilburn, Pat, and Peter Sullivan. 2002. *Good Questions for Math Teaching: Why Ask Them and What to Ask, K–6.* Sausalito, CA: Math Solutions. *19*

Lim, Kien. 2013. "A Collection of Lists of Mathematical Habits of Mind." El Paso: The University of Texas at El Paso. www.math.utep.edu/Faculty/kienlim/HoM _Collection.pdf. *31*

Litton, Nancy. 1998. *Getting Your Math Message out to Parents: A K–6 Resource.* Sausalito, CA: Math Solutions. *7*

Litton, Nancy, and Maryann Wickett. 2008. *This Is Only a Test: Teaching for Mathematical Understanding in an Age of Standardized Testing.* Sausalito, CA: Math Solutions. *19, 28*

Lobato, Joanne, and Frank Lester, eds. 2010. *Teaching and Learning Mathematics: Translating Research for Secondary School Teachers.* Reston, VA: National Council of Teachers of Mathematics. *22*

Lu, Adrienne. 2013, October 22. "Who Is an 'English-Language Learner'?" *Stateline: The Daily News Service of The Pew Charitable Trusts.* www.pewstates.org /projects/stateline/headlines/who-is-an-english-language-learner-85899514092. *2*

Lucas, George. 1977. *Star Wars.* Los Angeles, CA: Twentieth Century Fox. DVD. *37*

Martin, W. Gary, John Carter, Susan Forster, Roger Howe, Gary Kader, Henry Kepner, Judith Reed Quander, et al. 2009. Focus in High School Mathematics Series: Reasoning and Sense Making. Reston, VA: National Council of Teachers of Mathematics. www.nctm.org/standards/content.aspx?id=23749. *27, 33*

Marzano, Robert. 2003. "Using Data: Two Wrongs and a Right." *Educational Leadership* 60 (5). *25*

Masingila, Joanna O., ed. 2006. *Teachers Engaged in Research: Inquiry in Mathematics Classrooms, Grades 6–8.* Series edited by Denise S. Mewborn. Reston, VA: National Council of Teachers of Mathematics. *22*

Mason, John, L. Burton, and K. Stacey. 2010. *Thinking Mathematically.* New York: Pearson.

Mastrull, Sarah. 2002. "The Mathematics Education of Students in Japan: A Comparison with United States Mathematics Programs." Bristol Township, PA. www.gphillymath.org/ExempPaper/TeacherPresent/Mastrull/SMastrull.pdf. *18*

Mathematics Assessment Resource Service. "Mathematics Assessment Project: Assessing 21st Century Math." University of Nottingham, Shell Center for Mathematical Education, and University of California at Berkeley, http://map.mathshell.org. *19*

Mathematical Sciences Education Board and National Research Council. 1989. *Everybody Counts: A Report to the Nation on the Future of Mathematics Education.* Washington, DC: National Academies Press. *27, 28*

Mathematics INstruction using Decision Science and Engineering Tools (MINDSET). www.mindsetproject.org. *27*

Mcafee, Andrew. 2013. "What Will Future Jobs Look Like?" www.ted.com/talks /andrew_mcafee_what_will_future_jobs_look_like. *9*

McCallum, William, and Jason Zimba. 2011. "Common Core State Standards: A New Foundation for Student Success: The Importance of Mathematical Practices." James B. Hunt, Jr. Institute for Educational Leadership and Policy and the Council of Chief State School Officers. http://educore.ascd.org/Resource /Video/833668a0-77be-405f-8a5a-3b9cd442c3bb. *31*

McGatha, Maggie. 2009. "Mathematics Specialists and Mathematics Coaches: What Does the Research Say?" *Research Brief*, edited by Judith Quander Reed. Reston, VA: National Council of Teachers of Mathematics. *13*

Middleton, James A., and Amanda Jansen. 2011. *Motivation Matters and Interest Counts.* Reston, VA: National Council of Teachers of Mathematics. *10, 12*

Melanese, Kathy, Luz Chung, and Cheryl Forbes. 2011. *Supporting English Language Learners in Math Class, Grades 6–8*. Sausalito, CA: Math Solutions. *2, 37*

Middleton, James A., and Amanda Jansen. 2011. *Motivation Matters and Interest Counts*. Reston, VA: National Council of Teachers of Mathematics. *4*

Mirra, Amy, ed. 2005. *A Family's Guide: Fostering Your Child's Success in School Mathematics*. Reston, VA: National Council of Teachers of Mathematics. *7*

Mislevy, Robert J., Russell G. Almond, and Janice F. Lukas. 2004. "A Brief Introduction to Evidence-Centered Design: CSE Report 632." Los Angeles: National Center for Research on Evaluation, Standards, and Student Testing (CRESST) and Center for the Study of Evaluation (CSE), UCLA. *28*

Molina, Concepcion. 2012. *The Problem with Math Is English: A Language-Focused Approach to Helping All Students Develop a Deeper Understanding of Mathematics*. San Francisco, CA: Jossey-Bass. *2, 13, 37*

Moschkovich, Judit. 2009. "Using Two Languages When Learning Mathematics: How Can Research Help Us Understand Mathematics Learners Who Use Two Languages?" *Research Brief*, edited by Judith Quander Reed. Reston, VA: National Council of Teachers of Mathematics. *2, 37*

National Center on Education and the Economy. 2008. *Tough Choices or Tough Times: The Report of the New Commission on the Skills of the American Workforce*. San Francisco, CA: Jossey-Bass. *12, 27*

National Council of Supervisors of Mathematics. 2007–present Improving Student Achievement Series: Research-Informed Answers for Mathematics Education Leaders. Denver, CO: NCSM. www.mathedleadership.org/resources/position.html. *22*

———. 2007a. "Improving Student Achievement by Leading Effective and Collaborative Teams of Mathematics Teachers." Improving Student Achievement Series: Research-Informed Answers for Mathematics Education Leaders. Denver, CO: NCSM.

———. 2007b. "Improving Student Achievement by Leading Sustained Professional Learning for Mathematics Content and Pedagogical Knowledge Development." Improving Student Achievement Series: Research-Informed Answers for Mathematics Education Leaders. Denver, CO: NCSM.

———. 2008a. "Improving Student Achievement by Leading the Pursuit of a Vision for Equity." Improving Student Achievement Series: Research-Informed Answers for Mathematics Education Leaders. Denver, CO: NCSM.

———. 2008b. "Improving Student Achievement in Mathematics for Students with Special Needs." Improving Student Achievement Series: Research-Informed Answers for Mathematics Education Leaders. Denver, CO: NCSM.

———. 2009a. "Improving Student Achievement in Mathematics by Addressing the Needs of English Language Learners." Improving Student Achievement Series: Research-Informed Answers for Mathematics Education Leaders. Denver, CO: NCSM. *2*

———. 2009b. "Improving Student Achievement in Mathematics by Leading Highly Effective Assessment Practices." Improving Student Achievement Series: Research-Informed Answers for Mathematics Education Leaders. Denver, CO: NCSM.

———. 2010. "Improving Student Achievement in Mathematics by Promoting Positive Self-Beliefs." Improving Student Achievement Series: Research-Informed Answers for Mathematics Education Leaders. Denver, CO: NCSM. *1*

———. 2011. "Improving Student Achievement in Mathematics by Systematically Integrating Effective Technology." Improving Student Achievement Series: Research-Informed Answers for Mathematics Education Leaders. Denver, CO: NCSM.

———. 2012a. "Improving Student Achievement in Mathematics by Expanding Learning Opportunities for the Young." Improving Student Achievement Series: Research-Informed Answers for Mathematics Education Leaders. Denver, CO: NCSM.

———. 2012b. "Improving Student Achievement in Mathematics by Expanding Opportunities for Our Most Promising Students of Mathematics." *Improving Student Achievement Series: Research-Informed Answers for Mathematics Education Leaders.* Denver, CO: NCSM.

———. 2013a. "Improving Student Achievement by Infusing Highly Effective Instructional Strategies into RTI Tier I Instruction." *Improving Student Achievement Series: Research-Informed Answers for Mathematics Education Leaders.* Denver, CO: NCSM.

———. 2013b. "Improving Student Achievement in Mathematics by Using Manipulatives with Classroom Instruction." *Improving Student Achievement Series: Research-Informed Answers for Mathematics Education Leaders.* Denver, CO: NCSM.

National Council of Teachers of Mathematics. *Essential Understandings Series.* Reston, VA: National Council of Teachers of Mathematics.

———. 1970–present. *Journal for Research in Mathematics Education.* Reston, VA: National Council of Teachers of Mathematics.   22

———. 2007–2013. *Research Briefs.* Reston, VA: National Council of Teachers of Mathematics.   22

———. 1980. *An Agenda for Action.* Reston, VA: National Council of Teachers of Mathematics.   32

———. 1989. *Curriculum and Evaluation Standards for School Mathematics.* Reston, VA: National Council of Teachers of Mathematics.   27, 31, 39

———. 1991. *Professional Standards for Teaching Mathematics.* Reston, VA, National Council of Teachers of Mathematics.   4, 12, 15, 23, 32

———. 1995. *Assessment Standards for School Mathematics.* Reston, VA: National Council of Teachers of Mathematics.   19

———. 2000. *Principles and Standards for School Mathematics.* Reston, VA: National Council of Teachers of Mathematics.   23, 27, 31, 32, 33, 34, 35, 36, 37, 38, 39, 40

———. 2004. "Figure This!" Reston, VA: National Council of Teachers of Mathematics. www.figurethis.org/index.html.   32

———. 2005. "Assessing to Learn and Learning to Assess." *Professional Development Focus of the Year.* Reston, VA: National Council of Teachers of Mathematics. www.nctm.org/profdev/content.aspx?id=4420.   19

———. 2007. "Research on Students' Thinking and Reasoning About Averages and Measures of Center." *Student Learning Research Brief,* edited by Judith Quander Reed. Reston, VA: National Council of Teachers of Mathematics.

———. 2011. "Technology in Teaching and Learning Mathematics." *Position Statement.* Reston, VA: National Council of Teachers of Mathematics.   24

National Environmental Satellite, Data, and Information Service [NESDIS]. "NOAA National Operational Model Archive and Distribution System (NOMADS)." http://nomads.ncdc.noaa.gov.   35

National Governors Association Center for Best Practices and Council of Chief State School Officers. 2010. *Common Core State Standards for Mathematics.* Washington, DC: NGA Center and CCSSO. www.corestandards.org.   3, 7, 12, 14, 15, 16, 18, 19, 21, 22, 23, 27, 31, 32, 33, 34, 35, 36, 37, 38, 39, 40

National Oceanic and Atmospheric Administration and NOAA Satellite and Information Service. "Model Data." www.ncdc.noaa.gov/data-access/model-data. 35

National Numeracy Network. http://serc.carleton.edu/nnn/index.html.   33

National Research Council. 2001. *Adding It Up: Helping Children Learn Mathematics.* J. Kilpatrick, J. Swafford, and B. Findell, eds. Mathematics Learning Study Committee, Center for Education, Division of Behavioral and Social Sciences and Education. Washington, DC: National Academy Press.   3, 22, 23, 31, 34, 39, 40

Nielsen, Kim E. 2010. *Beyond the Miracle Worker: The Remarkable Life of Anne Sullivan Macy and Her Extraordinary Friendship with Helen Keller.* Boston, MA: Beacon Press.   5

Nolan, Edward C. *Using Multiple Representations in Algebra (6–12), E-Seminar ANYTIME*. Reston, VA: National Council of Teachers of Mathematics. Recorded seminar. www.nctm.org/catalog/product.aspx?id=14116. *32*

Noyce, Pendred E., and Daniel T. Hickey. 2011. *New Frontiers in Formative Assessment*. Cambridge, MA: Harvard Education Press. *19*

Oakes, Jeannie. 1987. "Tracking in Secondary Schools: A Contextual Perspective." *Educational Psychologist* 22(2) (March): 129–53.

Oakes, Jeannie, and Marisa Saunders, eds. 2008. *Beyond Tracking: Multiple Pathways to College, Career, and Civic Participation*. Cambridge, MA: Harvard Education Press. *7, 9*

Obama, Barack. 2009 (September 8). "Prepared Remarks of President Barack Obama: Back to School Event." www.whitehouse.gov/MediaResources /PreparedSchoolRemarks. *1*

Otten, Samuel. 2011. "Cornered by the Real World: A Defense of Mathematics." *Mathematics Teacher* 105 (1): 20–25. *12, 32*

Palmer, Parker J. 2007. *The Courage to Teach: Exploring the Inner Landscape of a Teacher's Life*. San Francisco, CA: Jossey-Bass. *11*

Papert, Seymour. 1972. "Teaching Children to Be Mathematicians Versus Teaching About Mathematics." *International Journal of Mathematical Education in Science and Technology* 3 (3): 249–62. *31*

———. 1993. *Mindstorms: Children, Computers, and Powerful Ideas, 2nd edition*. New York: Basic Books. *24*

———. 1994. *The Children's Machine: Rethinking School in the Age of the Computer*. New York: Basic Books. *24*

Papert, Seymour, and Cynthia Solomon. 1971. "Twenty Things to Do with a Computer." *Artificial Intelligence Memo #248*. Cambridge: Massachusetts Institute of Technology. http://dailypapert.com/?p=1058. *24*

PARCC. "Partnership for Assessment of Readiness for College and Careers." http://parcconline.org. *19, 28, 31, 34*

Parrish, Sherry. 2010, 2014. *Number Talks, Grades K–5: Helping Children Build Mental Math and Computation Strategies, Updated with Common Core Connections*. Sausalito, CA: Math Solutions. *3, 38*

Pearson, P. David, and M. C. Gallagher. 1983. "The Instruction of Reading Comprehension." *Contemporary Educational Psychology* 317–44. Boston: Allyn and Bacon, MA. *23*

Pink, Daniel. 2006. *A Whole New Mind*. New York: Riverhead Trade. *5*

Pollak, Henry. 2003. "A History of the Teaching of Modeling." In *A History of School Mathematics*, edited by George M. A. Stanic and Jeremy Kilpatrick, 647–71. Reston, VA: National Council of Teachers of Mathematics. *35*

———. 2014. "What Is Mathematical Modeling?" http://dese.mo.gov/divimprove /curriculum/documents/cur-math-comcore-what-is-mathematical-modeling.pdf. *35*

Polya, George. 1981. *Mathematical Discovery: On Understanding, Learning and Teaching Problem Solving, Combined Edition (Volumes I and II)*. New York: John Wiley and Sons. *32*

———. 2004. *How to Solve It: A New Aspect of Mathematical Method*. Princeton, NJ: Princeton University Press.

———. 2009. *Mathematics and Plausible Reasoning, Volume II: Patterns of Plausible Inference*. Princeton, NJ: Princeton University Press. *31, 33, 34*

Popham, W. James. "Ten 'Must-Know' Facts About Educational Testing." National PTA. www.pta.org/programs/content.cfm?ItemNumber=1724. *28*

Purcell, Kristen, Alan Heaps, Judy Buchanan, and Linda Friedrich. 2013. "How Teachers Are Using Technology at Home and in Their Classrooms." Washington, DC: Pew Internet and American Life Project. http://pewinternet.org/Reports/2013 /Teachers-and-technology.aspx. *36*

Ramirez, Nora, and Celedon-Pattichis. 2012. *Beyond Good Teaching: Advancing Mathematics Education for ELLs*. Reston, VA: National Council of Teachers of Mathematics. *2*

Rath, Tom, and Barry Conchie. 2008. *Strengths Based Leadership: Great Leaders, Teams, and Why People Follow*. New York: Gallup Press. *29*

Ravitch, Dianne. 2010. *The Death and Life of the Great American School System: How Testing and Choice Are Undermining Education*. New York: Basic Books. *23, 28*

Ray, Max. 2013. *Powerful Problem Solving: Activities for Sense-Making with the Mathematical Practices*. Portsmouth, NH: Heinemann. *32*

Reed, Judith Quander. 2008. "What Can We Learn from Research?" *Research Brief*, edited by Judith Quander Reed. Reston, VA: National Council of Teachers of Mathematics.

Reys, Barbara. 2006. *The Intended Mathematics Curriculum as Represented in State-Level Curriculum Standards: Consensus or Confusion?* Charlotte, NC: Information Age Publishing. *16*

Reys, Robert E., Mary Lindquist, Diana V. Lambdin, and Nancy L. Smith. 2010. *Helping Children Learn Mathematics, 10th edition*. New York: John Wiley and Sons. *3*

Rennie Center for Education Research and Policy. 2007. "Seeking Effective Policies and Practices for English Language Learners." Cambridge, MA: Rennie Center for Education Research and Policy. *2*

Rigsbee, Cindi. 2011 (March 16). "Reflections of a Dance School Dropout." *Teacher Leaders Network*, March 16. Bethesda, MD: Education Week Teacher. www.edweek.org/tm/articles/2011/03/16/tln_rigsbee_dance.html?tkn=UNCCuZrks DcNlFyQl8u4xnY2sHTJMVsmvCvy&cmp=clp-sb-ascd. *4*

Ripley, Amanda. 2013. *The Smartest Kids in the World: And How They Got That Way*. New York: Simon and Schuster. *1*

Robbins, Mike. 2007. *Focus on the Good Stuff: The Power of Appreciation*. San Francisco, CA: Jossey-Bass. *29*

Rogers, Fred. 2003. *The World According to Mister Rogers*. New York: Hyperion. *7*

Ronau, Robert N., Christopher R. Rakes, Sarah B. Bush, Shannon Driskell, Margaret Niess, and David Pugalee. 2011. "Using Calculators for Teaching and Learning Mathematics." *Research Brief*, edited by Karen King. Reston, VA: National Council of Teachers of Mathematics. *36*

Russell, Susan Jo, Deborah Schifter, and Virginia Bastable. 2011. *Connecting Arithmetic to Algebra*. Portsmouth, NH: Heinemann. *17, 38, 39*

Schifter, Deborah, Virginia Bastable, and Susan Jo Russell. 2008. *Patterns, Functions, and Change Casebook*. Developing Mathematical Ideas. Lebanon, IN: Dale Seymour Publications. *38, 39*

Schifter, Deborah, Virginia Bastable, Susan Jo Russell, and Stephen Monk. 2008. *Number and Operations, Part 3: Reasoning Algebraically About Operations Casebook*. Developing Mathematical Ideas. Lebanon, IN: Dale Seymour Publications. *38*

Schleppenbach, Meg. 2010. "How Can Teachers and Schools Use Data Effectively?" *Research Brief*, edited by Sarah DeLeeuw. Reston, VA: National Council of Teachers of Mathematics.

Schmidt, William H., Curtis C. McKnight, and Senta A. Raizen. 1997. *A Splintered Vision: An Investigation of U.S. Science and Mathematics Education*. Norwell, MA: Kluwer Academic Publishers. *14*

Schoen, Harold L., and Randall I. Charles. 2003. *Teaching Mathematics Through Problem Solving: Grades 6–12*. Reston, VA: National Council of Teachers of Mathematics. *32, 40*

SEDL. 2014. "Concerns-Based Adoption Model." www.sedl.org/cbam. *21*

Seeley, Cathy L. 2009. *Faster Isn't Smarter*. Sausalito, CA: Math Solutions. *3, 20, 25*

———. 2010a. "Navigating the Peaks and Valleys of Teaching." *New England Mathematics Journal* 42 (May): 64.

————. 2010b. "Walking the Walk." Presentation for the Key Curriculum Press's Ignite! session at the annual meeting of the National Council of Supervisors of Mathematics, San Diego, CA, April 19. www.youtube.com/watch?v=RwCc8po4Vkc. *30*

Shaughnessy, Michael. 2008. "What Do We Know About Students' Thinking and Reasoning About Variability in Data?" *Research Brief*, edited by Judith Quander Reed. Reston, VA: National Council of Teachers of Mathematics.

Sims, Peter. 2011. *Little Bets: How Breakthrough Ideas Emerge from Small Discoveries.* New York: Free Press. *5, 8*

————. 2013. "Five of Steve Jobs's Biggest Mistakes." *HBR Blog Network*, January. Boston, MA: Harvard Business Review. http://blogs.hbr.org/2013/01/five-of-steve-jobss-biggest-mi. *8*

Skinner, B. F. 1972. *Beyond Freedom and Dignity.* New York: Knopf. *8*

Smith, Aaron. 2014. "African Americans and Technology Use: A Demographic Portrait." Washington, DC: Pew Internet and American Life Project. http://pewinternet.org/Reports/2014/African-American-Tech-Use.aspx. *36*

Smith, Margaret Schwan, and Mary Kay Stein. 1998. "Selecting and Creating Mathematical Tasks: From Research to Practice." *Mathematics Teaching in the Middle School* 3 (February): 344–50.

————. 2011. *5 Practices for Orchestrating Mathematics Discussions.* Reston, VA: National Council of Teachers of Mathematics. *4, 34*

Smith, Stephanie Z., and Marvin E. Smith, eds. 2006. *Teachers Engaged in Research: Inquiry in Mathematics Classrooms, Grades Pre-K–2.* Series ed. Denise S. Mewborn. Reston, VA: National Council of Teachers of Mathematics. *22*

Sowder, Judith, and Bonnie Schappelle, eds. 2002. *Lessons Learned from Research.* Reston, VA: National Council of Teachers of Mathematics. *22*

Stanford Center for Opportunity Policy in Education. 2014 (January 28). "Student-Centered Learning Approaches Are Effective in Closing the Opportunity Gap." Stanford, CA: SCOPE. https://edpolicy.stanford.edu/news/articles/1137. *12*

Stanford University. "Supporting ELLs in Mathematics." Stanford, CA: Stanford University. http://ell.stanford.edu/teaching_resources/math. *2*

Stanley, Dick, and Jolanta Walukiewicz. 2004. "Delving Deeper: In-Depth Mathematical Analysis of Ordinary High School Problems." *Mathematics Teacher* 97 (4), 248–55. *12, 32, 40*

Steen, Lynn Arthur, ed. 1990. "Pattern." *On the Shoulders of Giants: New Approaches to Numeracy.* Washington, DC: National Academy Press. *27, 39*

Steen, Lynn Arthur. 2006. "Facing Facts: Achieving Balance in High School Mathematics." *Mathematics Teacher* 100 (5) 86–95. *27*

Stein, Mary Kay. 2007. "Selecting the Right Curriculum." *Research Brief*, edited by Judith Quander Reed. Reston, VA: National Council of Teachers of Mathematics.

Stenmark, Jean Kerr, ed. 1991. *Mathematics Assessment: Myths, Models Good Questions, and Practical Suggestions.* Reston, VA: National Council of Teachers of Mathematics. *19*

Stigler, James W., and James Hiebert. 1999. *The Teaching Gap: Best Ideas from the World's Teachers for Improving Education in the Classroom.* New York: Free Press. *23*

Stone, James R. III, and Morgan V. Lewis. 2012. *College and Career Ready in the 21st Century: Making High School Matter.* New York: Teachers College Press. *9*

Strutchens, Marilyn, and Judith Reed Quander, eds. 2011. *Focus in High School Mathematics: Fostering Reasoning and Sense Making for All Students.* Reston, VA: National Council of Teachers of Mathematics. *13*

Tate, William F., Karen D. King, and Celia Rousseau Anderson. 2011. *Disrupting Tradition: Research and Practice Pathways in Mathematics Education.* Reston, VA: National Council of Teachers of Mathematics. *22*

Thames, Mark Hoover, and Deborah Loewenberg Ball. 2010. "What Math Knowledge Does Teaching Require?" *Teaching Children Mathematics* 17 (4): 220–29. *40*

Thomas-EL, Salome. 2006. *The Immortality of Influence: We Can Build the Best Minds of the Next Generation.* New York: Kensington Publishing. *1, 6*

———. 2013. Closing keynote address of the Scholastic Math Summit in Miami, December 11, 2013. Thomas Edison Charter School. Wilmington, DE. *1*

Tobias, Sheila. 1995. *Overcoming Math Anxiety, revised edition.* New York: W. W. Norton. *10*

Transition to College Mathematics and Statistics (TCMS). www.wmich.edu/cpmp/tcms. *27*

Tugend, Alina. 2011. "The Role of Mistakes in the Classroom." *Edutopia.* www.edutopia.org/blog/benefits-mistakes-classroom-alina-tugend. *8*

U.S. Department of Labor. Bureau of Labor Statistics. 2013. The Employment Situation— October 2013. www.bls.gov/news.release/archives/empsit_11082013.pdf. *12*

Usiskin, Zalman. 2006a. "From the 1980s: What Should Not Be in the Algebra and Geometry Curricula of Average College-Bound Students?" *Mathematics Teacher* 100 (5): 68–77. *27*

———. 2006b. "Reconsidering the 1980s: A Retrospective after a Quarter Century." *Mathematics Teacher* 100 (5): 78–79. *27*

———. 2011. "Mathematical Modeling in the School Curriculum." http://ucsmp .uchicago.edu/resources/conferences/2011-09-11. *35*

———. 2012. "The Ethics of Using Computer Algebra Systems (CAS) in High School Mathematics." http://ucsmp.uchicago.edu/resources/conferences /2012-03-01. *28, 36*

Vanderbilt University Center for Teaching. "Teaching Topics: Interactions with Students: Motivating Students." http://cft.vanderbilt.edu/teaching-guides/interactions /motivating-students. *4*

Van de Walle, John, and Lou Ann Lovin. 2005a. *Teaching Student-Centered Mathematics: Grades K–3 Volume 1.* Teaching Student-Centered Mathematics Series. Boston, MA: Allyn and Bacon. *3*

———. 2005b. *Teaching Student-Centered Mathematics: Grades 3–5 Volume 2.* Teaching Student-Centered Mathematics Series. Boston, MA: Allyn and Bacon. *3*

———. 2005c. *Teaching Student-Centered Mathematics: Grades 5–8 Volume 3.* Teaching Student-Centered Mathematics Series. Boston, MA: Allyn and Bacon. *3*

Van de Walle, John, Karen Karp, and Jennifer M. Bay-Williams. 2009. *Elementary and Middle School Mathematics: Teaching Developmentally, 7th edition.* Boston, MA: Allyn & Bacon.

Van Dyke, Frances. 2012. *It's All Connected: The Power of Representation to Build Algebraic Reasoning, Grades 6–9.* Sausalito, CA: Math Solutions. *17*

Van Zoest, Laura R., ed. 2006. *Teachers Engaged in Research: Inquiry in Mathematics Classrooms, Grades 9–12.* Series ed. Denise S. Mewborn. Reston, VA: National Council of Teachers of Mathematics. *22*

von Rotz, Leyani, and Marilyn Burns. 2002. *Lessons for Algebraic Thinking: Grades K–2.* Sausalito, CA: Math Solutions. *17, 39*

Washington State Office of Superintendent of Public Instruction. "Smarter Balanced Assessment Consortia." Smarter Balanced Assessment Consortia, www.smarterbalanced.org. *19, 28, 31, 34*

Webel, Corey. 2010. "Connecting Research to Teaching: Shifting Mathematical Authority from Teacher to Community." *Mathematics Teacher* 104 (4): 315– 18. *12, 15, 34*

Weingarten, Gene. 2007. "Pearls Before Breakfast." *Washington Post*, April 8.

Weiss, Iris R., and Pasley, Joan D. 2004. "What Is High-Quality Instruction?" *Educational Leadership* 61 (5): 24–28. *1, 6*

———. 2006. "Scaling Up Instructional Improvement Through Teacher Professional Development: Insights from the Local Systemic Change Initiative." *CPRE Policy Briefs.* Philadelphia: University of Pennsylvania. *21, 26*

Wenglinsky, Harold. 2000. "How Teaching Matters: Bringing the Classroom Back into Discussions of Teacher Quality." Princeton, NJ: Educational Testing Service. *3*

Whitman, Carmen. 2011. *It's All Connected: The Power of Proportional Reasoning to Understand Mathematics Concepts, Grades 6–8*. Sausalito, CA: Math Solutions. *17*

Wickett, Maryann, Katharine Kharas, and Marilyn Burns. 2002. *Lessons for Algebraic Thinking: Grades 3–5*. Sausalito, CA: Math Solutions. *17, 39*

Wieman, Robert, and Fran Arbaugh. 2013. *Success from the Start: Your First Years Teaching Secondary Mathematics*. Reston, VA: National Council of Teachers of Mathematics. *11*

Wiliam, Dylan. 2007a. "Five 'Key Strategies' for Effective Formative Assessment." *Research Brief*, edited by Judith Quander Reed. Reston, VA: National Council of Teachers of Mathematics. *14, 19*

———. 2007b. "What Does Research Say the Benefits of Formative Assessment Are?" *Research Brief*, edited by Judith Quander Reed. Reston, VA: National Council of Teachers of Mathematics. *14, 19*

———. 2011. *Embedded Formative Assessment*. Bloomington, IN: Solution Tree. *19*

———. 2012. *Sustaining Formative Assessment with Teacher Learning Communities*. Amazon Digital Services, Inc. *19*

———. 2013. *Formative Assessment: The Bridge between Teaching and Learning in High School Mathematics*. PowerPoint presentation for NCTM High School Interactive Institute, August 2. www.dylanwiliam.org/Dylan_Wiliams_website /Presentations.html. *22*

Willingham, Daniel T. 2009. *Why Don't Students Like School?* San Francisco, CA: Jossey-Bass. *1*

Zickuhr, Kathryn. 2013. "Who's Not Online and Why." Washington, DC: Pew Internet and American Life Project. pewinternet.org/Reports/2013/Non-internet-users. aspx. *36*

Zimmerman, Gwen, John Carter, Timothy Kanold, and Mona Toncheff. 2012. *Common Core Mathematics in a PLC at Work, High School*. Reston, VA: National Council of Teachers of Mathematics. *23, 25, 27*

Index

About the Author

For more than forty years, Cathy Seeley has been a mathematics educator and change facilitator at the local, state, and national level. She is deeply committed to a high-quality mathematics education for every student. Among her experiences in education, she has worked as a mathematics teacher, district mathematics coordinator, and state mathematics director for grades K–12. After her return in late 2001 from teaching mathematics (in French) as a Peace Corps volunteer in Burkina Faso, Cathy was elected to serve a two-year term as President of the National Council of Teachers of Mathematics. In that role, she was awarded an EXCEL Gold Award from the Society of National Association Publications (SNAP) for her President's Message, "Embracing Accountability." Cathy has given presentations in forty-nine states, Mexico, Canada, Portugal, France, Germany, South Africa, and China. She has appeared on television and radio and authored or coauthored various publications including mathematics textbooks and her 2009 book, *Faster Isn't Smarter—Messages About Math, Teaching, and Learning in the 21st Century*, which received a 2010 AEP Distinguished Achievement Award honoring excellence and innovation in the educational community. Cathy recently retired as a Senior Fellow (now Emeritus) with the Charles A. Dana Center at the University of Texas, where she worked on state and national policy and improvement efforts, with a focus on prekindergarten–grade 12 mathematics education. For more information, see Cathy's website at www.cathyseeley.com.